Early hominid posture and locomotion [by] John T. Robinson. Chicago, University of Chicago Press [1972]

xi, 361 p. illus. 24 cm.

Includes bibliographical references.

1. Man - Origin. I. Title.

72-77306*

Early Hominid Posture and Locomotion

John T. Robinson

Early Hominid Posture and Locomotion

The University of Chicago Press, Chicago and London

The University of Chicago Press, Chicago 60637
The University of Chicago Press, Ltd., London
 Published 1972
Printed in the United States of America
International Standard Book Number: 0–226–72230–9
Library of Congress Catalog Card Number: 72–77306

For Sybil, Richard, and Peter
in appreciation

Contents

Preface

The primary purpose of this study is to provide a moderately detailed account of the postcranial remains of early hominids from the Republic of South Africa as seen through the eyes of one student. Some of the material that has been found in the East African region since 1959 has been included. I have seen some of the original specimens of this material, but have used only such information as has already been made public by the persons responsible for their study or is deducible from published information.

The secondary purpose of this work is to integrate the conclusions stemming from the study of the postcranial remains with those already arrived at on the basis of detailed study of the cranial remains, especially the parts which are most commonly preserved, the teeth and jaws. This integration is carried out in the final, rather lengthy chapter, which is different in nature from the rest of the book. It includes explanations of some of the basic aspects of my approach to interpretation of the early hominid material; a résumé of my interpretation of the cranial material, and a review of the main conclusions reached in the preceding fourteen chapters. These conclusions from the cranial and postcranial evidence are then used to provide a general interpretation of the main steps, and the factors that influenced them, in the transition from an ape grade of organization to the fully human grade. This involves as an integral element an interpretation of the course and causes of the achievement of erect posture. This final chapter can be read as a self-contained unit by those who do not wish to plough through the detail of the rest of the book.

Before embarking on this study it had seemed to me clear that *Australopithecus* and *Paranthropus* were distinctly different anatomically, behaviorally, and ecologically—a point of view that I had argued at some length. A major conclusion resulting from this study of the postcranial remains was that the two forms differed even more than I had previously supposed. *Paranthropus* evidently was even less manlike and *Australopithecus* more manlike than the cranial evidence alone had suggested. For this reason, after completing the manuscript, I reconsidered the section on nomenclature. As a result it was rewritten, and the suggestion I had made of sinking *Australopithecus* into *Homo* but retaining *Paranthropus* as a separate genus was put into practice.

It is becoming customary to use complex multivariate forms of statistical analysis in the study of fossil primate remains, and occasionally one meets the implication in the literature that unless such is the case a study cannot be regarded as modern, adequate, and genuinely "scientific." I do not subscribe to this view: statistical analysis is a most valuable aid to clarifying relationships that already exist in the material being studied. Statistical analysis, however, will not bring out relationships that are otherwise quite undetectable. Moreover, unless the statistical analysis is thoroughly grounded in an understanding of the biology of the situation it is more likely to be misleading than helpful—as many uncritical studies demonstrate only too readily. Statistical analysis does not possess biological judgment, nor is it a substitute for biological judgment—the final decisions on biological meaning have always to be made by the biologist, as Simpson, Roe, and Lewontin (1960), among others, have pointed out. Even avowed protagonists of the sophisticated statistical approach, after a study that used discriminant functions on some aspects of early hominid problems, state in their summing up, "For the most part our findings are similar to those reported by comparative anatomists with experience of the material, mainly on the basis of visual inspection" (Ashton, Healy, and Lipton 1957). In this study I have depended primarily upon anatomical analysis and have not used sophisticated statistics. Straightforward parametric statistical analysis has been employed, however, as well as a simple, primarily graphical method of dealing simultaneously with a number of characters to demonstrate patterns of similarity or difference; the logarithmic ratio diagram. These have proved to be useful additions to the primarily comparative anatomical analysis used. If my level of biological judgment has been inadequate for the task, then this work fails—but in that case it would not have been rescued had I employed more sophisticated statistics.

I wish to apologize to my colleagues for the length of time it has taken for this study to reach publication. The delay has resulted from many factors, chief among them being the necessity to obtain at first hand much of the metrical and anatomical background data needed for comparison—a most time-consuming task—and the fact that I was primarily concerned with research in South Africa, where this work was started, but substantial teaching commitments have greatly reduced research time during the eight years since I came to the United States of America.

Many persons or institutions have assisted me in various ways in the course of this work, and I extend to them all my most sincere thanks for their kindness and courtesy. Listing all to whom I owe a debt of gratitude would be a formidable task, and the list would run to a number of pages. I hope that those whose names do not appear will nevertheless accept my thanks as being no less sincere.

For financial assistance I wish to mention especially the Wenner–Gren Foundation for Anthropological Research, New York; the U.S. National Institutes of Health, Bethesda; the National Science Foundation, Washington, D.C.; the University of Wisconsin, Madison, including its Research Committee and the Wisconsin Alumni Research Foundation; the Transvaal Museum, Pretoria; the South African Council for Scientific and

Industrial Research, Pretoria; the Nuffield Foundation, London; the Boise Fund, Oxford University; and Mr. and Mrs. L. Donnelly, Johannesburg, whose friendship is valued and whose generosity has extended in many directions over many years.

For access to specimens or collections I wish to thank the following institutions or individuals: the Transvaal Museum, Pretoria, and especially its former Director, Dr. V. FitzSimons, and its present Director, Dr. C. K. Brain; the Department of Anatomy of the University of the Witwatersrand, Johannesburg, especially Professor R. Dart and Professor P. V. Tobias; Dr. L. S. B. Leakey and Dr. Mary Leakey, Nairobi; the British Museum (Nat. Hist.), London, especially the Subdepartment of Anthropology, and Drs. K. P. Oakley and D. Brothwell, the Department of Mammalogy and the Department of Geology; the Powell–Cotton Museum, Birchington, and its curator, Mr. L. Barton; the Anthropological Institute of the University of Zurich, its present Director, Professor J. Biegert, and its former Director, Professor A. Schultz; the Museum of Natural History, Paris, and the late Professor C. Arambourg and Dr. Y. Coppens, then of that institution; Professor R. von Koenigswald, then of the University of Utrecht, and Dr. J. Huizinga of the same university; Dr. D. Hooijer of the Natural History Museum, Leiden; the American Museum of Natural History, New York; the Smithsonian Institution, Washington, D.C.; the Museum of Comparative Zoology, Harvard University; Yale University and Drs. E. Simons and D. Pilbeam; the Field Museum of Natural History, Chicago, and Dr. C. Liem; Professor F. C. Howell, then of the University of Chicago.

For assistance with various phases of final preparation of the manuscript I wish to thank: Miss V. Gutgesell, who also assisted with some computation; Mrs. J. Williams and Mrs. K. Steudel. Special thanks go to Mr. J. Dallman and Miss C. Hughes of the Department of Zoology, University of Wisconsin, for preparing the final versions of some of the illustrations. It is a pleasure to acknowledge many helpful discussions with my friend and colleague Dr. L. Freedman, who also read much of the manuscript in an early stage.

1 Preamble

Introduction

Man is a curious animal and one of the directions in which his curiosity is especially active is in the investigation of his own origins. Such an inquiry is of vast scope since it involves broad philosophical questions concerning the nature of the universe and the place of man within it. In a sense, all seeking for knowledge is a part of man's attempt to understand himself.

Within this vast field of inquiry, one aspect is of especially direct relevance and that is the investigation of the evolutionary emergence of man as a distinct kind of animal: an animal with self-consciousness and the capacity for conceptual thought; an animal who can discover knowledge, who can use it, store it, and transmit it effectively to others of his kind whether near him or on another planet, whether coeval with him or dwellers in the future. Man is unique among living organisms, at least in this solar system, with potentialities that seem to be enormously greater than those of any other organism about which we have any knowledge. With the coming of man evolution attained a new major level, acquired a new dimension. Man is surely the dominant organism on this planet, even though it used to be thought that such a view represented nothing more than distasteful anthropocentric eccentricity or even downright anthropolatry. But man really is dominant, superior, or what you will, and is so because of a basic change in his behavior. This expresses itself in the manifold aspects of cultural activity, which is the most important single mechanism for adaptation possessed by man and not possessed by other animals except at best in the most rudimentry form.

The capacity for culture is apparently a consequence of the capacity for conceptual thought. Seemingly, therefore, the key change in the emergence of man from nonhominids was the effective development of the capacity for conceptual thought. But it is not clear what gives man this capacity. The assumption is often made that it is primarily a matter of an increased number of circuits in the larger brain of man that allows incoming information to be related in many different ways to the existing store of information. The fossil record seems to indicate that human cultural activity began while the brain was no larger than that of living apes. Perhaps, therefore, there was also some physiological change in the central nervous system so that man's cultural ability depends on both more complex circuitry than had previously existed and improved functioning of those circuits. Whatever the true nature of this change may turn out to be, it is unlikely to be discovered from the fossil record. But the record can indicate when the change was already in effect because its consequences can be found in the form of fossilized remains of material culture, such as recognizable artifacts in stone or bone. The presence of these demonstrates that the basic change from "ape" to "man" had already occurred.

Anatomical changes are associated with behavioral changes and are of considerable importance. If effective use is to be made of cultural activity, it is probable that an at least appreciably greater facility with erect posture will be necessary than is possessed by living pongids. This implies that cultural activity would not have been a real adaptive possibility capable of emerging under the control of natural selection unless erect posture had already existed to an appreciable extent as a prospective adaptation.

Fortunately, a not inconsiderable amount of fossil material is known that is of direct relevance to an investigation of the locomotor changes associated with the coming of man. The fossils concerned belong to the very early hominids, generally referred to as the australopithecines. These are known from various parts of the Old World, but the greatest number of specimens has been found so far in South Africa, most of them in the Sterkfontein Valley. Appreciable numbers of specimens are now being found in East Africa. The australopithecines have provided so far the earliest evidence known of cultural activity of a specifically human sort, and it seems clear that these creatures were closely associated with the actual emergence of man. Their posture and locomotor habit are thus of great interest. Only about 10 percent of the known fossil specimens belong to the postcranial skeleton. It is the purpose of this work to give a general account of the postcranial material from South Africa—some specimens being described here for the first time—and to discuss the light it throws on the posture and locomotor habit of the australopithecines. The pelvis is better represented in the known material than is any other part of the postcranial skeleton, and the major portion of this book will therefore concern this region of the body.

Nomenclature

The present tendency with respect to hominid taxonomic nomenclature is to lump all of the australopithecines together in a single genus, *Australopithecus* (see especially various authors in Washburn 1963). This tendency is reflected in discussions of the australopithecines in which the various characteristics are usually referred to as if they belonged in equal measure to all australopithecines. As an illustration of this, the authors of a recent textbook on human evolution state that the degree to which the skull vertex rises above the upper orbital margins is proportionately much greater in australopithecines than in pongids, the former agreeing much more closely with modern man. This statement is correct with respect to *Australopithecus* but is not true of *Paranthropus*; all of the evidence now available indicates that the latter is indistinguishable from pongids in this feature. Many such examples could be quoted. Instead, it seems to be the case that, in a series of significant features, what is true of one of the two major types of australopithecine is not true of the other.

The available evidence provides much support for the thesis that the two forms of australopithecine differed in important respects not only in anatomy but also in ecology and behavior in ways that are of especial significance with regard to the emergence of man (Robinson 1961, 1962, 1963, 1966, and 1967). The two types appear to have occupied quite different adaptive zones, and for this reason to lump them in a single genus does not seem to be a taxonomically sound practice. The reasons for this statement are discussed at length in the final chapter (see pp. 245–256). Because they are so different, lumping them in a single genus also leads to repeated errors in the literature when characteristics belonging to one form alone are attributed to the genus *Australopithecus* as a whole when this designation is used to include both forms.

In this work I shall continue to distinguish between the two forms of australopithecine at the generic level, as originally proposed (Robinson 1954; elaborated in 1962). The former paper is one of the very few in which the morphological evidence is evaluated at first hand from a taxonomic point of view as opposed to the many that include statements of opinion about australopithecine nomenclature in the absence of systematic taxonomic analysis of the relevant evidence. In the above system the more robust, low-vaulted form, which has a sagittal crest and small canine teeth compared to postcanine teeth, is placed in the genus *Paranthropus* Broom, 1938; the more gracile, higher-vaulted form, which does not usually have a sagittal crest but has relatively large canines for a hominid, is placed in the genus *Australopithecus* Dart, 1925.

Without exception, I have employed the above usage since 1954. It is my firm belief, however (Robinson 1965, 1966, and 1967), that the organisms included in the genus *Australopithecus* in the above sense belong to the same phyletic line or lineage as we do and represent an early phase in the establishment of the adaptive pattern characteristic of modern man. The reasons for this view are presented more fully in the last chapter. Therefore, it is logical to sink *Australopithecus* into the genus *Homo* as a junior synonym since they share the same basic adaptive pattern. Because *Paranthropus* appears to occupy a different adaptive zone, it should remain as a distinct genus. This classification thus includes the known hominids of the later Pliocene and Pleistocene in two genera only, *Paranthropus* and *Homo*.

This classification removes justification for the vernacular term "australopithecine" because its adoption leaves no taxonomically distinct group of early hominids to which such a term could apply. When it was believed that the Family Hominidae was horizontally divided into a later

Subfamily Homininae, which included "true" man, and an earlier Subfamily Australopithecinae, which included more primitive hominids that were not yet "true" man, the term australopithecine was entirely justifiable. But the whole burden of the analysis of the anatomical, ecological, and behavioral aspects of the fossil material that has led me to propose the *Paranthropus*—*Homo* generic scheme is that the dichotomy is much more nearly vertical than horizontal: *Australopithecus* was much more like the later *Homo erectus* and *H. sapiens* than it was like its contemporary, *Paranthropus*. The study of the postcranial skeleton reported on in this work makes this even more clear than it had previously been because of the apelike, power-oriented propulsive mechanism of *Paranthropus* and the modern, human type of speed-oriented mechanism possessed by *Australopithecus*. The implication is that *Paranthropus* was incompletely adapted to erect posture and possibly still spent much time up in the trees while *Australopithecus* was essentially fully adapted to erect bipedality in the fashion of modern man. This difference appears to be considerably greater than I had believed it to be when I embarked on this study of the postcranial skeleton.

If subfamily distinction is to be retained in the Family Hominidae, then on the basis of the above scheme the subfamilies would have to be Homininae, now including *Homo* plus "*Australopithecus*" rather than the former alone, as it previously did, and Paranthropinae. This follows from the provisions of the International Code of Zoological Nomenclature that require that a subfamily name consist of the name of the type genus, suitably modified, and the suffix "-inae." On the face of it the maintenance of subfamily distinction within the Hominidae would seem quite unnecessary, especially as both taxa would be monotypic with one genus apiece. Further reflection suggests that the scheme has merit. The more recently discovered Pliocene species of *Gigantopithecus*, *G. bilaspurensis* (Simons and Chopra 1969), is more hominidlike in mandibular morphology than is the later *G. blacki* and is probably geologically not much older than the earliest remains of *Paranthropus*. *G. bilaspurensis* seems to me to bear considerable resemblance to *Paranthropus*, especially with respect to increase of the occlusal surface area of the postcanine teeth and decrease of the size of the canines and incisors. In this respect *G. bilaspurensis* appears to occupy a position transitional between that of the great apes and of *Paranthropus* but is much less similar to *Australopithecus* and *Homo*. Moreover, there is some suggestion in the East African material of *Paranthropus*, as compared to that from South Africa, that the geologically more recent specimens tend to be smaller and less masssive than the older ones. The evidence is insufficient to prove this because at present the oldest material is very fragmentary and does not include substantial pieces of jaw or skull. But much of the East African material appears to be both older and more robust than the South African material. It is possible, therefore, that the ancestor of *Paranthropus* was a rather robust animal, which apparently was also a characteristic of *G. bilaspurensis*. To repeat, the evidence at present is much too scanty for the hypothesis to be treated as well-founded; but it is a reasonable possibility.

Based on a study of the gelada, Clifford Jolly (1970), has postulated that the initial trend away from the ape grade or organization toward the hominid grade involved a shift from a forest environment to a more open habitat and a shift from a more typical ape type of diet to what he calls a granivorous diet of grass seeds, stems, and roots. This, he postulates, would involve increased occlusal area size in the postcanine teeth and reduction of the anterior teeth, especially the canines. He believes that *Paranthropus* has characteristics consistent with this view and accepts my interpretation that *Paranthropus* represents a stage ancestral to that of *Australopithecus*. Pilbeam (1970) appears to accept Jolly's basic thesis; he believes also that *G. bilaspurensis* was adapted for foraging in relatively open country and that its dental and other mandibular features resulted from dietary specializations associated with this nonforest habitat. While I have some reservations about the diet suggested—it seems much too restricted as well as insufficient for a large animal of that type at all seasons of the year—the general trend of these arguments concerning dietary change is what I have propounded for many years as the basis of the difference between *Paranthropus* and *Australopithecus* (= *Homo africanus*). The shift from forest to woodland to grass savannah habitat, as I elaborate in the final chapter, appears logical and easily understood in terms of the adaptive features involved. Even though they now accept the basic dietary argument, it is interesting that Simons, Pilbeam, and Jolly appear to believe that the resemblances between *G. bilaspurensis* and *Paranthropus* are parallelisms, not consequences of a closeness of relationship. One may speculate that they believe the ancestral hominid has already been found in the form of *Ramapithecus*. After study of the material, it is my opinion, discussed in detail elsewhere, that *Ramapithecus* does not possess one known character that is typical of hominids alone and does have some that are inconsistent with its being a hominid or even trending clearly in a hominid direction. On morphology alone, but especially if the whole analysis of early hominids is taken into account, it seems that the much larger and more *Paranthropus*-like *G. bilaspurensis* is a more suitable ancestor of the hominids. It may be that *G. blacki* and *Paranthropus* are both representatives of a stock, of which *G. bilaspurensis* is an earlier representative, that diverged from a pongid stage (perhaps *Dryopithecus indicus* as Pilbeam [1970] has suggested) as a result of adaptation to woodland conditions. More than one lineage seems to have been involved in the later stages. *Gigantopithecus* is at present known only from dental and mandibular material and *G. bilaspurensis* from a single mandible; clearly much more information is needed before such hypothesizing can be securely based. At present it seems a reasonable course to place *Gigantopithecus* and *Paranthropus* together in a sort of parahominid group as distinct from the fully erect, bipedal, culture-bearing hominids "*Australopithecus*" plus *Homo*. For this purpose the Subfamily Paranthropinae would serve admirably.

Because it has long seemed justifiable to sink "*Australopithecus*" into *Homo* (Robinson 1965, 1966, and 1967), and because the study here reported adds significantly to that justification, I propose formally to sink *Australopithecus* Dart, 1925 into *Homo* Linnaeus, 1758—as was in effect already done in Robinson (1967). The following scheme will be employed in this work:

Family: Hominidae Gray, 1825
Subfamily: Paranthropinae
Genus: (1) *Gigantopithecus* von Koenigswald, 1935
(2) *Paranthropus* Broom, 1938
This includes *Zinjanthropus*, *Meganthropus* from Java, and *Paraustralopithecus*
Subfamily: (2) Homininae
Genus: *Homo* Linnaeus, 1758
This includes *Australopithecus*, *Meganthropus* from Africa, *Telanthropus*, *Sinanthropus*, *Pithecanthropus*, *Atlanthropus*, and other supposed genera that are now generally treated as falling into either *H. erectus* or *H. sapiens*

"*Tchadanthropus*" is not assigned a position in this scheme since the one specimen on which it is based is too fragmentary and too damaged through sand-blasting for its affinities to be determinable. *Ramapithecus* and *Kenyapithecus* I regard as pongids, not hominids, and hence they do not have a place in this scheme. *Hemanthropus* I believe to be an invalid taxon since it appears to be composed of specimens belonging to more than one already known taxon.

Because this formulation has not been used previously, to the best of my knowledge, definitions are provided here of the two subfamilies.

The Subfamily Paranthropinae is defined as consisting of a group of higher primates having cheek teeth with proportionately large occlusal surfaces of low relief associated with anterior teeth, especially canines, which are proportionately small to extremely small; the canines wear in such a manner that the major wear facet either starts on the apex or soon incorporates the apex; the mandible is very robust with proportionately tall and more or less vertical ascending ramus; the face is robust and flat with prominent cheekbones so that the face is commonly dished; the frontal area is narrow with no more than slight convexity so that no forehead exists, and the vertex is relatively low as in pongids; a sagittal crest in the region of the vertex is usual; some evidence of adaptation to erect posture is already present (i.e., more than in pongids), but adaptation is incomplete and does not include a proportionately shortened ischium and lengthened lower limb—the propulsive mechanism is thus, as in apes, essentially power-oriented.

In explanation, it seems to me that these purely morphological criteria reflect adaptation to a grassy woodland type of habitat as distinct from forest conditions, on the one hand, and open plains conditions on the other. The way of life was essentially apelike rather than manlike, involving neither hunting nor cultural activity of significance. There was some modification of diet to a somewhat differently specialized herbivorousness than that of forest apes, placing much stress on crushing and grinding. Much time was spent on the ground using erect posture of moderate efficiency, but some time was still spent in the trees for shelter and possibly also for feeding. The locomotor habit was thus a compromise between erectness and facility in quadrupedal climbing.

The Subfamily Homininae is defined as consisting of higher primates with spatulate canines that do not protrude significantly beyond the adjacent teeth and in which wear starts on the apex; the cheek teeth occlusal surfaces are not strongly enlarged and specialized for crushing and grinding as in the paranthropines and consequently there is more harmony of proportion between anterior and cheek teeth as contrasted to the paranthropine condition. The mandible has a relatively lower and less vertical ascending ramus. There is always some convexity of the frontal region to produce at least a moderate forehead development, and the vertex is relatively high as compared to pongids; a sagittal crest is very rarely present. The face is less robust and is never dished because the cheekbones are less well developed and the nasal area better developed than in paranthropines. Adaptation to erect posture is always well developed and includes a relatively short ischium and long lower limb to produce a speed-oriented propulsive mechanism. Significant cultural activity is present—tool-using in the earliest forms, with tool-making added at a relatively early stage.

These morphological criteria relate to a different basic adaptation than that of the paranthropines. Hominines have culture as a highly significant aspect of their adaptation. This originated with a change to a hunting way of life on dry plains. The hominine adaptation involved significant changes in diet (to omnivorousness) and behavior (cultural activity as well as other changes). As part of these changes, erect posture was developed to a level of efficiency where quadrupedal activities became so difficult as to no longer be of practical significance and tree-climbing ceased to be a normal part of their way of life. With this change to a basically cultural adaptation, a wholly new dimension was added to evolution, representing a major step in the evolution of evolution itself. Because the essence of this adaptive mode lies in behavior, and because it leads to an increasing degree of interference with the process of natural selection by substituting conscious, direct adaptation, physical criteria become increasingly useless for defining hominines. Indeed, it is questionable whether hominines should be included in the classificatory system that encompasses paranthropines and all other living organisms.

Sinking *Australopithecus* into *Homo* raises the question of the correct trivial name to be used with *Homo* to designate the species now widely known as *Australopithecus africanus*. The basically correct procedure is to substitute the newly assigned generic name for the old: *Homo africanus*. Some authors have done this; see for example Simpson (1963) and Simons (1967). When a species is transferred from one genus to another, however, the problem of homonomy must always be considered. In this instance it was pointed out some years ago (Robinson 1967) that the trivial name *africanus* is preoccupied in the genus *Homo* and is thus unavailable in this case under the priority rule of the International Code of Zoological Nomenclature. Unfortunately, owing to a *lapsus calami* on my part, the wrong reason (prior use by Broom for Boskop Man in 1918) was given for its unavailability. This was brought to my attention by Oakley (private communication) and later by Simons (1967). The Boskop case related to another nomenclatural problem with which I was dealing at the time of writing the paper here concerned. This other problem involved the availability of the trivial name *capensis* for the genus *Homo* in the event that "*Telanthropus*" *capensis* Broom and Robinson, 1949, is sunk into *Homo* but retained as a separate species of which *Homo* "*habilis*" might be a synonym. Some authors discussing the latter species believed this course of action might be desirable, although it does not appear to be the case to me, and I had long since sunk "*Telanthropus*" into *Homo erectus* (Robinson 1961). This case, of course, has no relevance to the availability of *africanus*, as I was fully aware; it was a simple oversight on my part not to notice until after the article had appeared in print that the grounds for the nonavailability of *capensis* had been given instead of those for *africanus*.

The name *Homo africanus* was used in that exact form as long ago as 1904 by Sergi. In a later work he modified this usage so that *africanus* became the equivalent of a lower level in the species-group (Sergi 1908). Basing himself on this latter usage, von Eickstedt (1937) incorporated the name *H. sapiens africanus* in a very lengthy classification of man. It was accepted by

Peters (1937) in a paper dealing directly with hominid nomenclature, the purpose of which was to determine which names were valid under the provisions of the International Code of Zoological Nomenclature. Later von Eickstedt's entire classificatory scheme was reproduced in Saller's third edition (in three volumes) of Martin's well-known *Lehrbuch der Anthropologie* (Martin and Saller 1957). As von Eickstedt's works, as well as the text of Martin and Saller, are well known and represent part of the fundamental literature of physical anthropology, the use of *H. sapiens africanus* for some living African peoples has been readily accessible to workers in this field for decades. These facts would suggest that *africanus* is not available as a trivial name in *Homo* for the species *Australopithecus africanus*. It is irrelevant that the prior usage of *africanus* was at the subspecies level since the Code declares that a name used validly at any level within the species-group makes it unavailable for any other taxon in that genus at any level within the species-group. Valid use of *africanus* at the subspecies level therefore precludes its subsequent use in *Homo* at either the species or the subspecies level.

If *africanus* is not available in this case, the proper procedure is to select the next most senior synonym if such exists. The next most senior synonym is *transvaalensis*, stemming from Broom's *Australopithecus transvaalensis* from Sterkfontein erected in 1936. Broom subsequently erected a new genus for this species, *Plesianthropus*. This genus, along with its trivial name, was sunk into *Australopithecus africanus* as a synonym (Robinson 1954), and there is no evidence that anyone disagrees with that action. Thus *Plesianthropus transvaalensis* became *Australopithecus africanus* and would now become *Homo transvaalensis*.

As nomenclatural problems so often are, this one is complex. In 1858 Linnaeus distinguished more than one species in *Homo*, and many authors since that time have written about the classification of man. Sifting through this literature is an enormous task. I cannot categorically declare that *africanus* was not used in a species-group sense prior to its use in 1904 by Sergi. Apparently it was not, though some workers appear to believe that Blumenbach used it in a subspecies sense for African peoples in 1775 in his famous published doctoral thesis, *De Generis Humani Varietate Nativa*. This view does not appear to be correct; he did not use a Linnean type of nomenclature here or in later writings, but always referred to African peoples as the Ethiopian division of mankind. This is true also of the English translations of some of his works, including the above, published in 1865.

Assuming that the first proper nomenclatural use of the name *africanus* in *Homo* was that by Sergi, there are still two avenues to be investigated before accepting the name *Homo transvaalensis*, which at this point appears to be the correct one. The first concerns the statute of limitations built into the Code, Article 23(b), which states in principle that a name that has not been used for fifty years is to be regarded as a forgotten name and thereby becomes available for another usage. Mayr (1969), a member of the International Commission for Zoological Nomenclature, has explained the intent of this rather cryptic provision of the Code as providing a basis for preserving a commonly used name against the claim of priority of an older name that either was never much used or that has fallen into desuetude during at least the past fifty years. Manifestly the use of *africanus* is much more widespread as referring to an early fossil hominid than it is with reference to some living African peoples. But the use of the name in the latter context in a number of major texts and papers by a number of authors through much of the last fifty years eliminates the possibility of invoking Article 23(b) to take the desirable step of preserving *africanus* in the genus *Homo* to refer to "*Australopithecus*" *africanus.*

The other avenue to be investigated is the validity of the earlier usage in terms of the provisions of the Code. It may seem strange to put this consideration last since clearly the claim to priority of a name depends on its being a valid name and this point should thus be the first to be investigated. In this case it seemed evident that the claim to validity of *H. sapiens africanus* was well founded since it had been used a number of times in a Linnean manner for defined organisms by a number of authors, including one whose avowed purpose was to investigate claims to validity under the Code (Peters 1937). Pursuing this trail back to its origin, however, has led me to a conclusion different from the one that at first had seemed correct. The name appears to be valid from 1937 on from when von Eickstedt took it over from Sergi and incorporated it in a very comprehensive classification that is very different from that of Sergi. The usage is Linnean, and there is no doubt about what the name refers to. One may quibble over the fact that von Eickstedt often uses a quadrinominal nomenclature, dividing some subspecies into named subdivisions. Clearly such fourth terms have no standing under the Code. Von Eickstedt specifically quotes Sergi (1908) as the source from which he derives his use of *H. sapiens africanus.* In a paper that follows immediately on that of von Eickstedt in the same journal, Peters gives *H. sapiens africanus* as a valid name under the Code and attributes it to Sergi (1908) without involving von Eickstedt in the validation, though apparently he had had access to the latter's paper in manuscript. But when one examines Sergi's 1908 classification, it is obvious that his usage is by no means Linnean and cannot be regarded as conforming to the requirements of the Code. He has three genera, referred to as such, *Homo europaeus*, *Homo afer*, and *Homo asiaticus.* The "genus" *Homo afer* is divided into four species (his term), one of which is *Homo eurafricus.* This "species" is divided into eight varieties (his term), one of which is *Homo africanus*. From his classification it becomes clear that in this case the full name from the genus level down is *Homo afer Homo eurafricus Homo africanus*. To demonstrate that the name *Homo* is not needlessly repeated in this series and should be ignored after its first use, Melanesian pygmies are classified as *Homo afer Pygmaeus africus Pygmaeus melanesiensis*. He was evidently using the three levels, genus, species, and subspecies, but making each binominal in its own right. This usage is not in conformity with the Code, and Sergi's names therefore do not have validity. Moreover, as far as I can determine, Sergi does not use the name *sapiens* at all. Consequently, even if his names were valid, he cannot be used as the source of the name *Homo sapiens africanus*. The only other possible source of validation of this name is his 1904 work, though both Peters and von Eickstedt unequivocally quote his 1908 work as their authority.

The peculiar classificatory system used in the 1908 work does not occur in Sergi's 1904 book, which does not set out to provide a real classification. After discussing the major divisions of living man in very general terms and without illustrations, and devoting a very brief paragraph only to the African Negro, he proposes a classification that he describes as provisional. He simply lists the genus *Homo* with seven species, one of which is *H. africanus*. Here also *sapiens* is not used. The names are given without definition; in the case of *africanus* one must assume that he intended it to apply to the African Negro—this is not stated directly. This treatment, it seems to me, can hardly be regarded as sufficiently explicit to lend validity to the names used, nor can it be a source for the name *H. sapiens africanus*.

This leaves the validity, if such exists, of the name *H. sapiens africanus* as deriving from von Eickstedt's use of it; his use has page priority over that of Peters. Since their work was published in 1937 but *Australopithecus africanus* was established in 1925, the latter has priority. In the circumstances the name of the latter species when transferred to the genus *Homo* must be *H. africanus*. Thus I, and Mayr (1950) before me, was in error in using *Homo transvaalensis*.

In this work, therefore, the name *Homo africanus* will be used in place of *Australopithecus africanus*. Because the nomenclatural scheme adopted above removes justification for the vernacular term "australopithecine"—a convenient term but one that often leads to error and confusion—I will no longer use it. Where a vernacular approximation is needed "early hominids" will be used instead.

2 The *Homo africanus* Innominate Bone

Material

The known specimens of *H. africanus* innominate bone are in two collections, that of the Transvaal Museum, Pretoria, and that of the Department of Anatomy, University of the Witwatersrand, Johannesburg, both in the Republic of South Africa.

The material in the Transvaal Museum collection, identified by catalogue number, is as follows:

Sts 14 a virtually complete pelvis of an adult individual. It consists of almost the entire right innominate bone and most of the left innominate as well as almost half of the sacrum. A number of other bones of the postcranial skeleton are present. This is the most complete single individual of an early hominid known at present.

Sts 65 the greater part of the right ilium of an adult individual, with part of the pubis still attached.

Both of these specimens came from the Lower Breccia at Sterkfontein. This breccia is currently believed to be of Lower Pleistocene age.

The material in the University of Witwatersrand collection is as follows:

MLD 7 an almost complete left ilium of a juvenile individual.

MLD 8 the major portion of a right ischium of a juvenile individual.

Both of these specimens are from immature individuals in which the parts of the innominate had not yet fused. The specimens were found together and belonged to individuals of similar age; they may, therefore, have come from the same individual.

MLD 25 the major portion of a left ilium of an immature individual.

Since MLD 7 and 25 are both from the left side, two different individuals are represented. All three of these specimens are from Makapansgat.

The general features of the *H. africanus* innominate bone are well known from a number of publications based on study of the original material (e.g., Dart 1949a and b; Broom, Robinson, and Schepers 1950; Broom and Robinson 1950; Le Gros Clark 1950, 1955, and 1957; Dart 1957 and 1958).

The best known specimen from Sterkfontein is the right innominate of Sts 14. When this was first described, it was known that the block from which it had come also contained at least part of the left innominate of the same individual. This appeared to be a small piece from the neighborhood of the anterior superior iliac spine. There was also evidence of the presence of some vertebrae and the considerably damaged proximal end of a femur. The natural cast of the lateral face of the right innominate was intact in the block. For this reason, and because of the fragmentary, damaged, and delicate appearance of the remaining pieces in the block, the late Robert Broom decreed that no preparation was to be done on this block of breccia. Years later, after the death of Broom and when considerable experience with the acetic acid technique of preparation had been acquired, I had a cast made of this block and then prepared it. In it was found most of the left innominate, much of the sacrum, all of the lumbar and about two-thirds of the thoracic vertebrae, some ribs, and the proximal half or so of the left femur as well as some more fragmentary bone.

Almost the whole pelvis of this adult individual is thus available and has suffered relatively little damage. The middle portion of the right iliac crest has suffered slight crushing and displacement. The area of damage extends for some 6.5 cm along the crest and for a distance of roughly 2 cm down the blade from the crest. The region of the anterior superior iliac spine is missing. Fortunately, all of the area that is damaged or missing on the right innominate is intact and undisturbed on the left, except for some very superficial damage to the crest in the region of the anterior superior spine. The surface bone is missing over a small area of the lateral surface of the ilium just above the greater sciatic notch and extending toward the posterior superior iliac spine. Most, though not all, of the bone surface of the ischial tuberosity is missing, but the surrounding surface is intact and well preserved. The ischial ramus is present but has been distorted to some extent. The body of the pubis is intact and apparently undistorted, but the superior ramus has been damaged by being broken across not far from the acetabular margin and, as well, a part of the distal end is missing. The inferior ramus has suffered distortion also. No part of the pubic symphyseal surface is present, and the surface bone in the region of the pubic tubercle is missing. The break through the superior ramus of the pubis resulted in the part distal to the break adopting a quite unnatural angle to the body. The displaced portion was cemented into this unnatural position by matrix. Whatever caused this break also caused the ischial ramus to be broken near the tuberosity. The parts of the ischium and the pubis that lay between the two breaks were thus arranged in a quite unnatural relation to the remainder of the innominate. In the illustration of the bone published while Broom was alive, this portion of the bone is shown in this unnatural position. Subsequently I removed the cementing matrix, using acetic acid, and was thus able to restore the displaced portion approximately to its original position. This could not be done with complete accuracy because of a little warping of the part concerned by the forces that caused the breaks. Many of the illustrations now in the literature thus give a wrong impression of the lower end of this innominate bone.

The left innominate has suffered more damage than has the right, but fortunately in such a way that what is damaged on it is in almost every case intact on the right innominate. This is also the case with the pubis: the superior ramus is complete except for a small patch of bone that is missing from precisely the region of the pubic tubercle, but the symphyseal surface is undisturbed and well preserved and much of the inferior ramus is intact. Unfortunately, the left ischium, like the right, has the surface bone missing from the tuberosity. The acetabulum has suffered some damage but has been restored from the intact right side. A band of bone from the region of the anterior inferior iliac spine to the posterior part of the iliac crest is missing but has been restored from the intact right side. The anterior iliac spine itself is missing but has also been restored from the intact right side.

The sacrum is represented by the left halves of the first two sacral vertebral elements: actually slightly more than half of each is present, therefore the midline is present; hence the right halves can be restored satisfactorily from the left.

With the exception of the inferior portion of the sacrum, the entire pelvis is thus known and can be assembled into almost its exact original condition. The inferior distorted portion of the right innominate precludes a proper fit at the pubic symphysis.

Sts 65 is a specimen that has not been described hitherto. It consists of the greater portion of the ilium of a right innominate bone of which much of the crest is missing. The posterior superior spine is present, and a part of the crest itself has been detached and is present though there is no satisfactory contact between it and the rest of the specimen. The region of the posterior superior spine has been displaced slightly. Not much depth of bone is missing anywhere along the region of the mostly missing crest. The whole of the auricular surface is present. Much of the anterior inferior spine is missing, and only a small portion of the iliac part of the acetabulum is present. Part of the body of the pubis is present, but damaged, and part of the superior ramus is present in a reasonable state of preservation. The specimen is generally well preserved but evidently suffered a fair amount of weathering before fossilization. This fact suggests that the specimen had been lying on the surface outside the cavern for a substantial period of time before it fell in and became incorporated in the deposit. As a consequence many fine cracks are present, and these are especially instructive since the split-line pattern of the bone is thereby admirably displayed.

The known pelvic material listed above demonstrates that the adult *H. africanus* innominate bone resembles the modern human innominate in its general features. Compared to the pongid innominate, as seen in living pongids, that of *H. africanus* is short and broad with an iliac blade that projects so far behind the lower half of the innominate that a strongly developed, deep greater sciatic notch is formed. Such a well-developed greater sciatic notch is invariably present in modern man while in pongids it is either absent or at best poorly developed because the iliac blade is never strongly produced in a posterior direction. This is true even of the gorilla, which has the broadest iliac blade of any of the pongids; even here the greater sciatic notch is poorly developed and quite unlike that in man and in *H. africanus.*

The acetabulum is very similar in its characteristics in pongids and in hominids. *H. sapiens* and the gorilla have similar-sized, proportionately large acetabula; the chimpanzee, orang, and Sts 14 have proportionately smaller, similar-sized acetabula, and the gibbon has proportionately and absolutely the smallest acetabulum.

Anatomically, the ischium in *H. sapiens* and the pongids differs primarily in that the ischial tuberosity in the former covers a proportionately larger area of the ischium than it does in the latter. *H. africanus* resembles the pongids more than man in this respect. With regard to ischium length (i.e., functional length from the center of the acetabulum to the distal end of the tuberosity), the pongids have proportionately long ischia, while that of modern man is proportionately somewhat shorter. Sts 14, however, has an ischium that is even shorter than that of modern man.

On direct comparison of actual specimens of innominate bones of man and of the various pongids, it is obvious that the pongid innominates are very similar in their general characteristics and are very different from that of man. There is no difficulty distinguishing the two groups and no instances will be found of specimens that raise doubts as to the proper category in which to place them. Such direct comparison also demonstrates that the innominate of *H. africanus* cannot be confused with that of any living pongid; to the naked eye it manifestly and without doubt sorts with modern man rather than with the pongid group.

Comparative Features

The statement above, that naked-eye inspection allows immediate differentiation between pongid and hominid innominates, is less easy to document than appears probable at first sight. The reason for this, curiously enough, is that the differences between pongid and hominid innominates are extensive and far-reaching so that there is practically no unchanged feature that can act as a standard of comparison. Even where an obvious similarity appears to be present, it may in fact be spurious. This is the case with iliac breadth: at first sight it seems that pongid innominates are mostly narrow and the iliac of modern man is broad, but the gap between the others is bridged by the gorilla innominate. The apparent similarity between gorilla and modern man is spurious, however, since the proportionately great iliac breadth is reached quite differently in the two forms. In the gorilla the expansion of the blade is primarily an expansion of the anterior portion whereas in man the expansion is primarily in the back portion of the blade. This can readily be seen in that man has a markedly developed greater sciatic notch and the gorilla has not and that the gorilla has a strongly protuberant anterior superior iliac spine region whereas man has not.

When ratios are used for comparison—that is, where variation in one dimension is being measured against the standard of another dimension—similar ratios may cloak a considerable amount of anatomical and functional change if both dimensions have altered in similar ways.

In the present case it has seemed to me that an illuminating technique of analysis is the use of the logarithmic ratio diagram proposed by Simpson (1941). It allows direct comparison of size relationships of the same dimension in different taxa without having to relate the dimension to some other dimension to provide the comparison. Furthermore, plotting a series of dimensions in this manner allows direct comparison of the pattern of ratios for the taxa being compared. I have used a slight modification of this technique to illumine further the nature of the differences between pongid and hominid innominates. It consists of attempting to cancel out as much as possible differences due to size, thus throwing into relief the differences due to changes in proportion.

In the ratio diagram as typically constructed, the various dimensions of one taxon are used as the base of comparison; hence for each dimension the various taxa are compared using the chosen taxon as standard. This taxon will then appear in the ratio diagram as a straight line representing zero deviation from the standard. Another taxon that has identical proportions but is either smaller or larger than the standard will show up in the diagram as a straight line parallel to the standard (line of zero deviation), but to the left of it if smaller and to the right if larger. To discount size difference, this second line simply has to be shifted to coincide with the zero line of the standard. Taxa whose relevant proportions differ from those of the standard will appear in the diagram as a series of zigzag lines connecting the points for each dimension, which will fall at varying distances on either or both sides of the standard zero line. To cancel out the effects of size difference is not now so easy since the zigzag lines cannot be made to coincide with the zero line. Therefore a suitable dimension has to be chosen that will be used to make the lines on the diagram as nearly coincident as possible with each other and with the zero line but without in any way altering their shape.

Two criteria will assist in the choice of the most suitable dimension:

(1) it should be one in which there is the least amount of absolute size variation from one taxon to another in the group being compared, and (2) it should be as stable as possible within each of the taxa concerned—i.e., it should have as small a coefficient of variation as possible.

In the present case, acetabular width seems to meet these criteria best and has thus been employed to reduce the effects of size difference. The lines representing each taxon in the ratio diagram as already constructed are now copied onto tracing paper in such a way that the point representing acetabular width in each case coincides with the zero line but otherwise maintains the same orientation with regard to the zero line that it had in the original diagram. The lines that varied fairly widely in the diagram will now be much more compactly arranged about the zero line, with those of similar pattern approximating coincidence as they approach identity of ratio pattern. Differences between the various lines will now be reflecting differences of pattern of proportions much more obviously than on the original diagram.

Figures 4, 5, and 6 are ratio diagrams for twelve dimensions of the innominate bone of the gibbon, orang, chimpanzee, gorilla, *H. sapiens*, and *H. africanus* as represented by Sts 14. Figure 4 uses *H. sapiens* as the standard, Figure 5 uses the chimpanzee as standard, and Figure 6 uses Sts 14 as standard. The conclusions are the same regardless of which form is used as the standard of comparison. Figures 7 and 8 are modified ratio diagrams based on acetabulum width as the standard for coincidence; the former uses *H. sapiens* as zero line and the latter the chimpanzee.

Size differences are not completely eliminated in the modified type of diagram, and this can readily be demonstrated by constructing several diagrams using different dimensions for coincidence. If the selection of the dimensions used is carried out carefully according to the criteria suggested, the conclusions drawn from different diagrams agree very closely indeed.

Any of the diagrams illustrated here shows clearly that, though the pongids differ among themselves, there is a well-defined pongid pattern of variation in these twelve dimensions that is distinct from the *H. sapiens* pattern. It is also obvious that although *H. africanus* (Sts 14) has a pattern that is not identical with that of *H. sapiens*, the differences between them are minor and the differences from the pongid pattern are very much greater. The diagrams thus objectively demonstrate the validity of the subjective conclusion from naked-eye observation that the *H. africanus* innominate resembles that of *H. sapiens* very much more closely than it does that of any of the living pongids.

The modified ratio diagrams are very illuminating. Apart from the above conclusions, they clearly show the differences that exist between the various pongids and also indicate at a glance not merely in which features pongids and hominids are most alike or least alike, but also which ratios between dimensions will most clearly demonstrate difference between the two groups. For example, dimensions *c* (anterior margin of ilium length) and *a* (length of posterior margin of ilium between posterior spines) both have markedly different proportions in the two groups, but an index constructed from them would not distinguish pongids from hominids because the differences between the values for *a* and *c* in the two groups are of the same magnitude and sign. Hence two dimensions that each give differences of significance would give an index that showed no significant difference. On the other hand, it is obvious from the modified diagrams that an index using *h* (acetabulum width) and either *a* or *c* would distinguish sharply between pongids and hominids—as can be seen indeed in Figure 9.

It is worth enumerating briefly the more significant points that emerge from these ratio diagrams.

One of the somewhat unexpected conclusions is that iliac breadth is useless as an indicator of difference between pongids and hominids. This follows from the fact that the amount of variation in proportional relationship in this dimension is not very great and from the further fact that *H. sapiens* and *H. africanus* (Sts 14), whose values coincide, fall in the middle of the range of variation of the pongids in a position virtually indistinguishable from that of the chimpanzee and orang. A similar conclusion follows in the case of the diagonal dimension from the posterior inferior iliac spine to the anterior superior iliac spine (*d*), which is moderately correlated positively with iliac breadth. It is also clear that while the chimpanzee, orang, *H. sapiens*, and *H. africanus* (Sts 14) all have proportionately very similar iliac widths, the proportionate width of the gorilla ilium is relatively great and that of the gibbon relatively small.

Ilium height, on the other hand, is proportionately different in pongids and modern man. Pongids form a rather compact group having a proportionately longer ilium, that of the chimpanzee and gibbon being proportionately a little longer than that of the gorilla and orang. *H. africanus* (Sts 14) falls distinctly closer to the *H. sapiens* position than to the pongid, but at least this one individual had a proportionately longer ilium than is average for modern man although well within the standard population range of variation (3 s.d. limits) of the latter (see also Figure 10).

Two other dimensions that are moderately well correlated with ilium height are the length of the anterior margin of the ilium measured in a direct line between the anterior superior iliac spine and the nearest edge of the acetabulum at the inferior spine (*c*) and the posterior iliac margin length measured between the posterior iliac spines (*a*). In both of these there is a very marked proportional difference between *H. sapiens* and the pongids. In both cases the pongids fall in a fairly compact group a long way outside the range of variation (3 s.d. limits) for man. Of the two *c* is the more significant, probably, even though the proportional difference is actually slightly greater in the case of *a*. The reason for drawing this conclusion is that, though *a* has a reasonable coefficient of variation in the pongids (10.7–13.3), it is a very variable dimension in *H. sapiens* (24.2). This latter great variability does not appear to be due to sexual dimorphism since the coefficient of variation seems to be of this order in both sexes. Part of this variation may in fact be spurious, possibly the result of measuring error, because the two points involved are often much less well-defined in *H. sapiens* than they are in the pongids and therefore less easy to locate with accuracy. In the case of *c*, the coefficient of variation for *H. sapiens* is 11.2 and that for the great apes ranges between 7.4 and 11.3, the gibbon having a value of 14.5 (see also Fig. 9).

In both *c* and *a* the *H. africanus* (Sts 14) value falls very close to that of *H. sapiens* so that it differs from the pongids to virtually the same extent in either case as does *H. sapiens*. These two dimensions thus differentiate very clearly indeed between pongids on the one hand and *H. sapiens* and *H. africanus* on the other.

Another dimension in which there is a clear difference in proportion between pongids and modern man is the direct distance between the posterior inferior iliac spine and the ischial spine (*e*). This is not closely correlated with iliac height but is more nearly a reflection of the extent to which the posterior inferior portion of the iliac blade is bent backward and downward. The pongids vary among themselves in this respect; the gibbon has this distance proportionately the longest, thus showing the least effect of backward and downward expansion of the iliac blade, and the chimpanzee is a close second. Next comes the orang, and close to it the gorilla, with the most obvious influence of changes in the posterior inferior part of the blade. The pongids thus tend to separate into two groups in this respect, but this differentiation is not strong and there would be much overlap in the individual ranges of the taxa concerned. All of the pongids, however, have *e* proportionately very long compared to that of *H. sapiens*. Once more, *H. africanus* (Sts 14) agrees very closely indeed with *H. sapiens*, thus differing sharply from all of the pongids. The closest pongid (mean value) to modern man falls well outside the range of variation (3 s.d. limits) for the latter. Indeed, calculating confidence intervals for the means of *H. sapiens* and the nearest pongid (gorilla) at the level of $p = .001$, and plotting these on the modified diagram in the appropriate manner, indicates that the difference between the two is significant at that confidence level. In fact, at that confidence level the zone of nonoverlap between the confidence intervals is greater than the combined lengths of the two confidence intervals themselves, indicating that p is actually much smaller than .001 (see also Fig. 11).

The situation with regard to ischial length revealed by the modified ratio diagrams is also interesting. In this work ischial length is measured from the center of the acetabulum to the end of the ischial shaft. This agrees closely with ischial length as commonly used in the literature as measured from the contact point of the three innominate elements in the acetabulum, which is not quite in the center of the latter. The reasons for using the center of the acetabulum are, first, that measured in this way ischial length agrees most closely with the maximum length of the moment arm of the hamstring muscles and, second, in the adult the fusion point of the three innominate elements is usually difficult or impossible to locate with certainty. Either of these usages, which are virtually the same, differs sharply from the usage of Washburn in the context of evolutionary changes in the innominate bone and their functional significance (e.g., Washburn 1963). In this paper he writes of ischial length as being synonymous with the distance from the ischial margin of the acetabulum to the nearest edge of the ischial tuberosity; that is to say, the acetabular portion of the ischium is ignored as well as the tuberosity itself and only the shank portion between these is regarded as ischial length. It is difficult to see what functional significance this particular piece of the ischium has.

To return to relative proportions: there is less variation in relative ischial length in the forms being compared than might be expected. For example, all but the gorilla fall just within the standard population range of variation for *H. sapiens*, and even the gorilla falls only a short distance outside that range. It should be noted here that the sample of 10 used for the gorilla (in this particular instance taken from data provided by Schultz [1930]) does not appear to reflect parent population characters quite as well as do the others, and it may be that the proportionate value obtained for the gorilla is slightly too high. Schultz's sample seems to have included a number of unusually large individuals, hence the mean is probably slightly high. If this is so, the pongid means have a very compact distribution, occupying the upper end of the standard population range for *H. sapiens*. This demonstrates that there is a real difference between pongids and modern man in this respect: pongids have ischia that proportionately are longer appreciably than that of *H. sapiens*.

This proportionate difference is considerably less obtrusive than those in some of the iliac height dimensions already discussed. Nevertheless, that it is highly significant can readily be demonstrated by constructing an index using acetabular width and ischial length (Fig. 12). It is directly comparable to the ratio diagram values since in both cases the effect of body size differences is eliminated by using acetabular width as a standard of comparison. Using such an ischio-acetabular index shows that the difference between the pongid with proportionately the shortest ischium (orang) and modern man is highly significant even with $p = .001$. Moreover, it will become apparent later that evaluation of the functional significance of ischial length differences also involves considerations other than those used here. Such analysis demonstrates that ischial length differences play a very important role in major adaptive differences between pongids and modern man.

What is of great interest at this point is that the ratio diagram shows that proportionately *H. africanus* (Sts 14) had the shortest ischium of any of the forms being compared. This contrasts sharply with Washburn's already quoted statement that *H. africanus* had a long ischium. Instead the value for Sts 14 falls on the lower extreme of the *H. sapiens* standard population range and about as far below the average for man as the pongids fall above that range.

There is remarkably little variation in average relative length of the pubis in pongids; they all fall in an extremely compact group just outside the upper end of the standard population range for modern man. *H. sapiens* thus has a shorter pubis than have the pongids. In this instance *H. africanus* (Sts 14) agrees very closely with the pongids, falling just on the upper end of the standard population range for *H. sapiens*. Indeed, this is the only instance among the measurements used in which *H. africanus* diverges significantly from *H. sapiens* in the direction of the pongids (see also Fig. 13).

The length of the pubic symphysis proportionately is appreciably less in *H. sapiens* and *H. africanus* (Sts 14) than it is in the pongids, the mean values for the three great apes falling just around the upper end of the standard population range for *H. sapiens*. The gibbon, however, proportionately has a rather long symphysis as compared to that of the other pongids.

It would seem, therefore, that the significant differences or similarities in proportion between the innominate of *H. sapiens* as compared to that of the pongids are as follows:

(1) There is a reduction of relative iliac height (*m*), which is accompanied by an even greater reduction in the heights of the anterior (*c*) and posterior (*a*) margins of the iliac blade.

(2) A second significant change is the proportional reduction in the distance from the posterior inferior iliac spine to the ischial spine (*e*). This apparently simple change actually reflects a complex of changes that involve expansion of the posterior part of the iliac blade and much clearer definition of the greater sciatic notch. This expansion is followed by reduction of the amount of protrusion of the anterior

part of the iliac blade. Hence there is a proportional expansion in posterior iliac blade width that is obvious to the naked eye; in spite of this, the relative width of the whole iliac blade of modern man is within the range of variation of that of living pongids. The dimension *e* is thus especially valuable in this case since it does reflect a real difference between man and pongid that is directly related to this modification of iliac shape and proportion, which does not show up as a change in the proportional width of the ilium because of compensating changes in other parts of the ilium.

(3) There is proportional reduction to a smaller extent in pubic symphysis length in man.

(4) There has been a relatively slight reduction in proportionate length of the ischium and also of the pubis.

(5) There has been no change in the relative proportion of over-all iliac breadth or in dimensions closely correlated with it.

(6) The relative size of the acetabulum has remained especially stable. This appears from the standard ratio diagrams, not from the modified ones, which make use of this acetabular stability.

There are other anatomical changes that will emerge in the course of this work; here I am concerned to isolate the changes—or lack of change—of proportion of a major nature related to the bone as a whole and reflected in major dimensions.

It is proper to point out here that I do not believe that metrical features are the most significant in comparing forms as different as are pongids and hominids in pelvic morphology and function. Metrical features certainly must be examined and analyzed as one of the many aspects of the study of a group or groups of organisms; but similarities in size or proportion need not necessarily be of significance nor need differences indicate different functions. Understanding of an organism can be achieved only by taking into account all aspects of structure, function, and relationship to the environment, and this understanding can be as much hindered as helped by overemphasis on metrical features and too heavy reliance on the calculating machine for one's judgments. Such procedures are much more likely to be helpful where comparisons involve closely related organisms, as in comparing demes of one species or the species of one genus. Building up a broad picture of the nature of the adaptation of an organism is the most generally useful way of understanding it and its relationship to others. This is the approach that I have attempted to employ in this work.

As appears in the foregoing section, it seems incontrovertible on direct inspection that the *H. africanus* pelvis is very much more similar to that of man than to that of any pongid in which it is known. This conclusion is confirmed by metrical comparisons of the sort made above as well as later in this work. Since every other known part of the skeleton of this form supports this conclusion, I shall not be concerned to present large amounts of evidence to show that *H. africanus* is not a pongid. The view held by most students of the subject that *H. africanus* is a hominid obtains massive support from the metrical and other evidence presented in this volume. Consequently, I am more concerned here to analyze the similarities and differences between *H. africanus* and the later species of *Homo* than to make detailed comparisons between it and each of the pongids. Comparative information is indeed presented, but as a background to the comparisons between *H. africanus* and later forms of *Homo* rather than as an end in itself.

Original metrical data are presented in the Appendix and will be referred to in the text where pertinent. Many of the measurements used are not standard ones and often cannot be compared directly with published measurements taken by others. For this reason similar measurements were taken on samples of various pongids and of *H. sapiens* to allow meaningful comparisons to be made. The various dimensions used are shown in the key on p. 264.

Ilium

In pongids the iliac blade is thin over most of its extent and is either flat or more or less uniformly curved so that the gluteal surface is mildly convex and the iliac fossa concave. A thickened column extends from the main shaft of the innominate in the acetabular region, up the posterior part of the iliac blade to the posterior superior iliac spine. This column is the only marked thickening of the ilium found in all of the pongids and causes substantial enlargements at the posterior end of the iliac crest. The crest runs in a shallow curve, with lateralward convexity, or is straight, from the posterior to the anterior superior spine. The thickening at the posterior end of the crest is more pronounced on the lateral surface; consequently the latter is not uniformly convexly curved but has a slight concavity near the posterior superior spine. In the orang the iliac crest is more or less straight, while it is most convex in the gorilla.

In pongids there is also a slight thickening of the anterior margin of the iliac blade between the acetabulum and the anterior superior spine. This is usually poorly developed in all but the gorilla. In the latter, where there is considerable sexual dimorphism in body size, there may be quite substantial thickening in males. It does not coincide with or parallel the anterior iliac margin. Instead, it starts above the acetabular margin a short distance in from the edge of the ilium and trends toward the anterior margin of the latter, which it meets a little more than halfway up to the anterior superior spine. This thickened region is presumably needed for additional mechanical strengthening in the case of the gorilla males because of the large size of the iliac blade and the extent of the forward projection of the anterior superior spine region, coupled with the great bulk and muscularity of adult males.

In *H. sapiens* the ilium is more complex and differs considerably in thickness in different places. Besides the strongly developed posterior column on the gluteal surface, which is equivalent to that in the pongid ilium, there is also a well-developed thickening anteriorly. This is in the form of a somewhat ill-defined pillar extending from the acetabulum to the crest at a point about one-third of the distance from the anterior superior spine to the posterior superior spine. At this point the crest is considerably thickened to form the iliac tubercle. Mednick (1955) has shown that this acetabulo-cristal pillar is formed by thickening of the outer table only, not by a general thickening of the bone as a whole in this region. Because of this pillar the gluteal surface does not have a more or less uniformly convex curvature from one side to the other; instead, the surface is concave posteriorly and convex anteriorly, thus contributing to the sigmoid curvature of the crest when viewed from above. Furthermore, the crest is considerably thickened from a short distance posterior to the tubercle to the anterior superior spine. A second thickening of the crest, also not found in pongids, occurs where the iliac fossa meets the sacro-pelvic surface. This thickening extends medialward, not lateralward as does the tubercle. As a result of these thickenings of the crest, as well as that at the posterior end on the gluteal surface—which also occurs in pongids—the crest in man has a well-developed sigmoid curve when seen from above.

There is a great deal of difference, however, between the thickening near the anterior margin of the pongid ilium and the acetabulo-cristal pillar of man. The latter is well back from the anterior margin of the ilium and runs directly to the iliac tubercle and not at all toward the anterior margin or the anterior superior spine. The sort of poorly developed thickening near the anterior margin commonly seen in pongids is also usually present in man and occasionally even the more definite thickening seen in gorilla males may be found in man, trending diagonally toward the anterior superior spine. The human type of acetabulo-cristal pillar does not occur in any of the living pongids.

The ilia of pongids and that of *H. sapiens* thus differ very markedly from each other in these respects as well as the dimensions already discussed. The ilium of *H. africanus* closely resembles that of *H. sapiens.*

The *H. africanus* iliac crest has a fairly well-developed sigmoid curve when seen from above, and the double curvature of the gluteal plane, convex anteriorly and concave posteriorly, is also present. There is also an anteriorly placed thickening of the outer table corresponding in part to the acetabulo-cristal pillar of man. Natural breaks in the bone demonstrate clearly that this is indeed a thickening of the outer table. This thickened area, however, is not exactly the same as that commonly found in *H. sapiens*; it occupies a greater area proportionately than in the latter, and the line of greatest thickening is more anteriorly placed than it is in the latter. In *H. sapiens* the pillar or zone of greatest thickening runs up more or less in the long axis of the innominate and reaches the crest well back from the anterior superior spine in the area of the tubercle, as already noted. In *H. africanus* the pillar of greatest thickening trends from the upper acetabular margin on a gentle curve forward and upward in the direction of the anterior superior spine. This acetabulo-spinous column is well-developed and quite unmistakable. But there is another, less obvious one further back. This one corresponds in location to the one found in *H. sapiens* and is part of the relatively large general area of thickened outer table. In Sts 14 there is comparatively little differentiation into two distinct columns of thickening; in Sts 65, on the other hand, the two columns are better differentiated. The development and distribution of areas of thickened outer table is therefore more complex in *H. africanus* than is usual in *H. sapiens*, though there is an obvious similarity. This matter will be discussed in greater detail later in this work (see Figs. 19 and 20).

It is perhaps a little misleading to speak of a "pillar" or "column" in the above connection. What is involved is a moderate degree of thickening of the outer table of the bone—which is not very thick to start with—over a moderately large area of the gluteal surface. Moreover, the zone of thickening is not sharply demarcated from the thinner parts since one grades insensibly into the other. The term "buttress" will be used for these thickenings.

It is not easy to be sure about the presence or absence of an iliac tubercle in *H. africanus*. In Sts 65 the relevant part of the crest is missing. On the left innominate of Sts 14 a well-defined protuberance is present in a position similar to that occupied by the tubercle in *H. sapiens*. This protuberance is very well defined: the iliac crest suddenly thickens from 6.4 mm to 8.9 mm, an increase of 39 percent. Unfortunately, a small area of damage to the outer table makes it impossible to determine just how far back along the crest this protuberance reached. The protuberance appears to be quite natural, not pathological. It thus appears to be a genuine tubercle, though it differs from that usual in *H. sapiens* in that it protrudes rather abruptly. Unfortunately, most of the corresponding region of the iliac crest of the right innominate of this specimen is damaged and displaced sufficiently for it to be impossible to see whether a sharply jutting protuberance was also present there. Fortunately, the area of damage ends in just about the position corresponding to the small area of damage on the left side where the protuberance is present and obscures the nature of the posterior end of the tubercle. At this point on the right side the iliac crest begins to thicken (from 4.9 mm to 5.6 mm), and there is evidence of organization of the bone of the sort suggesting that much increased tendinous insertion occurred at this point. It seems very probable, therefore, that this is the posterior portion of a protuberance similar to that present on the left iliac crest. If this is true, then it seems highly probable that the structure on the left side is a true iliac tubercle.

The reason for the *H. africanus* tubercle arising too abruptly from the lateral surface of the iliac blade appears to be as follows. In *H. sapiens* the most strongly developed portion of the thickened outer table, which forms the acetabulo-cristal buttress, reaches the iliac crest just at the point where the tubercle occurs. The latter thus seems to be an extension of the pillar. This is not true in the *H. africanus* specimen in which the column of greatest thickening reaches the iliac crest well forward of the tubercle. In this case the tubercle does not blend into the most thickened portion of the pillar, even though there is thickening below it, and thus appears to protrude more abruptly (see Figs. 21 and 22).

The different placement of the column of greatest thickening in *H. africanus*, as compared to *H. sapiens*, gives the ilium a characteristic look. The column is more sharply defined in some respects than it is in *H. sapiens*—in Sts 65 it actually forms a low crest—and runs obliquely forward toward the anterior superior iliac spine. At the most prominent part of the column, the surface of the ilium changes direction sharply, forming a flat slope that passes rapidly down to the anterior margin of the ilium. The above statements apply to the column of maximum thickening; as already mentioned, however, the column of thickening in *H. africanus* is not confined to a rather narrow zone, as it is in *H. sapiens*, but spreads over a wider area with the less-prominent portion occupying the same position as the whole buttress usually does in the latter. Remnants of the oblique components of the buttress, however, still may be found in the modern human ilium.

Of the two juvenile ilia from Makapansgat described by Dart (1949a and b), one (MLD 7) does not have this clearly defined column of greatest thickening, though the general thickening of the outer table is present and the natural split-lines in the bone have a direction coinciding with that shown by the acetabulo-spinous component in the other ilia, including the juvenile specimen MLD 25 from Makapansgat. Thus MLD 7 appears to have the same sort of strengthening on the anterior part of the gluteal surface as have the other specimens, but without the extra prominence of the acetabulo-spinous component. This may be a reflection of the youth of the individual concerned, the youngest (ontogenetically) represented in the pelvic material.

In all four individuals represented by the five known ilia of *H.*

africanus, the region of the anterior superior spine differs from that found in *H. sapiens*. In the latter the distance from the anterior superior spine to the acetabular margin, that is, the anterior margin, is short compared to the distance from the acetabular center to the highest point on the iliac crest. In modern man the mean of the former dimension ($N = 40$) is 46 percent of the latter as compared to values of 66 percent for the chimpanzee and 65 percent for the orang ($N = 10$ in both). The iliac crest drops quite sharply from the highest point forward to the anterior superior spine. Seen in side view, the upper margin of the ilium is thus strongly convex as a rule, with the maximum curvature near the middle, or slightly anterior of the middle, of the distance from the posterior to the anterior superior iliac spines.

This strongly arched iliac crest is not characteristic of the *H. africanus* specimens. In these the iliac crest rises quite sharply upward from the posterior superior spine and then passes flatly across to a point close to the anterior superior spine, after which it curves very sharply down to the latter. As a result the iliac crest of *H. africanus* is not nearly so arched in the middle, when seen in side view, and is higher at either end than is the case in *H. sapiens*. The relative lengths of the anterior margin and the iliac height are similar to those in man and different from those in the pongids; in the case of Sts 14, the former dimension is 44 percent of the latter. The strongly arched iliac crest in modern man is not, therefore, the result of the iliac height being proportionately higher than in *H. africanus* but rather is primarily the result of the latter form having proportionately more iliac blade above a line connecting the highest point on the crest with the anterior superior spine. This causes the region of the anterior superior spine in *H. africanus* to have a more massive look than is the case in *H. sapiens* (see Fig. 1).

Not only is the *H. africanus* anterior superior spine region more massive than the equivalent in modern man, it is also more protuberant. This is readily apparent in visual inspection and can also be demonstrated by comparing the distance between the anterior superior spine and the posterior inferior spine (d) with the distance between the latter and the anterior inferior spine where it touches the acetabular margin (g). Using the average dimensions for a sample of 50 of *H. sapiens*, the latter dimension is 76 percent of the former, while in Sts 14 the equivalent figure is 69 percent. This percentage relationship is given by the index $g \times 100/d$ rather more accurately than by comparing mean dimensions. From Table 31 it will be seen that the mean value for this index in *H. sapiens* is 77.5 with a calculated standard population range of 60.4–94.6. The value of 69.2 for *H. africanus* (Sts 14) is therefore within the standard population range but well away from the mean and toward the lower end of the range. Since this represents a single observation, it may not necessarily be significant. Fortunately, it is possible to check this value because the other available specimens of *H. africanus* ilium indicate that the Sts 14 specimen is probably typical in this respect. The index cannot be determined accurately for the other specimens, but the morphology of the anterior part of the ilium is so similar that it is probable that the Sts 14 value lies in the middle portion of the distribution rather than toward either end. The single Sts 14 value is thus smaller than average for *H. sapiens* and probably, with the corroborative support of the other specimens, reflects a lower mean for *H. africanus*.

A similar situation is found among the great apes. The chimpanzee and orang have ilia in which the region of the anterior superior spine is not prominent, while visual inspection indicates that in the case of the gorilla this region is protuberant. The above index supports this conclusion, as shown in Figure 23. The means fall very close indeed to that of modern man—though this aspect of iliac morphology is very different in the two groups—while that of the gorilla is distinctly lower, the mean being slightly lower than the value for Sts 14 (*Pan*, 76.7; *Pongo*, 75.8; and *Gorilla*, 61.6). In fact, values equivalent to the means for the chimpanzee and orang could be expected only very rarely in gorillas on the basis of the present sample. It is important to note here that the validity of this argument depends upon comparing similarly shaped ilia. This index does not necessarily give a good idea of relative degree of protuberance of the anterior superior iliac spine in ilia that are quite differently shaped. Hence the comparisons between chimpanzee, orang, and gorilla ilia are valid, as are comparisons between *H. africanus* and *H. sapiens*; comparisons between the latter two, on the one hand, and the three pongids on the other, would be much less meaningful in this instance.

When seen from above, the *H. africanus* iliac crest does not have as strongly developed a sigmoid curvature as does *H. sapiens*. This is due primarily to the fact that in the latter the crest swings quite markedly medialward as it approaches the anterior superior iliac spine, while in *H. africanus* this deviation is less strong. Possible reasons for this difference will be discussed below (pp. 54–56).

The region of the anterior superior spine of the ilium is not only more protuberant in *H. africanus* than in *H. sapiens* but also projects more strongly in a lateral direction than does that of the latter. The pelvic basin (pelvis major) is more open at the front than is that of *H. sapiens*. The entire iliac blade is actually somewhat differently orientated in *H. africanus*, but this point can be more effectively dealt with when the pelvis as a whole is considered.

The iliac blade of the living pongids differs very considerably from those of *H. africanus* and *H. sapiens* in the above respects. In the pongids the highest point of the iliac crest is in the immediate neighborhood of the posterior superior spine and in virtually all cases the crest drops more or less regularly downward to the anterior superior spine, which is quite a bit lower down than is the posterior end of the crest.

A feature in which the *H. africanus* ilium agrees very well with that of *H. sapiens* is the presence of a well-developed anterior inferior spine. This spine is never powerfully developed in pongids and is frequently totally absent. In *H. sapiens* it is always well-developed and serves for the attachment of the straight head of rectus femoris, which is the main proximal tendon of this muscle, as well as for a stengthened portion of the ilio-femoral ligament, the Y-shaped ligament.

Evidently similar functions were served in *H. africanus* because: (a) there is very great similarity in the nature and degree of development of the spines in the two forms, and (b) in Sts 14 the right innominate has the spine so well preserved that the two distinct and dissimilar areas for attachment of the tendon and the ligament can clearly be seen.

In Sts 65 the anterior inferior spine is incomplete but enough is preserved to show that it was also large in this individual. Of the juvenile ilia from Makapansgat, MLD 7 has a well-developed spine, but it is not now possible to see how large the spine was in the other, less-complete, specimen.

The scar for attachment of the reflected head of rectus femoris is visible in Sts 14 and is especially clear in Sts 65 on the outer slope of the acetabular rim in a similar position to that of *H. sapiens.*

With regard to general features, the medial surface of the ilium is very similar in structure to that of *H. sapiens*. But there is a difference between the two with respect to the relative proportion between the auricular region and the iliac fossa. An approximate measure of the relative sizes of these two regions may be obtained by comparing the width of the sacral surface (measured in the middle of and at right angles to the long axis of the surface) with the width of the iliac fossa (measured from the anterior superior iliac spine to the nearest point on the sacral surface), as indicated in Schultz (1930: 346–47).

In modern man the auricular area occupies a substantial area of the mesial surface of the ilium. This is reflected in the index consisting of iliac fossa width × 100/sacral surface width, which has a mean value of 152.3 (N = 15) according to Schultz (1930). In *H. africanus* the sacral surface proportionately is considerably smaller than this, Sts 14 (right) having a value for the index of 268. That is to say, in modern man the iliac fossa width is 152 percent of the sacral surface width but in Sts 14 it is 268 percent of the sacral surface width. The latter figure is much more similar to that pertaining to pongids than it is to that pertaining to modern man. The figure for Sts 14 is exactly that given by Schultz for two individuals of the orang, while two chimpanzees gave a value of 304.8. The value for gorillas is 387.2 ($N = 13$) and for the gibbon 224.1 ($N = 23$). Unfortunately, these sample sizes are very small, but the general pattern is very clear in spite of that: *H. sapiens* has a relatively broad auricular area for its general body size. The range of this index overlaps in the case of the latter and the gibbon, so that even on fairly small samples the observed ranges overlap extensively. The similarities between modern man and the gibbon with regard to this index spring from different causes, however. In man the ilium has increased in width during its phylogenetic history, but the size of the auricular surface has increased to an even greater extent. In the gibbon, on the other hand, the ilium has undergone relatively little expansion in width, and the index is low primarily because the ilium is narrow. In this respect the gibbon forms a bridge between the great apes and monkeys, both cercopithecoid and ceboid, in which the ilium is relatively narrow. This is apparent from the iliac breadth index (breadth × 100/length), which usually ranges between 30 and 50 in monkeys, between 50 and 100 in pongids, and is invariably over 100 in modern man. The gibbon index usually lies somewhere in the 40–60 range, it being more monkeylike in this respect than are the great apes (Straus 1929; Schultz 1930; and Chopra 1962). The few values of the index consisting of iliac fossa width × 100/sacral surface width for monkeys given by Schultz range from about the mean for modern man (around 150) down to 2.5 for a single specimen of *Oedipomidas*.

The proportionately large iliac fossa breadth compared to auricular region breadth in *H. africanus* as compared to modern man is a consequence of two factors. Not only is the auricular area itself narrow, but also, as we have already seen, the region of the anterior superior iliac spine is relatively more protuberant in *H. africanus* than it is in *H. sapiens*. Each of these contributes to the inflated value of the index.

The relatively small size of the auricular region is found in all four individuals of *H. africanus* represented by the known pelvic material, and therefore it would appear that this is a normal characteristic (see Figs. 15 and 24).

As in the case of *H. sapiens*, there evidently was a good deal of variation in the morphology of the auricular region. In Sts 14 (right) the iliac tuberosity is rather sharp and protuberant and is separated from the auricular surface by a fairly deeply excavated hollow. The auricular surface itself has a moderately large protuberance near the middle with a tuberosity that is almost as well-defined and large as the iliac tuberosity. The auricular surface is thus almost saddle-shaped. In Sts 65 the auricular surface is almost flat with only a very slight humped effect near the middle. The region of the iliac tuberosity is a little damaged, but it seems that the tuberosity was as well-developed as it is in Sts 14. In MLD 7 the auricular surface is almost flat, as it is in Sts 65, but the iliac tuberosity was evidently more poorly developed and the space between it and the auricular surface not as deeply excavated as is the case in both Sts 14 and Sts 65. Since MLD 7 is not fully adult, however, it may be that the adult anatomy would have been closer to that of the other two (see Figs. 24 and 25).

The basic anatomy of the auricular region in *H. africanus* seems to be the same as that of *H. sapiens*, differing only in a tendency for a pronounced, spinelike iliac tubercle and somewhat higher relief of the whole area than is usually the case in the latter, as well as in the relatively small size of the auricular area compared to the ilium as a whole. Virtually the whole pelvis is present in Sts 14, including more than half of the first two sacral vertebral elements as well as fourteen vertebrae, almost all of which are reasonably complete. From this material it is clear that though the innominates are of small *H. sapiens* size, the vertebrae—including those composing the sacrum—are significantly smaller than those of *H. sapiens*. Judging from the relatively small size of the auricular area in all four individuals of *H. africanus* represented by the known innominates, this proportionate smallness of vertebral size to ilium size was a normal feature of this taxon.

Acetabulum

The morphology of the acetabulum of the *H. africanus* innominate is detailedly similar to that of *H. sapiens* except for two features. The first is that in over-all size it tends to be smaller than is usual in *H. sapiens*, both in absolute size and in proportionate size as compared to, say, the height of the ilium. In neither respect is the *H. africanus* acetabulum outside the standard population range for *H. sapiens*, but in both it falls close to the lower end of the range. This also reflects a smaller femoral head size. Thus far one proximal end of a femur is all that is known, and it belongs to the same individual as the Sts 14 pelvis. In this specimen the head itself is missing. Part of the neck and 178 mm of the shaft are preserved, and the shaft is much more slender than is the case in modern *H. sapiens* individuals who have innominates comparable in size to Sts 14. The femoral head, judged from acetabulum size, was of adequate size for the femur but small for the size of the innominate. Thus it is again apparent that the pelvis of this individual was large in proportion to the rest of the body.

The second difference concerns the margin of the acetabulum adjacent to the anterior inferior iliac spine. At this point the margin is somewhat indefinite, merging with the lower portion of the anterior inferior spine. The cause of this appears to have been the encroachment onto the acetabular margin of the area of attachment of the ilio-femoral ligament. This feature is clearly developed on both innominates of the Sts 14 pelvis and also seems to have been present on MLD 7. This specimen belonged to a juvenile, and the three elements of the innominate had not yet fused. The area of the acetabulum that is best preserved in this specimen is the one relevant in the present connection. In the cases of both Sts 65 and MLD 25, damage to the relevant area makes it impossible to judge whether these individuals also had the character described above.

The acetabulum is discussed further in the section on the pelvis as a whole.

Ischium

The ischium has the same general structure and orientation in *H. africanus* as in *H. sapiens*. The posterior border is continuous with that of the ilium below the greater sciatic notch, the two together forming a short, straight border between the curve of the greater sciatic notch and the ischial spine. The latter proportionately is slightly better developed in Sts 14 than is usual in *H. sapiens* but has the same characteristics as in the latter. Between the spine and the tuberosity of the ischium is a smooth area of bone in the lesser sciatic notch with a reasonably well-defined upper border, which is very closely similar to the *H. sapiens* equivalent. It is reasonable to conclude, therefore, that this region in *H. africanus*, as in *H. sapiens*, was covered with cartilage over which was situated a bursa separating the tendon of obturator internus from the ischium.

The ischium, however, is not detailedly similar to that of *H. sapiens* in all respects. For example, the nature and extent of the ischial tuberosity differs from that of the latter. Most unfortunately, no specimen is available in which the whole surface of the tuberosity itself is intact; therefore it is not possible to compare the morphology of the surface in these two forms. Enough of the tuberosity is available for it to be clear that its surface proportionately is considerably smaller than is that of *H. sapiens*. The area for muscular attachment does not extend so far up the ischium toward the acetabulum.

Chopra (1962) has raised a question about the ischial tuberosity of *H. africanus*; he points out that Broom and Robinson (1950) described the edge of the tuberosity as being sharp and that sharp-edged tuberosities always appear to be associated with a gluteus maximus that passes lateral to and not over it. This is not the conclusion reached by Broom and Robinson at that time. Chopra's question is a legitimate one, since indeed a sharp-edged tuberosity does imply a nonsapient arrangement of the tuberosity and gluteus maximus. And yet further preparation of the specimen and careful examination under the microscope has convinced me that Broom and I were mistaken in believing that the tuberosity on the right innominate of Sts 14 (the left one was not then known) was intact. Most of the actual surface is missing and cancellous bone is visible. Only the "small rounded knob" (Broom, Robinson, and Schepers 1950) described as being present near the upper end is in fact intact tuberosity surface. At this point the edge of the tuberosity is not sharp, as can be seen in Figure 27. The sharp edge described for most of the tuberosity therefore is a sharp edge formed because tuberosity surface bone is missing. Where the tuberosity surface is intact the edge is not sharp; hence it seems reasonable to conclude that the tuberosity as a whole did not have sharp edges. Furthermore, because of the short iliac height, the expanded and bent back posterior portion of the iliac blade and consequently low position of the sacrum in relation to the acetabulum, as well as the known position of the gluteal tuberosity on the femur of this individual, it seems unlikely that gluteus maximus could have been arranged in any manner other than passing over the top of the ischial tuberosity. The position of the sacroiliac and femoral attachments of this muscle in relation to the tuberosity seem to leave no other alternative.

As has already been noted, where the ischium is referred to in the literature, the statement is usually made that "the australopithecines" have a long ischium (e.g., Washburn 1963; Napier 1964 and 1967). What is usually being referred to in such cases is the relatively great length of the ischial shank between the acetabulum and the nearest edge of the tuberosity. No comparative dimensions are given to substantiate the view that "the australopithecine" ischium is long. It is an error to speak of "the australopithecines" here without distinction since the proportionate lengths of the ischia of *H. africanus* and *Paranthropus* appear to be quite different on present evidence. The statement is erroneous for a further reason when applied to *H. africanus* because the ischium of the latter is not merely shorter than that of *Paranthropus* but is short even when compared to that of modern man.

Before ischial length can be discussed meaningfully it is necessary to decide on a definition. It has already been indicated that the definition used in this work is in fact the maximum length of the ischium measured from the center of the acetabulum to the end of the shaft measured on a straight line passing down the center of the shaft from the center of the acetabulum. This represents the nearest possible approximation to the maximum length of the moment arm of the hamstrings that is available from the bare innominate bone. This would seem to be the most reasonable criterion of ischial length since the most obvious functional effect of variation in ischial length is to alter the mechanical advantage of the hamstrings. Other measures, such as the shank length used by Washburn, are clearly much less directly related to functional aspects of the ischium. For example, the distance between the tuberosity and the acetabulum is only a part of the moment arm length, and spreading or contracting the nearer margin of the tuberosity toward or away from the acetabulum without altering the total length of the ischium would not alter the maximum mechanical advantage possessed by the hamstrings. Variations in the distance between acetabulum and tuberosity could be taken as an index of functional ischial length only if the lengths of the acetabular and tuberous portions of the ischium bore a constant proportion to the length of the shank in between them in all individuals of one taxon and also from one higher primate taxon to another. No one has presented evidence that this is true, and simple inspection demonstrates effectively that it is not true.

Using the maximum length of the ischium, then, it transpires that the absolute length of the *H. africanus* ischium is short compared to that of *H. sapiens* and the gorilla, of about the same length as the average for the chimpanzee and orang, and very long compared to that of the gibbon (see Tables 21–26 and Fig. 28). From the modified ratio diagrams (Figs. 7 and 8), however, it is evident that there is little proportional difference in ischial length among the pongids, that all pongids have relatively long ischia compared to *H. sapiens*, and that *H. africanus* (Sts 14) has an appreciably shorter ischium than the average for *H. sapiens*. Indeed, the Sts 14 value coincides with the lower limit of the standard population range (mean ± 3 s.d.) for *H. sapiens*. This evidence clearly indicates a proportionately short ischium for *H. africanus*, even though based on a single individual. The only corroborative evidence so far available is that from the juvenile ischium of MLD 8 from Makapansgat. All that can be said about it in this respect is that it is consistent with an adult ischium of the sort found in Sts 14—but this does not mean a great deal.

It would seem to me that another, perhaps more illuminating, way to judge relative ischium length is to compare it with femur length; that is to say, compare the length of the moment arm of the hamstrings with the length of the lever that is moved by them. It would actually be more proper to use the distance from the acetabulum to the ground rather than femur length (see Smith and Savage 1956), but there is no way of accurately determining this distance. This comparison is even more directly related to function and the maximum mechanical advantage of the hamstrings than is ischium length alone. Using mean dimensions, the following values are obtained of femur length as a percentage of the length of the ischium:

Man	551♂ 567♀*	Orang	339♂	Gibbon	504♂
Chimpanzee	366♂	Gorilla	308♂	Sts 14	684♂

* (Australian Aborigine: data from Davivongs 1963a and b.)

The value given above for Sts 14 was obtained by using what appears to me to be the most probable estimate of the original length of what evidently was more than half of a femur of Sts 14 itself. The evidence on which the estimate is based is dealt with at some length in the section describing the femur.

Assuming for the moment that the original length was no more than that of the preserved piece, then its percentage length of the ischium length of that individual would be 440. This is well above the mean values for all of the pongids except the gibbon. Since the preserved piece of femur is broken at a point before the expansion leading to the distal condylar region had begun to manifest itself, it is obvious that the femur in fact must have been longer appreciably than the preserved piece; hence a value of 440 is distinctly too low.

Alternately, one may calculate what the femur length should have been to give a value equal to the above mean for *H. sapiens* ♀: a figure of 255.2 mm results. This compares with 198 mm for the length of the preserved piece. It is totally impossible to complete the Sts 14 femur, so that it has the anatomical features of the known distal ends in two specimens from the Sterkfontein Lower Breccia, by adding 57 mm only to the length. It would appear necessary to add at least twice that distance onto the Sts 14 specimen to complete it in conformity with the known femoral distal end anatomy of this form. Clearly, therefore, the index value for Sts 14 must have been considerably higher than the average for *H. sapiens.*

These facts argue in favor of a proportionately long femur in Sts 14. In other words, compared to femur length the ischium is relatively short. Hence, on these grounds the same conclusion is reached as we reached on the basis of ischium length alone; namely, that proportionately *H. africanus* (Sts 14) has the shortest ischium of all the forms here being considered.

The relative shortness of the gibbon ischium compared to femur length is of interest. At first sight it may appear out of character in this respect, but this is not so. The gibbon is a highly specialized brachiator, and in this form of locomotion the lower limbs are used relatively little. When not brachiating the gibbon will often be using bipedal posture in moving along the upper surface of larger branches, in which case it does not need a lot of power from the hamstrings so much as capacity to move the femur rapidly. Speed of movement at the end of the lever is favored by a relatively short moment arm and a relatively long lever. Specialization for power use of the hamstrings—as in the heavier quadrupedal climbers—is associated with a relatively long moment arm and a relatively short lever. Man, *H. africanus*, and the gibbon tend in one direction (speed), and the others tend in the opposite direction (power).

One might also argue that to some extent the action of the hamstrings relates to weight bearing or, at any rate, to the weight of the trunk that has to be supported and moved. Some dimension representative of trunk weight could be used as a means of estimating relative length of the ischium. For this purpose I selected the maximum transverse diameter of the first sacral body (centrum) as having a relationship to the weight of the trunk supported by the pelvis. Using an index consisting of ischium length × 100/transverse diameter of first sacral body gives the following values, using mean dimensions:

Man	183	Gorilla	265
Chimpanzee	215	Sts 14	167
Orang	210		

This again indicates that *H. africanus* (Sts 14) proportionately has a short ischium, that that of modern man is longer, and that those of the great apes are longer still.

In view of the above discussion, it would seem that from any meaningful point of view *H. africanus*, as represented by Sts 14, had a relatively very short ischium. This analysis also suggests that it is misleading to concentrate on the distance between the acetabular margin and the edge of the tuberosity as reflecting ischial length. By doing so one overlooks a significant part of the functional role and evolutionary dynamics of the ischium. In *H. sapiens*, as compared to *H. africanus*, the functional length of the ischium has increased by a relatively small amount, but the area occupied by hamstring attachment has increased by a greater amount, thus leaving short distance between the two.

It is important to note, however, that the shift in *H. sapiens*, as compared to *H. africanus*, appears to involve an expanded area of attachment of the hamstrings and not a shift of the whole area of attachment toward the acetabulum. This does not affect the maximum length of the moment arm, which is actually longer in *H. sapiens* than in *H. africanus*, as has already been demonstrated. It is quite possible that no more is involved than an increase in the absolute size of the hamstrings, and thus of their area of attachment, resulting from a proportionate increase in the size of the thighs and rest of the trunk as compared to the pelvis.

The ischial ramus in Sts 14 is similar to that in *H. sapiens* except that it appears to have been more slender, relatively, near its origin from the body and proportionately more robust where it becomes continuous with the pubis. The ramus, however, is missing in Sts 14 on the left side and is damaged on the right side; hence all the details of its morphology are not clear.

The entire ischium is missing in Sts 65, as it is also in MLD 25. In MLD 7 much of the ischium and its ramus is present, but since the specimen was immature at death it does not give much information other than to confirm the main features seen in Sts 14.

Pubis

The pubis is almost complete on the left side of Sts 14 and has suffered a small amount of distortion only. Most of the right pubis is also present but has been damaged to a considerable extent. In Sts 65 much of the superior ramus of the pubis is present but is not very well preserved. Neither of the two immature innominate bones from Makapansgat has any part of the pubis preserved.

The absolute length of the pubis of *H. africanus* (Sts 14) is a little less than the average length for *H. sapiens* and much greater than that of the gibbon. The average pubic length for the chimpanzee is very slightly greater than that for *H. sapiens*, that for the orang is even longer, and that for the gorilla is the longest (see Tables 21–26 and Fig. 29). The modified ratio diagrams (Figs. 7 and 8), however, indicate that relative pubis length is almost identical in all four pongids and that of Sts 14 agrees very closely indeed with the latter so that they form a single compact group on the diagram. All of these have relatively long pubes compared to *H. sapiens* and fall just about on the extreme of the standard population range for the latter. The difference therefore is not especially great, and there is considerable overlap in all of the standard population ranges. Of the twelve dimensions used in the ratio diagrams, however, pubis length is the only one in which *H. africanus* (Sts 14) agrees more closely with the pongids than it does with *H. sapiens*.

In Sts 14 the body of the left pubis is well preserved and relatively large. There evidently has been a slight distortion of the body by pressure near the margin of the obturator foramen, and a small area of the surface in the region of the pubic tubercle is missing. Because of the latter damage it is not now possible to see whether a pubic tubercle was present. The entire symphyseal surface is present and measures 28.4 mm by 12.1 mm. The width of the body from the symphyseal surface to the obturator margin is 27.2 mm. This relatively large pubic body for an innominate that is a little small in comparison with that of *H. sapiens* suggests that this specimen belonged to a female individual. This conclusion is supported by the nature of the greater sciatic notch, which is relatively wide open. It is not known, however, whether these criteria, which apply to the innominate of *H. sapiens*, also apply to *H. africanus*. All that can be said at this stage is that the Sts 14 specimen has a series of characters that, if encountered in a specimen of *H. sapiens*, would be regarded as strong indication that the individual concerned had been female.

Except for the symphyseal area, the body is not thick: in the center of the body the total thickness is 5 mm. The superior ramus is also slender and more delicately constructed than is the case in *H. sapiens* innominate bones of roughly similar over-all size to those of Sts 14. The pubic crest is relatively well defined. Most of the inferior ramus is missing on the left side but is present on the right side, though considerably distorted and pushed out of position. This leaves some doubt as to the exact original position occupied by it and the ischial ramus. As preserved, it gives the impression that the obturator foramen was more nearly rounded, instead of roughly triangular in shape, as is usually the case in *H. sapiens*. This impression, however, seems to be due largely to the distortion, and if one uses what is present of the pubis on the left side and takes into account the nature of the buckling and displacement of the ischial ramus and pubic inferior ramus, it seems that the original position and structure was probably the same in this specimen as is usually the case in *H. sapiens*, except for the somewhat greater robustness of the latter (see Figs. 16, 17, 30, and 31).

Damage and warping of the left innominate of Sts 14 has been sufficient, however, to preclude reconstruction of the parts that are preserved into precisely the correct position. For example, the pubic symphysis is not now vertical and in the midsagittal plane, where it ought to be. Hence, in the photographs of the complete pelvis the whole pubis is a little too low in relation to the rest of the pelvis.

In Sts 65 most of the superior pubic ramus is preserved but none of the rest of the bone. What is preserved is not especially instructive, though it does confirm that the pubis of the left side of Sts 14, as reconstructed, is a little too low. Sts 65 exhibits a feature that is not obvious in Sts 14: a well-developed and quite sharp crest. It is best developed at about the point of junction between ilium and pubis and is therefore partly developed on the ilium and is arranged exactly on the line where the iliac surface of the innominate changes to the pelvic surface. It appears to correspond with what in *H. sapiens* is known as the pecten pubis but extends further round toward the auricular area in Sts 65 than is usually the case in *H. sapiens* (see Fig. 24).

After examination of Sts 65, it is possible to identify a trace of this crest on the right innominate of Sts 14—the left one has suffered damage in this area. There is no actual crest or pecten pubis, but in the area equivalent to the maximal development of the crest in Sts 65 there is an incipient ridge in the form of a more abrupt change in direction of the bone surface than is usually the case at this point in *H. sapiens*. Since the crest is so well developed in Sts 65 that it partly obstructs the inlet of the true pelvis, it is possible that this specimen belonged to a male. There is no reason whatever to suppose that this crest is a pathological development since the bone composing it is completely normal in appearance and traces of a similar structure are present in Sts 14 and in *H. sapiens.*

There are some other indications, none of them strong, that suggest that Sts 65 may have belonged to a male individual. For example, the greater sciatic notch is flexed, that is, less widely open, than is that in Sts 14; but the difference is not great. Also, the thickened column of bone on the gluteal surface of the ilium that extends in the direction of the anterior superior iliac spine is more strongly developed than is that in Sts 14. This suggests, but does not prove, greater muscularity and perhaps greater weight than in the case of Sts 14. The possibility that this feature may reflect greater weight results from the conclusion that at least part of the function of the thickened column is to withstand strain applied by the inguinal ligament when it is under tension. Such tension comes in part from support of the abdominal viscera. Thickening of the column can thus be construed as perhaps being in part due to relatively heavier abdominal viscera.

These features, taken together, give a slight preference to maleness rather than femaleness for Sts 65, though this conclusion cannot be regarded as strongly indicated. The very probably female Sts 14 innominates and the less complete innominate of Sts 65 are of very similar size. If they belonged to individuals of opposite sex and if there was a noticeable degree of sexual dimorphism in body size in *H. africanus*, one would not expect such close similarity in size. And yet only two individuals are represented and, since male and female size ranges would overlap appreciably unless dimorphism was very great, little significance can be attached to the similarity in size, especially as the diagnosis of sex in the case of Sts 65 is very uncertain. Hence the available material does not serve to convey much information with regard to sexual dimorphism in body size. Somewhat clearer information is available from the vertebral column.

The Innominate Split-line Pattern

The direction of organization of the bone usually can be seen on well-preserved specimens. Sts 65 had undergone an appreciable amount of weathering before being incorporated into the fossil deposit and, as a result, fine cracks developed over the outer surface of the specimen. These serve the same purpose as do the artificially induced splits produced in partly decalcified bone by the split-line technique: they indicate the direction of the internal organization of the bone. These directions are indicated in Figures 20 and 32.

The well-defined cracks, or natural split-lines, on the gluteal surface of Sts 65 have a number of directions. A strongly directional set of lines passes upward and forward from the region behind the posterior margin of the acetabulum. This system fans out over the fairly broad area of gluteal surface that consists of thickened outer table bone. It is at least partly divided into two systems. One of these is centered on the oblique thickened column extending up to the anterior superior iliac spine. This system coincides with the greatest thickening of the outer table. The second system extends upward in the direction of the iliac tubercle and fans out widely as it approaches the iliac crest. There is a zone of thickened outer table coinciding with this system. It is not as well-developed as the thickening associated with the first, the acetabulo-spinous, system and is not entirely separated from the latter, though there is a slight hollow separating the two. Near the crest the cracks of the acetabulo-cristal system curve over sharply to parallel the crest and are thus more or less at right angles to their direction further away from the crest.

A third clearly defined system of split-lines curves away from behind the acetabulum and passes backward in the direction of the posterior superior iliac spine. This system coincides mainly with the much thickened bone that passes round the curve of the greater sciatic notch. The area of greatest concavity on the gluteal surface, in which the bone is thinnest, does not appear to have directional orientation, and natural split-lines are here conspicuous by their absence. The posterior superior iliac spine system (acetabulo-sacral system) and the acetabulo-cristal system divide sharply from each other just below the area of thin, splitless, unorganized bone.

This pattern of split-lines is also readily discernible, even though less obvious than in the weathered Sts 65, on both innominates of Sts 14, confirming that the system that radiates upward on the anterior portion of the gluteal surface is partly divided in two. The separation between the acetabulo-cristal and acetabulo-spinous systems is here also marked by either a slight flattening of the surface or a slight concavity, as in the case of Sts 65.

The split-line pattern of the gluteal surface of the *H. sapiens* ilium is similar to, but not quite the same as, that of *H. africanus*. As Mednick (1955) has shown—and my studies confirm—in *H. sapiens* there is a thickened column of outer table bone that passes upward from the acetabular region toward the iliac crest. This is fairly narrow and passes upward directly in the long axis of the innominate bone to the region of the iliac tubercle on the crest. This corresponds exactly to the posterior portion of the broader zone of thickening found in *H. africanus*, the acetabulo-cristal system. Usually *H. sapiens* does not have a clearly defined anterior system corresponding with the more anteriorly situated acetabulo-spinous system of *H. africanus*. And yet this is not always the case, as is demonstrated most effectively by a modern innominate of *H. sapiens* in the collections of the Transvaal Museum, Pretoria. In this specimen there is a very clearly defined anterior system with a thickened ridge developed on the outer table, which trends directly toward the anterior superior spine. This column starts from the position of the scar of attachment of the reflected head of rectus femoris—which is also the position of origin of the anterior column in *H. africanus*. In such a specimen of modern *H. sapiens*, which has two divergent directions of organization and columns of outer table thickening with a slight concavity of surface between the two, the resemblance to the *H. africanus* condition is remarkably close. The main difference in this case is due to the relatively reduced anterior superior iliac spine region and the greater medialward trend of this region as compared to the *H. africanus* specimen.

The modern pongid ilium has a relatively simple pattern on the gluteal surface. The main thickening, the acetabulo-sacral buttress, passes upward and backward to the area of the sacroiliac articulation. Split-lines follow this buttress and then curve over in a broad zone and meet those that pass up the anterior edge of the blade and curve over below the crest toward the back portion of the blade. Usually the anterior column is weakly developed and is far less clearly defined than is the acetabulo-sacral buttress. As already noted, however, in the gorilla the iliac blade is considerably expanded forward and in males the anterior column or buttress may be quite well developed but trends up the anterior margin of the ilium. Although this condition resembles that of the hominids a little more than does that of the other pongids, the similarity is not great. It seems probable that the thickening of the anterior margin is needed for added mechanical strength in a very large animal with so large an innominate in which the anterior margin is absolutely longer in males than it is in the other pongids.

In *H. sapiens*, *H. africanus*, and modern pongids the split-line pattern and organization of bone of the medial iliac surface (iliac fossa and pelvic surface) are more nearly similar than is the case with respect to the gluteal surface. In pongids the medial surface has the same simple pattern that is found on the gluteal surface. In *H. sapiens* there is one significant deviation from this simple pattern in that the anterior split-lines passing upward from the back of the acetabular region do not simply curve over into those that pass horizontally across the top of the blade. Instead they converge strongly, along with the horizontal upper ones that parallel the crest, toward the anterior superior iliac spine. There is a small subsidiary pattern of split-lines that branches from the main anterior set and converges on the anterior inferior iliac spine. Although Mednick (1955) discusses the patterns as though they are the same, she illustrates this difference between the pongid and *H. sapiens* medial iliac patterns, though less clearly than is usually the case on actual specimens.

The *H. africanus* ilia have the same type of anteriorly convergent pattern seen in *H. sapiens*, and it is very clearly developed. Although the appropriate areas of the innominate are not suitably preserved in all cases to allow detection of the small set of lines that converge on the anterior inferior spine, what evidence is available suggests that this was present also. The significance of these differences in structural organization will be discussed in a later chapter.

3 The *Homo africanus* Sacrum

The sacrum of *H. africanus* is represented in the known collections by a single incomplete specimen. This belongs to the same individual, Sts 14, from which came the two nearly complete innominate bones that were discussed in the previous chapter. A little more than half of the left side of each of the first and second sacral vertebrae or vertebral elements is preserved and it is therefore a comparatively simple matter to reconstruct the missing portions of these as mirror images of the preserved portions. Upon reconstruction the two innominates can be joined successfully in order to produce a virtually complete pelvis (see Figs. 33–36).

Although there is a very great deal of anatomical similarity between the sacra of the pongids and of *H. sapiens*, nevertheless there are clear differences in proportion and orientation of the sacrum in relation to the innominates and the vertebral column between the two groups.

To compare the absolute size of the sacrum, maximum breadth and length in the midsagittal plane may be used. Maximum breadth presents no difficulties, and the measurement here used is the same as that normally used in the literature. As for length (f), since the *H. africanus* specimen includes the first two sacral elements only, the first two vertebral elements were measured in the midsagittal plane on the anterior (pelvic) surface of the sacrum for samples of the three great apes as well as for modern man for purposes of comparison. The means for the chimpanzee and orang fall very close together, while that for the gorilla is appreciably larger. The mean for *H. sapiens* falls roughly midway between the two, although there is considerable overlap of both the observed ranges and the standard population ranges. The single value for *H. africanus* (Sts 14) is very low compared to any of the others, falling below the standard population ranges for all but the orang (see Fig. 37).

With regard to breadth (e), on the other hand, the mean for modern man is well above those for all three of the great apes, which fall fairly close together. The *H. africanus* value is a little above the mean for the chimpanzee, coincides with the mean for the orang, and is a little below the mean for the gorilla (see Fig. 38).

In absolute size, then, the *H. africanus* (Sts 14) sacrum is very small compared to that of any of the three great apes or that of modern man.

Comparing the length and breadth by means of the length/breadth index ($f \times 100/e$) is very instructive. This shows that the three great apes have relatively long and narrow sacra with means falling between 60 and 70, while modern man has a proportionately broad sacrum with a mean of 48. The overlap between the standard population ranges of the three great apes is very extensive, and the differences between the means of the samples used are not significant at the $p = .001$ level. On the other hand, the difference between the mean of modern man and that of the most nearly similar great ape (orang) is significant at a level where p is very much smaller than .001. The value for *H. africanus* (Sts 14) is 46.8; this is close to, but smaller than, the mean for modern man and is thus even further from the nearest pongid mean. Hence the difference between this value and the mean of the nearest great ape (orang) is also highly significant statistically (see Fig. 39).

Modern man thus has a very significantly broader sacrum on the average than has any of the great apes, and the one available *H. africanus* specimen agrees very closely indeed with the mean for modern man and falls outside the standard population ranges for all three of the great apes.

The orientation of the sacrum in the pelvis is different in pongids as compared to man. Schultz (1930: 354) has shown that if the pelvis of a pongid is so orientated that a line drawn from the center of the acetabulum to the center of the auricular surface is vertical, then the long axis of the sacrum makes a small angle only with this line. Thus the long axis of the sacrum lies almost in the long axis of the pelvis and does not make a significantly large angle with the lumbar portion of the spine. In modern man, on the other hand, the long axis of the sacrum is almost at right angles to the line joining the center of the acetabulum and that of the auricular surface. It also makes a distinct angle with the long axis of the lumbar portion of the vertebral column. The orientation of the *H. africanus* (Sts 14) sacrum is like that in modern man and differs correspondingly from that of the pongids (see Fig. 40).

Although the sacrum of *H. africanus* agrees so much more closely in the features already considered with modern man than with the pongids, it is not identical in its characteristics with that of *H. sapiens*. The difference springs largely from facts that have already been mentioned: the size of the vertebrae in Sts 14 is small for an individual with innominate bones of the size found in this individual, and the size of the bodies of the sacral vertebrae is small compared to the lateral mass. This is especially true of the first sacral vertebra. As has already been noted, the auricular surfaces of the innominates of Sts 14, like those of the other known specimens, are relatively small for the size of the innominates. Hence even though the lateral masses of the first two sacral vertebrae are relatively large for their bodies, nevertheless they are rather small compared to the innominate size in comparison with the condition in *H. sapiens*. This can readily be seen by comparing the condition seen in *H. africanus* with that in the Bush pelvis, which is commonly of roughly comparable size—in the latter the sacrum is appreciably larger. The *H. africanus* sacrum is small compared to that of *H. sapiens*, but the bodies of the sacral vertebrae are disproportionately smaller.

This can readily be seen by referring to Figure 41, a modified ratio diagram using *H. sapiens* as standard and employing the length of the first two sacral bodies as the basis for coincidence. The maximum sacral breadth (*e*) for the three great apes is proportionately small compared to that of modern man, a conclusion already reached by a different route. The Sts 14 values are virtually coincident with the corresponding ones of modern man. The other two dimensions are the maximum transverse diameter of the upper face of the first sacral vertebra (*a*) and the diameter at right angles to this (*b*). Here also the pongid values are low compared to the *H. sapiens* equivalents. The Sts 14 values fall among the pongid values, well away from *H. sapiens*. *H. africanus* therefore agrees closely with *H. sapiens* and differs noticeably from the pongids with respect to the proportionate size of the length and breadth dimensions. On the other hand it differs from *H. sapiens* and agrees closely with the pongids in the proportionate size of the dimensions of the upper articular surface of the first sacral body. These facts demonstrate again that for the over-all size of the sacrum the sacral bodies are proportionately much smaller than they are in either *H. sapiens* or the great apes. The implication is that the pelvis is proportionately large for the trunk size in comparison with either the pongids or modern man.

The auricular surface on the Sts 14 sacrum occupies most of the end of the lateral mass of the first sacral vertebra and only a part of that of the second. In *H. sapiens* it usually extends over the second and partway onto the third. The fossa on the dorsal part of the lateral mass, which serves for the attachment of the powerful sacroiliac ligament, consists of a single fossa on the lateral mass of the first sacral vertebra. In *H. sapiens* it is usually a compound fossa with a well-defined depression on the lateral mass of the first sacral as well as another, connected with the first, on the lateral mass of the second sacral vertebra.

Although all the usual criteria show that the Sts 14 individual was a fully mature adult at the time of death, both the vertebral elements now present in the sacrum are completely unfused as far as the bodies are concerned. The lateral masses are fused completely so that the portions contributed by the two vertebrae are not in any way distinguishable. The bodies, however, are not only not fused, but the gap between them is large enough (1.5 mm) to indicate that a thin cartilaginous intervertebral disc was present. Moreover, the middle portions of both the upper and lower ventral margins of the second sacral centrum, as well as the lower margin of the first sacral centrum, are hollowed with the depression extending onto the intervertebral face of the centrum. There is no indication of this phenomenon on the promontory, however. It is not clear to me what the significance is of these little hollows; they have the appearance of being areas of attachment for ligaments.

In *H. sapiens* the centra of the sacral vertebrae are usually entirely fused in the older mature adult, though traces of the intervertebral disc may remain that are not visible from outside. Complete fusion of adjacent bodies on the pelvic face of the sacrum may be delayed until relatively late, and therefore it is not uncommon to find sacra belonging to individuals who otherwise have all the skeletal evidence of full maturity that nevertheless have some incompletely fused junctions. Where this is the case, it is the junctions nearest to the lumbar region that are unfused. I myself have not seen specimens of *H. sapiens* that otherwise appear to be fully mature and yet have the degree of lack of fusion seen in *H. africanus* Sts 14. It is possible that the degree of fusion in the *H. africanus* sacrum was not as great as that in modern man. But having for study only one sacrum of this form, there is no way of knowing whether it is modal or not and, if not, on which side of the mode it belongs. In view of the fact that a complete and reasonably substantial intervertebral disc was still present between the first and second vertebrae and that there is no evidence of any greater degree of fusion in the case of the second and third, it would seem reasonable to conclude that the tendency toward fusion is greater in modern man than it was in *H. africanus* of the time period represented in this case.

In conclusion it can be said that the *H. africanus* sacrum, as represented by this incomplete specimen, is more nearly like that of modern man than it is like that of pongids, but is not identical with the former. Anatomically there is a great deal of similarity between sacra of pongids and of modern man, the more significant differences being concerned with the relationship of breadth to length. The sacrum of modern man is relatively broad and that of *H. africanus* also has this characteristic. This is an important feature since it concerns the nature of the pelvis as a whole and of the birth canal in particular. This aspect will be discussed at greater length later in this work.

The differences between the *H. africanus* and *H. sapiens* sacra seem to concern the size relationship of the pelvis to the rest of the trunk. The small cross-sectional area of the *H. africanus* lumbo-sacral articular surface suggests a lightly built trunk. This is confirmed by the small and lightly built lumbar and thoracic vertebrae as well as the small ribs. On the other hand, the pelvis is relatively large, being within the observed range, albeit near the lower extreme, in size for modern man. The need for a wide sacrum as part of the process of achieving a large birth canal in a pelvis having short innominate bones, coupled with a small and lightly built trunk, has produced the unusual *H. africanus* sacrum with its small bodies but very long lateral masses.

Because the first two sacral elements alone are present in this specimen, it is not possible to judge whether the strong concave curvature of the pelvic surface of the sacrum, seen in *H. sapiens*, was present in *H. africanus*. Pongids usually have very little or no curvature of the pelvic surface.

4 The *Homo africanus* Pelvis

In the light of the discussion so far it appears that the bony pelvis of modern man differs from that of the pongids in the following major features:

(1) The ilium is considerably shorter in proportionate length.

(2) While there is no obvious difference in the relative breadth of the ilium, that of modern man is expanded backward much more than is true in pongids. In the latter, breadth of ilium is achieved more by forward than by backward expansion. The two sorts of expansion, from the relatively narrow monkey type of ilium, can readily be distinguished because the hominid expansion results in a strongly developed greater sciatic notch, whereas the pongid type is associated with a greater sciatic notch that is poorly developed at best.

(3) The iliac blade is orientated in a more lateral position than it is in the pongids, where the blade is more nearly posterior.

(4) The iliac blade, when seen from above, has a sigmoid curvature that primarily is due to the anterior portion of the blade being bent quite markedly medialward. The iliac fossa is thus markedly concave and the gluteal surface has two distinct planes.

(5) The outer table of the iliac blade has a well-defined acetabulo-cristal buttress or thickening for added mechanical strength. This relates to erect bipedality and the necessity to maintain lateral balance during locomotion.

(6) The anterior inferior iliac spine is large and protuberant in response to the added strains placed on rectus femoris and on the ilio-femoral ligament by the regular use of erect posture.

(7) The ischium is relatively short.

(8) The pubis is relatively short.

(9) The sacrum is proportionately wide. This not only serves to increase the size of the true pelvis but also pushes the iliac blades further apart, thus assisting in the achievement of the more lateral position of the blades.

(10) The sacrum has a different orientation in the pelvis so that the anterior part of its pelvic surface is roughly horizontal when the individual is standing erect. This is approximately the orientation of the pongid sacrum when the individual is in the quadrupedal position but quite different from that in a pongid standing in an erect position.

(11) The auricular surface of the sacrum (and on the ilium, therefore) is proportionately large.

The *H. africanus* pelvis has most of these characteristics developed in the hominid fashion—either completely so or at least very much more so than is true in the pongids. Of the points listed above, 1, 2, 3, 6, 7, 9, and 10 are as much true of *H. africanus* as they are of *H. sapiens*. In the case of 4 the resemblance is much closer to *H. sapiens* since there is a sigmoid curvature of the iliac crest and two gluteal planes, but the medialward flexure of the anterior portion of the blade is less well developed in *H. africanus* than it is in *H. sapiens*. In the case of 5, *H. africanus* again resembles *H. sapiens* very much more than it does any pongid since it does have an acetabulo-cristal buttress. The situation, however, is complicated by the fact that there is also a well-developed acetabulo-spinous buttress, which is usually not present in *H. sapiens*. With regard to 7, the ischium is reduced even further in *H. africanus* than is usually the case in *H. sapiens*, though it is not outside the standard population range of the latter and, therefore, is even less like the pongids than is usual in *H. sapiens*. In the case of 8 we find one instance (of two) in this list of characters where *H. africanus* clearly has greater resemblance to pongids than to *H. sapiens*; the pubis, though not actually outside the standard population range for modern man, is relatively somewhat longer than is that of *H. sapiens* and is very similar in proportional length to that of pongids.

The other instance of similarity concerns 11; *H. africanus* has a rather small auricular surface, and this the pongids also have. This is not a simple case of similarity to the pongids; as has been made clear, the expansion in sacral width characteristic of *H. sapiens* is very clearly present in *H. africanus*. Owing to the fact that the vertebral column—and evidently the trunk in general—is very small compared to the size of the pelvis, and because the anterior superior iliac spine is relatively protuberant, the auricular area is rather small compared to the width of the ilium, even though the sacrum has expanded in width proportionately quite as much as has that of *H. sapiens* and the lateral masses proportionately are even larger than those of the latter. The similarity between *H. africanus* and pongids with respect to 11 is therefore a consequence of a marked disparity in general trunk size as compared to pelvis size in *H. africanus*, a characteristic in which it resembles neither modern man nor the pongids.

Of the more obvious pelvic differences between pongids and *H. sapiens*, in only one (pubis length) does *H. africanus* clearly fall in the pongid category, and that is not one of the more significant differences. The resemblance to *H. sapiens* is thus very detailed and the differences from pongids and other higher primates very great. It is clear that the *H. africanus* pelvis is basically hominid in structure (see Figs. 42–46).

Unfortunately, only a single, essentially complete pelvis (Sts 14) is known at present. This does not prevent us from drawing conclusions about the nature of the complete pelvis from incomplete specimens since the main characteristics of the bony pelvis are determined chiefly by the characteristics of the innominate. At first sight it might appear that the most serious shortcoming in the known material is the presence of only a part of a single sacrum. Since the specimen is broken in such a way that the whole of sacral elements 1 and 2 can be determined accurately, and since the sacroiliac joint here involves these two elements only, little is lost by the absence of the other elements of the sacrum. Furthermore, in the absence of the sacrum itself, the orientation of the sacrum in the pelvis can be judged easily from the orientation of the auricular surface on the innominate. The size of the auricular surface also gives a rough measure of the size of the sacrum. For example, it is clear from the other innominate specimens that the relatively small, though wide, sacrum of Sts 14 is not unusual but evidently was characteristic of *H. africanus* of the Lower Pleistocene in South Africa.

A more serious lack, in fact, is that three of the five innominates known have the pubis largely or entirely missing, while the other two (Sts 14) have it either quite severely damaged (right side) or warped out of position (left side). It is fortunate that in at least one adult, probably female, the whole pubis is known. Though it is out of position due to distortion, the well-preserved symphyseal area indicates clearly enough what the original orientation must have been. This is confirmed by the damaged right side where the upper portion of the superior ramus is undamaged and apparently unwarped, as well as by Sts 65, which has much of the superior ramus apparently in position. So even this lack is not especially serious as far as we are concerned in determining the features of the entire bony pelvis. Obviously, it would be very useful to have at least one more intact pubic region.

Similarly, there is no adult ischium available in which the entire surface of the tuberosity is intact and well-preserved. From Sts 14 and MLD 7 it is clear what was the extent and orientation of the tuberosity surface, but it would be instructive to have the actual surface well-preserved in order to see just where and how the hamstrings attached.

In the case of the iliac tubercle, the only intact and well-preserved example is on the left side of Sts 14. Indications from the right side of that specimen, and also from Sts 65, are that this one tubercle is quite normal. These indications consist of the nature of the crest near the position of the tubercle and the presence of a thickened column of outer table bone, the acetabulo-cristal buttress, on the gluteal surface extending up toward the position occupied by the one preserved tubercle. But in view of the somewhat unusual nature of this tubercle, it would be valuable to have at least one more well-preserved example from a different adult individual.

As has already emerged from the description of the various parts of the pelvis, the *H. africanus* pelvic morphology agrees very closely with that of *H. sapiens*. In general, therefore, it differs from the pelvises of nonhominids in the same way as does that of *H. sapiens*. Associated with the shortening of the innominates, which we have already discussed, has gone a series of changes apparently concerned with the retention of a large true pelvis.

One of the zones in which change occurred is the thickened column of bone extending from the acetabular area to the region of the sacroiliac articulation. In nonhominid higher primates the sacroiliac articulation is relatively far above, or cranialward of, the acetabular region, and the plane of the pelvic inlet thus makes a relatively small angle with the long axis of the innominate bone. As the innominate length was reduced in the course of developing the characteristics seen in the hominid pelvis, the size of the pelvic inlet was maintained by the backward expansion of the ilium, among other things. The well-developed greater sciatic notch of the hominids bears witness to this change. Because of this change the pelvic inlet makes a relatively greater angle with the long axis of the innominates in hominids. Thus the relatively wide iliac blade in hominids came about in a manner unlike that of the iliac blade of the gorilla, where most of the expansion is forward under the influence of factors quite different from those operating in the case of the hominid pelvis.

As part of the process of maintaining the size of the true pelvis the sacrum expanded in width, as we have already noted. This expansion resulted in the ilia being pushed apart and, consequently, carried further to the sides of the trunk and reorientated to be more nearly parallel to each other. Since the acetabula are orientated to face laterally and this orientation was maintained, the reorientation of the ilium altered the relationship of the plane of the iliac blade to the acetabulum.

In the modern human female—and it is in the female that the size of the true pelvis is critical—the above process is assisted by the relatively greater length of the pubis, especially the body.

In other words, there has been modification all the way around the pelvic inlet, usually involving some expansion, in order to (1) maintain a large enough true pelvis in a pelvis in which the innominates were reducing in length, and (2) increase the absolute size of the true pelvis in later stages of hominid evolution to allow for increasing head size at birth as the brain expanded in volume.

Another factor involved was the orientation of the sacrum. If, as erect posture developed, the sacrum had rotated in relation to the ilium as the latter expanded backward and downward in order to maintain the same relationship to the spinal column as it has in quadrupeds, this tendency would have avoided strong curvature of the lumbar region, especially at the promontory. It would have opposed at the same time, however, the tendency to maintain or expand the size of the true pelvis by bringing the coccyx and lower end of the sacrum closer to the pubic symphysis. Not only has this rotation of the sacrum to reduce the need for lumbar curvature not occurred, but indeed rotation in the opposite direction, to increase the size of the true pelvis, appears to have taken place (Schultz 1930), thus increasing the degree of lumbar curvature. As a result, the ventral surface of the sacrum has come to form the roof of the true pelvis, which is not the case in nonhominid primates (see Fig. 40).

In the course of all these changes from the quadrupedal higher primate type of pelvis to the hominid type, the sacrum has come to lie almost opposite the pubic symphysis. This can be seen from the fact that a line from the upper end of the pubic symphysis to the promontory makes only a small angle with the plane of the upper articular surface of the first sacral body. In pongids and in monkeys this is not true since the pubic symphysis is well below or behind the sacrum and the angle referred to is relatively large.

In all of these features the *H. africanus* pelvis does not differ from that of *H. sapiens*. As we have already noted, however, there are some differences in detail between the two, and these will now be considered.

Careful comparison indicates that if the acetabula of these two forms are orientated in exactly the same way, for the most part the iliac blades are orientated similarly; that is to say, the reorientation of the iliac blade in relation to the acetabulum, referred to above, has occurred in both to the same extent. This is not true of the anterior portion of the iliac blade to quite the same extent as it is of the rest. We have already noted that the region of the anterior superior spine has a less strong medialward flexure in *H. africanus* than in *H. sapiens*. Also, this region is more strongly developed in the former, comprising a proportionately larger part of the blade. This is readily apparent even to the naked eye if one observes that the iliac crest curves quite sharply downward to the anterior superior spine in *H. sapiens* as a rule, whereas this is not so in *H. africanus*. The anterior margin of the ilium is also a little shorter proportionately in *H. sapiens*. This suggests that the reorientation of the iliac blade was not quite complete in all respects in *H. africanus*. One may speculate on the reason for this reorientation.

The expansion of the sacrum seems to be of most importance in relation to the expansion of the pelvic inlet and thus to size of the birth canal in relation to size of the head of the newborn. As I have already suggested, in the early stages of expansion of the sacrum the purpose of the expansion may not have been to *increase* the size of the true pelvis so much as to *maintain* its size in a pelvis in which the iliac blades were reducing in height. The effect of this expansion, however, is to push the iliac blades further apart. In an animal using the erect position, the basin of the false pelvis serves an important function in assisting in the support of the abdominal viscera. This function is of small importance in a quadruped in which the iliac blades are dorsal to the abdominal viscera, not below them, for most of the time. Thus reorientation of the iliac blades in part may have been related to need for support of the viscera in an animal frequently using erect posture. Furthermore, if the sacral widening had not been accompanied by reorientation of the iliac blades, the greater width of the false pelvis would have made an awkward structure that was not well suited either to the support of the abdominal viscera or to the support by muscular action of the trunk in the erect position.

In the circumstances it is readily possible to see why the iliac blades came to have a more lateral orientation. One may ask, however, why the general region of the anterior superior iliac spine would need to bend further around, as is indicated by the difference between the *H. sapiens* pelvis as compared to that of *H. africanus*. It seems to me that this could be explained in terms of the function of the inguinal ligament, which serves as an extension of the bony pelvis anteriorly to close partially the anterior gap in the false pelvis. It is an extremely powerful ligament, an important structural element of the lower abdomen. The necessity for good support in this region is clearly shown by the frequency with which rupture occurs, both inguinal and into the femoral triangle, calling for surgical correction. The necessity for support of this sort in this region is much greater in an erect form than in an animal that is primarily quadrupedal.

In *H. africanus* the inguinal ligament would have been attached so that the direction of tension was to some extent sideways onto the anterior superior spine region, which flares outward more than is the case in *H. sapiens* and is relatively more protuberant. Even in *H. africanus* there is some medialward bending of this region. In *H. sapiens*, however, the increased degree of medialward bending brings the anterior end of the ilium around to such an extent that the inguinal ligament is more nearly a direct continuation of the direction of the anterior part of the iliac crest. This condition is more efficient mechanically since it would produce a much smaller bending force in the anterior portion of the iliac blade. Furthermore, the fairly sharp downward slope of the crest in *H. sapiens* also brings the upper surface of the crest more nearly into line with the direction of the ligament. Hence this reduced and medially more flexed anterior portion of the iliac blade would appear to form a more efficient and effective combination with the inguinal ligament as supporting structures for a creature with an erect trunk than would have been the case in *H. africanus*.

The inguinal ligament, however, is not the only structure attaching in the region of the anterior superior iliac spine. Both sartorius and tensor fasciae latae also attach in this region. The *H. sapiens* condition in this respect would seem to provide a slightly more efficient direction of operation for tensor fasciae latae in an erect form, especially in relation to the operation of gluteus maximus and its effect on the ilio-tibial band. But it would seem that the inguinal ligament is the most important of these structures in relation to medialward bending of the anterior superior spine region. This is suggested, for example, by the strong trend of split-lines on the medial surface of the ilium toward the anterior superior spine, converging strongly onto this region from the whole of the anterior half of the iliac fossa. The structural organization in the bone of the medial surface, as indicated by this well-defined split-line pattern, would seem to be due to stresses resulting from tensional strain applied to the immediate region of the anterior superior spine. The only structure that is essentially confined, in its origin, to the immediate region of the spine itself is the inguinal ligament.

Napier (1964) has suggested that the disappearance of the protuberant anterior superior spine region in hominid evolution is related to the development of lumbar curvature and a medially facing iliac fossa. Lumbar curvature, however, is already well developed in Sts 14, and the iliac fossa differs only very slightly indeed from that in modern man.

It seems to me that none of these functional improvements is of such significance that very rapid evolutionary changes are to be expected in them. But they are significant enough for the present iliac structure to be more efficient than that of *H. africanus*, and therefore selection will have favored this more detailed adjustment to the basic features of the erectly bipedal pelvic type.

The anterior fibers of gluteus medius and minimus can act as medial rotators of the femur in man. It has been argued that this function is important in erect bipedal walking and that the relatively poorly developed medialward bending of the anterior portion of the *H. africanus* ilium would preclude this function; hence this form would not have been able to walk efficiently as a biped. Even assuming the validity of the first part of this argument, it seems to me that the latter part is invalid. The relatively weak degree of medialward bending in *H. africanus* is compensated for by the greater protuberance of the anterior part of the ilium. The small degree of medialward bending plus the greater degree of forward projection would place the anterior fibers of gluteus medius and minimus in the same relation to the greater trochanter of the femur in *H. africanus* as is achieved in *H. sapiens* by a greater degree of bending but less forward extension. The same functional result seems thus to have been achieved in the two by slightly different means.

As far as I can see, therefore, it appears that the difference in structure and orientation of the anterior part of the ilium in *H. africanus* as compared to *H. sapiens* does not reflect functional differences. The latter appears merely to represent a mechanically slightly more efficient version of functionally identical structures.

The difference in split-line tracts and in columns of thickened bone between the *H. africanus* and *H. sapiens* ilia anteriorly seems to be consistent with this view. The functionally important structures attaching to the region of the anterior superior spine stress it appreciably. Being mechanically less efficient in its orientation in *H. africanus*, the anterior part of the iliac blade needs relatively more support; hence the development of the anterior acetabulo-spinous buttress and the strong orientation of the split-lines toward the anterior superior spine on the iliac surface. With the reorientation of this region in *H. sapiens* the extra gluteal surface strengthening and anterior protuberance are not needed. Hence there is now only one thickened column, the acetabulo-cristal buttress, and this corresponds to the posterior one in *H. africanus*.

Mention may be made here that some authors have referred to the significance of various angles and torsions related to the pelvis. Much of this work has been only mentioned at meetings and published in abstracts related to meetings and cannot be considered here. Chopra (1962) has published some data on what he refers to as "the angle of pelvic torsion" (see also Chopra 1958), which is the angular relation between the iliac and ischio-pubic planes as measured in a pelvimeter designed by the author for that purpose. His results are set out in Figure 47, modified slightly so that standard population ranges are given rather than the 95 percent limits that he uses. Values for the gibbon are not given since the author measured two specimens only and thus did not give a value for the standard deviation. From the figure it is clear that the values for the great apes are high (including the two gibbon values) compared to that for modern man. A clear separation exists between the standard population ranges for the two groups. The value for *H. africanus* (Sts 14)—measured on a cast—is in the zone of nonoverlap between the great apes on the one hand and man on the other, though very slightly closer to the latter. Since this is a value based on a single individual, interpretation of these data is not easy. If Sts 14 represents a modal value for *H. africanus*, its position on the chart would mean something different than if it were a high value or a low value for the genus, shifting the resemblance from approximate neutrality to either greater resemblance to man or to the great apes, respectively. As Chopra points out, however, these data seem to have little if any significance in terms of locomotor function and posture since the three cercopithecoid monkey samples that he analyzed in this respect either bridge the gap between the great apes and man (*Papio*) or are distinctly more like man than they are like the great apes (*Cercopithecus* and *Macaca*). This is perhaps not surprising since the great apes are unusual in the degree to which the iliac blade has come to lie in a dorsal position. The orientation of the cercopithecoid iliac blade reflects a generalized condition common in mammals. Rather than man's ilium having reorientated into a more lateral position from a pongidlike one, it may have changed relatively little from the more generalized mammalian condition. Probably pongids independently have diverged significantly from a more monkeylike condition in this respect, having more dorsally situated ilia that lie almost in the same plane to form a flat bony plate across the lumbar region dorsal to the abdominal cavity. Only in the gorilla do they curve around to have, in part, a slightly more lateral position because of the proportionately great width of the blades. This extreme positioning of the pongid iliac blades is probably associated with the reduced length of the lumbar spine. It is improbable that hominid ancestors had a short pongidlike lumbar spine in which the number of vertebrae subsequently increased. One may speculate that the ilia of the Miocene pongids probably were orientated more nearly in the cercopithecoid fashion than are those of modern pongids.

One may note also that the measurement as devised by Chopra is not measuring quite the same thing in all cases. Since the measurement of the plane of the iliac blade depends on the position of the anterior and posterior superior spines, the plane being measured is not necessarily the same as the actual blade of the ilium. There will be close coincidence of the two in, say, the orang or the chimpanzee, although not in either the gorilla or modern man, though for different reasons. From this it is evident that the lesser degree of medialward flexion of the anterior portion of the *H. africanus* iliac blade is relevant here since it will give the innominate in this form a torsion value greater than that of man, even though most of the ilium and the rest of the innominate is similarly orientated in the two. We have seen that functional analysis of this region suggests no obvious difference in function between the two forms. Therefore, similarity of function appears to be associated with a real difference in degree of torsion as measured by Chopra. As we have seen, the conclusion that different degrees of torsion do not correlate closely with major locomotor patterns has already been demonstrated by the relationship of the values for man, pongids, and cercopithecoid monkeys. Though there is no obvious and simple relationship between orientation of the iliac blade and locomotor pattern, nevertheless it seems to me that there must be some reason for these differences, although they may not be obviously related to the function of the muscles. For example, the unusual orientation of the great ape ilium seems to be part of a pattern of alteration of the lower trunk that appears to involve an appreciable reduction in flexibility of the lower trunk. This can be seen in the upward expansion of the ilia and the reduction of the number of lumbar vertebrae so that the top of the iliac crest almost reaches the rib cage.

The wide open false pelvic basin of *H. africanus* in which the anterior superior iliac spine is protuberant and bent only a little medialward suggests that the abdomen may have been relatively large and protuberant, a feature not usual in man. This is also suggested by the fact that the iliac blades as a whole are a little less vertical than is usual in *H. sapiens*; they flare out sideways to a slightly greater extent than is usual in modern man—in whom there is a slight difference in this respect between the sexes. This is not the same feature that has been discussed in relation to the anterior portion of the iliac blade, but involves the whole ilium; that is to say, as the two iliac blades pass upward from the acetabulum, they diverge more from each other in *H. africanus* than is the case in *H. sapiens*. This provides a relatively wider false pelvic basin in the former than is the case in individuals of *H. sapiens* with innominates of comparable size. The lateralward bending of the iliac blade in *H. africanus*, which results in this outward flaring, occurs just above the acetabulum.

As has already been noted, the iliac tubercle in *H. africanus* is a little different from that in *H. sapiens*. The position of the tubercle in relation to other structures is the same in both. For example, in both the tubercle lies on a line running up from the most protuberant part of the ischial tuberosity past the posterior margin of the acetabulum up to the iliac crest.

In *H. sapiens*, however, the tubercle does not arise abruptly as a protuberant structure, but rather is a generalized thickening of the crest that blends with the thickened column, the acetabulo-cristal buttress, which reaches the crest in this position. In *H. africanus* the one preserved tubercle arises abruptly, and several factors appear to contribute to this condition. Since the anterior superior iliac spine region is more everted and elongate than in *H. sapiens*, this general region of the iliac blade is flatter and does not make so sharp a bend along a line running from the acetabulum to the region of the tubercle. This region is rendered even flatter by the fact that there is not simply one localized buttress of outer table thickening here, as in *H. sapiens*, but two that lie close together and diverge as they pass toward the iliac crest. Thus there is a broader zone of thickening in *H. africanus*, and the greater amount of thickening occurs in the anterior column, the acetabulo-spinous buttress. As a result, what in *H. sapiens* is a well-defined thickening, coinciding with the position where the gluteal plane changes direction and also the position where the tubercle occurs on the crest, is less compact in *H. africanus*; the main thickening does not coincide with the tubercle position, nor is there so sharp a change of plane of the gluteal surface.

It seems to me highly probable that if a pelvis of the *H. africanus* type underwent a slight reduction in degree of protuberance and increase in degree of medialward flexure of the anterior superior spine, the consequence would be to narrow the column or zone of thickened bone on the gluteal face to correspond with the posterior column, the acetabulo-cristal buttress, of the two now present. This would be a consequence of the improved mechanical efficiency of the anterior part of the iliac blade, especially in relation to the tensions emanating from the inguinal ligament. Hence there would be no need for the anterior, acetabulo-spinous buttress. This change would put the one column of thickening remaining directly in line with the tubercle—the second or posterior buttress is already in this position in *H. africanus*—and the gluteal surface morphology would be the same as that in *H. sapiens*. Furthermore, if the blades moved into the slightly more vertical position seen in the latter, then the direction of pull of the ilio-tibial band on the tubercle would be more nearly parallel to the gluteal face. These relatively minor adjustments, it seems to me, would produce exactly the *H. sapiens* condition.

Consequently, it appears that the changes in the area of the iliac tubercle between the *H. africanus* and *H. sapiens* stages do not concern any functional change in relation to the tubercle itself or the related area. Rather, the improved efficiency in relation to the line of tension of the inguinal ligament and associated structures on the anterior part of the ilium was responsible for modifying the region near the tubercle. The evidence for strong attachment of ligaments on the tubercle in Sts 14 is very clear. This suggests that, as in *H. sapiens*, the most powerfully developed fibers of the ilio-tibial band near its origin are those just posterior to tensor fasciae latae, many of which insert onto the femoral shaft in a common tendon with some of the fibers of gluteus maximus.

The acetabular region also differs in some minor ways from that of *H. sapiens*. In the description of the acetabulum it was pointed out that its morphology is in detail like that of *H. sapiens* with the exception that it is relatively slightly small and the margin is somewhat ill-defined at the base of the anterior inferior iliac spine.

On comparing the acetabulum size of *H. africanus* (Sts 14) with that of *H. sapiens*, the proportionately slightly smaller size is not only evident from comparative measurements but also from the fact that in the latter the upper margin coincides more or less in level with the upper edge of the greater sciatic notch and the lower margin with the level of the ischial spine, while in *H. africanus* it does not reach as far as either of these points.

The *H. sapiens* acetabulum is deeper than that of *H. africanus* because of its larger size, though presumably there is no difference in proportionate depth in view of the rather accurate nature of the ball-and-socket joint. Also, being larger, the bony walls are thicker. In *H. sapiens* the margin near the anterior inferior iliac spine reaches to the height of, or higher than, the spine, but in *H. africanus* the margin is well below the upper edge of the spine. There is a relatively greater buildup of the margin of the acetabulum in this general region in *H. sapiens*; this is the area of acetabulum most concerned with weight-bearing when the body is in the erect position. It is clear that the well-developed anterior inferior iliac spine serves as an important buttress in this area to help support the weight of the trunk on the femur head. The greater depth of the acetabulum results also in an increase in its distance from the posterior margin of the innominate passing down from the greater sciatic notch to the ischial spine.

None of these differences appear to concern functional differences between *H. africanus* and *H. sapiens*. They do appear to relate to improved mechanical efficiency with respect to weight-bearing by an increase in the absolute and relative size of the femoral head, hence also of the acetabulum. This distributes the weight over a larger bearing surface and so reduces the stress per unit area.

The most obvious pelvic differences between *H. sapiens* and *H. africanus* are, in the latter, the somewhat larger and less medially flexed anterior area of the iliac blade; the relatively small auricular region; the greater outward bending of the iliac blade relative to the lower portion of the innominate; the slightly weaker development of the acetabular margin in the region of the anterior inferior iliac spine; the usual presence of two well-developed buttresses on the anterior part of the gluteal surface of the ilium and the proportionately longer distance between the acetabular margin and the nearest edge of the ischial tuberosity even though the ischium is actually proportionately a little shorter than that of modern man.

As has already been noted in discussing the auricular area and the sacrum, the vertebral column in Sts 14 is slender compared to the size of the pelvis of the same individual. This is reflected in the relatively small auricular surface area, the proportionately small bodies of the sacral vertebrae for the size of the sacrum itself and of the iliac blades, and also in the relatively very small lumbar and thoracic vertebrae—as will be more obvious when these are described later in this work. The bodies of all of the vertebrae are small, the few available ribs of this individual are slender and indicative of a small rib cage, and the incomplete left femur is slender. All of these suggest a very lightly built, small individual. The pelvis, on the contrary, is more nearly the size of that of modern *H. sapiens*, that is, about the size of that of a small individual. Certainly it can be matched easily among pygmies.

This would suggest that the adult female of Sts 14 had a relatively wide and substantial pelvis for her body size. The outwardly flexed iliac blades that have well-developed and laterally flaring anterior regions suggest a well-developed and protuberant abdomen. The protuberant abdomen was probably also characteristic of males since all of the innominate bones known at present have the above iliac characteristics, and it is unlikely that only female pelvic remains have been preserved. Therefore the adult individual of *H. africanus* was evidently "pot-bellied" and, at least in the females, with rather wide hips. This feature of a protuberant abdomen is common in pongids but is not characteristic of modern man in good physical condition. It will have disappeared later in human evolution as the false pelvis was modified from the *H. africanus* condition to that of modern man.

5 Functional Analysis of the *Homo africanus* Pelvis

Some authors have argued that the pelvic remains of early hominids indicate that these forms were fully and habitually erect-walking (e.g., Broom and Robinson, Dart, and Le Gros Clark). Other authors have argued that there is evidence that this was not the case, that, at best, the level of adaptation to erect posture was incomplete and locomotion in this position was not efficient. This point of view is represented by, among others, Washburn and Mednick. Mednick has discussed the matter at some length (1955), and this work will be used as a basis for examining the thesis that early hominids were, at best, no more than partly adapted to erect posture and that their use of this locomotor habit must have differed a great deal from that of *H. sapiens.*

I have warned that error is an almost inevitable consequence of treating all of the early hominids as though they had the same characteristics, as Mednick does. In this discussion attention will be confined to *H. africanus* since the present study has shown that *Paranthropus* differs considerably from the former with regard to posture and locomotion. Therefore *Paranthropus* will be dealt with in a later chapter.

The basis of Mednick's argument is that some of the features of the gluteal surface of the *H. sapiens* ilium, such as the well-developed iliac tubercle and associated buttress of thickened bone, the consequences of the action of gluteus medius and minimus in balancing the body in the erect position while walking, are absent in early hominids. She correctly believes that the stabilizing effect of these two muscles on the pelvis in preventing the slumping of the pelvis to the unsupported side during walking is an important part of efficient erect-walking. Reynolds (1931) long ago pointed out the very great significance of accurate control of lateral balance for posture and locomotion in man (see also Waterman 1929). The conclusion that the absence of evidence of this stabilizing action in *H. africanus* must indicate relatively inefficient erect-walking is reasonable if it is true that such evidence is indeed absent. From the account already given of the relevant material, however, it is evident that the conclusion that such evidence is absent is incorrect—the error presumably a consequence of the fact, which Mednick makes clear, that she did not work from the original specimens. As we have seen, the specialized strengthening of the gluteal surface is present in *H. africanus* as well, apparently, as a well-defined iliac tubercle. The definition of two gluteal planes and the associated sigmoid curvature of the iliac crest, which Mednick believed to be absent also in the "australopithecines," are in fact present in *H. africanus.* As has already been noted, the latter are not quite so well developed in *H. africanus* as they are in *H. sapiens*, owing to the greater outward flaring of the anterior part of the ilium in the former.

It is not obvious whether the thickened buttress running up to the area of the iliac tubercle in *H. sapiens*, which does not occur in the pongids (as

Mednick points out), is due primarily to forces exerted by the ilio-tibial band or to the action of gluteal muscles. Possibly the enlargement of the iliac crest into a tubercle and the formation of the acetabulo-cristal buttress have different causations and need not necessarily be as closely associated as they are in *H. sapiens*. Certainly the condition in *H. africanus* is more complex than is usual in *H. sapiens* since there are two buttresses on the anterior part of the gluteal surface. The posterior one, the acetabulo-cristal buttress, is closely associated with the tubercle but the acetabulo-spinous buttress is not.

It would seem clear that in *H. africanus* this complex was not due to a single factor such as forces exerted by the gluteal muscles. As we have seen, an important factor in the case of the acetabulo-spinous buttress appears to have been tension exerted by the inguinal ligament on the end of the relatively long anterior superior iliac spine region of the ilium. This pull was reinforced by that of sartorius, a flexor of the thigh at the hip and a medial rotator of the thigh, which attaches just below the attachment of the inguinal ligament. But these were not the only influences responsible for the anterior buttress. It is likely that the relatively long and large anterior region of the ilium also was stressed by the weight of abdominal viscera, by the pull of lateral trunk muscles, and by the pull of anterior fibers mainly of gluteus minimus but also of medius.

For reasons already discussed, it seems probable that the reorientation of the anterior portion of the iliac blade was most directly influenced by the inguinal ligament. This would appear to have been the primary, but not the sole, factor responsible for the anterior and better-developed buttress.

The second, posterior, and less-well-developed acetabulo-cristal buttress is associated with the iliac tubercle in that the position of what appears to be a true iliac tubercle is directly above the end of the acetabulo-cristal buttress. As we have noted, however, the buttress does not expand into the tubercle in so obvious a manner as is usual in *H. sapiens*, though even here there is a good deal of variation. It is probable that the chief factor responsible for the thickening of the iliac crest in the form of the tubercle is the concentration of powerfully developed tendinous fibers of the ilio-tibial band that attach in this region. These run between tensor fasciae latae and the anterior margin of gluteus maximus, and some of them join fibers from the latter and insert onto the gluteal tuberosity of the femur. These fibers attaching to the tubercle thus come heavily under the influence of the action of the above two muscles and constitute the most powerful attachment of the ilio-tibial tract other than through these muscles. That there is important stressing of the ilio-tibial tract in erect-walking is suggested by the very fact that it is so powerfully developed as a specialization of the thigh fascia in man but is at best very poorly developed in other primates. Pauwels (1948) has demonstrated that this tract sharply reduces the stresses in the femur in man (see also Alexander 1968). The pull of the ilio-tibial tract may also contribute to the need for an acetabulo-cristal buttress. It seems more likely, however, that the continual contraction and relaxation of gluteus minimus and the anterior fibers of medius during walking contributes most strongly to the need for strengthening of the ilium in this region.

The posterior, acetabulo-sacral buttress is rather clearly associated in early hominids, *H. sapiens*, and pongids with weight-bearing. This is the obvious route in the ilium for the generation of the major stresses resulting from the support of upper trunk weight at the sacroiliac joints and trunk weight as a whole by the lower limbs. The line of maximum stress in the ilium is thus between the auricular surface and the upper part of the acetabulum. The function of the posterior buttress is thus the same in pongids and in hominids, although the morphology of the region is more or less identical in *H. africanus* and *H. sapiens* and rather different in pongids for reasons that do not directly concern the function of weight-bearing.

It seems clear, therefore, that all of the structural elements that Mednick identifies in the *H. sapiens* ilium as being products of functions directly associated with erect bipedalism are also present in *H. africanus*.

As we have seen, however, there are some differences between the condition seen in *H. africanus* and that in *H. sapiens*, and the question is whether these differences reflect differences of function. The relatively minor differences appear to relate directly to two factors: (1) there is a relatively long and protuberant upper anterior portion of the ilium, and (2) this region of the ilium is not recurved medialward to as great an extent in *H. africanus* as it is in *H. sapiens*.

The buildup of thickened outer table bone buttressing the anterior part of the ilium and the strong definition of the iliac tubercle suggest strongly that the powerful stabilizing action on the pelvis of the glutei and the stabilizing influence on both the pelvis and the knee of the gluteus maximus-tensor fasciae latae-ilio-tibial tract complex that are typical of erect bipeds were also characteristic of *H. africanus*. Furthermore, there appears to be no evidence that suggests that these functions were any less well developed than in modern *H. sapiens*.

This being so, the two differences referred to above—greater degree of protuberance and different orientation of the anterior portion of the iliac blade—may be interpreted in one of two ways. First, it could be that the individuals were relatively heavily built with short, stout torsos and protuberant abdomens, for which reasons the iliac blades extended further forward and the false pelvis was more open anteriorly. This does not seem a very probable explanation since the evidence at hand indicates, on the contrary, very lightly built individuals who were slender both above and below the pelvis. Second, these pelvic features might be interpreted as representing a stage in adaptation to erect bipedalism and that at a later stage these will have given way to the conditions seen in later forms of *Homo*.

In the early stages of remodeling the pelvis during adaptation to habitual erect posture, the innominates would come to lie more laterally in the pelvis as a result of the expansion of the sacrum. This, however, would not cause the anterior portion of the iliac blade to bend medialward. At this stage selection may well have favored increased length of the iliac crest both to increase the length of the zone of attachment of trunk muscles to help hold the trunk erect and to place the anterior fibers of gluteus medius and minimus in a more advantageous position to stabilize the pelvis and effect medial rotation of the thigh. The stresses in the lower front abdominal region resulting from frequent or regular use of the erect position would result in a more clearly developed inguinal ligament to provide support. The tension of this ligament on the anterior end of the iliac blade could have been the prime factor that caused the medialward bending of this region. Medialward bending would remove the necessity for the protuberance of this region. The resulting improved mechanical efficiency would reduce the need for the acetabulo-spinous buttress, and this would disappear as a separate entity from the acetabulo-cristal thickening. As we have seen, this anterior buttress still occurs occasionally in modern man—which may be an indication that formerly it was more common.

It would seem, therefore, that the fact that the anterior region of the *H. africanus* ilium had already bent medialward to some extent is an indication that the pattern of selection that brought about the *H. sapiens* condition in this region of the pelvis was already established then and was in the process of remodeling the *H. africanus* pelvis. If this interpretation is correct, then *H. africanus* was functionally no different in this respect than is *H. sapiens*. The same functions were being achieved in a slightly less efficient way. The difference between the two would represent no more than refinements of detail in the bipedal pattern rather than significant differences resulting from adaptation to a locomotor pattern different from erect bipedality or to so poor a level of adaptation to erect bipedality in the former that significant parts of the latter pattern were still missing.

There is no evidence, therefore, that supports Mednick's conclusion that the early hominids could not maintain lateral balance efficiently and were, at best, still in a transitional stage of adaptation to erectly bipedal posture. Dart (1958), in a brief discussion of the two juvenile ilia from Makapansgat, draws a similar conclusion: "In addition the natural split-lines in the fossils do not show an anthropoid pattern, as Mednick (1955) imagined possible, but a typically human pattern. These juvenile ilia show that the adolescents of *A. prometheus*, both male and female, simulate *Homo sapiens* (especially the Bush variety) in their iliac structure so closely that it would be unreasonable to imagine that bipedalism was any less distinctive of their gait than it is of living human beings."

Similarly, the medial face of the ilium shows the typical *H. sapiens* pattern and not that of pongids. The split-lines of the posterior half of the iliac fossa are much the same in pongids and in *H. sapiens*, but in the anterior half the pattern of the latter is quite different, with the major lines converging on the anterior superior iliac spine. The *H. africanus* ilia also have this pattern, the convergence evidently being caused by stresses emanating primarily from the inguinal ligament and sartorius. Since it is not possible to distinguish between the *H. africanus* and *H. sapiens* ilia structurally (except for the slightly more protuberant and less medially flexed anterior region already dealt with) on the medial side, there are no grounds for postulating differences of function here either. The auricular surface is relatively small in *H. africanus*, but this does not appear to reflect functional differences so much as imply that the pelvis was relatively more robust compared to the rest of the torso than is the case in *H. sapiens*. This conclusion is confirmed in this instance by the relatively very small lumbar and thoracic vertebrae.

The well-developed anterior inferior iliac spine of *H. africanus* agrees very closely in size and shape with that of *H. sapiens* but differs widely from the equivalent region of the nonhominid primate pelvis. This suggests that, as in *H. sapiens*, *H. africanus* had a relatively well-developed rectus femoris—and probably also quadriceps femoris as a whole—and also an ilio-femoral ligament strengthened and functioning in the manner of those of *H. sapiens*. A well-defined attachment scar indicates clearly that rectus femoris had a reflected head similar to that of *H. sapiens*. The powerful anterior thigh musculature and an ilio-femoral ligament functioning to prevent hyperextension of the pelvis on the thigh are integral parts of the adaptation to erect bipedality. At least one prosimian form is reported to have a well-developed anterior superior iliac spine. This does not have the peculiar features of the *H. africanus*, *H. sapiens*, and (as will be seen later) the *Paranthropus* spine, and its existence does not vitiate the conclusions drawn above. *Oreopithecus* also apparently had this structure well developed, but the only known pelvis is severely crushed—and this considerably obscures the picture; even if genuine, it does not appear to resemble closely the hominid spine, although Straus (1963) believes that it does.

In the description of the innominate bone morphology it was noted that the relationship of the *H. africanus* ilium to the rest of the innominate differed from that in *H. sapiens*. First, the ilium in the former is bent outward a little more than is usual in *H. sapiens*, in which the ilium is more nearly vertical. Second, the angle of torsion between the iliac plane and the ischio-pubic plane, as measured by Chopra (1962), is greater in *H. africanus*. The question is whether either of these differences reflects differences of function.

With regard to the former, the slightly greater degree of widening of the false pelvis in relation to the pelvic inlet, which makes it more basinlike, would mean a slightly increased capacity to support the abdominal viscera. The relatively wide false pelvis, coupled with the tendency for the anterior parts of the iliac blades to flare outward more than in modern man, means that there would have been less support anteriorly. This, as in the case of the gorilla, might mean that the abdomen was protuberant—especially, perhaps, in the early stages of erect posture before the abdominal wall became better adapted to coping with the stresses and strains involved. On the other hand, the relatively wide false pelvis in an otherwise lightly built and slender animal would provide a wider base from which the muscles of the lateral abdominal walls could operate to keep the trunk vertical. Furthermore, the relatively wide false pelvis brim would probably have afforded slightly better mechanical advantage for the muscles of lateral balance during walking, thus allowing slightly greater efficiency in this function, although the difference is hardly likely to have been significant. All of these points concern slight differences of efficiency in performing the same functions; there is nothing to suggest that the slight difference of orientation of the ilium involved a function different from that in *H. sapiens*.

The matter of pelvic torsion, measured by Chopra, has already been discussed. First, as Chopra himself appreciated, no clear relationship appears to exist between the degree of torsion and a particular locomotor pattern. Second, in this instance the difference between *H. africanus* and modern man simply reflects again, because of the measuring technique involved, the fact that the anterior part of the iliac blade flares outward more obviously in the former than it does in the latter. Consequently, this is not yet another difference but is another way of defining a difference that has already been dealt with.

The only other part of the *H. africanus* pelvis that is noticeably different from that of *H. sapiens*, and therefore may reflect functional differences, is the ischium. The interpretation of the ischium is, however, a complex problem.

First, there is no wholly intact adult ischium available at the present time. While both sides of the Sts 14 pelvis give a very good indication of almost all of the features of the ischium, most of the tuberosity surface is missing in each case and therefore it is not possible to be completely certain of the surface shape of the tuberosity. The evidence in this connection has already been discussed, and it appears that the apparently moderately sharp edge is an artifact due to the actual surface being gone. In one area only, on the right ischium, is the true surface present, and here the tuberosity margin is well-rounded, not sharp-edged. The tuberosity therefore seems to have had a somewhat rounded surface more nearly like that of *H. sapiens* and less like the flattened tuberosity with well-defined edge seen in the pongids.

In the description of ischial morphology, evidence was presented that demonstrates that *H. africanus* (Sts 14) has a shorter ischium, both absolutely and proportionately, than has *H. sapiens*. Contrary to general opinion, therefore, the ischium is shorter than that of modern man and considerably shorter than that of any of the pongids. This conclusion emerges from comparisons of ischial length by the ratio diagram method, as well as from comparisons of ischial length with femur length or with the transverse diameter of the articular surface of the sacrum with the last lumbar vertebra.

With respect to locomotion, an especially significant aspect of the ischium is its functional length (as defined in this work) since this approximates the maximum length of the moment arm of the hamstrings in extending the thigh at the hip joint. This dimension alone does not allow evaluation of its functional significance; at least one other dimension is necessary, and that is the total length of the lever being moved. In this case the latter distance is from the center of the acetabulum to the ground with the animal in its normal walking position. If the ratio between these two dimensions is large, then contraction of the hamstrings translates into a movement at the end of the lever that is relatively slow but powerful. If the ratio is small, on the other hand, then the movement at the end of the lever will be fast but relatively weak (Gregory 1912; Smith and Savage 1956). In pongids the ratio is rather large, not only because the moment arm is relatively long but also because the lower limb is short. For the chimpanzee the ratio is of the order of 1:5 but varies from about 1:4 to 1:5.5 during walking. These figures are misleading to some extent, however, since they are based on the lower limb length rather than on the distance from the acetabulum to the ground. The pongid lower limb is usually flexed at knee and hip at all stages of the stance phase, hence the lever length is distinctly less than the full length of the limb; thus the ratio is in practice larger than values calculated from limb length. In modern man the ischium is proportionately shorter than in the chimpanzee; as well, the lower limb is longer and, especially in walking, flexed so little that lever length and limb length are very similar. The ratio of moment arm length to lever length thus is appreciably smaller, being of the order of 1:10 but ranging up to about 1:13 during walking. In walking the hamstrings of modern man are functioning in a context in which speed of action is more emphasized than power, or alternately, energy is being conserved, whereas in the pongids there is more emphasis on power than on speed of action.

Naturally this conclusion does not take into account the degree of development of the hamstring muscles themselves, and this does affect the end result also. What one is concerned with in the above case is the architecture of the skeletal framework that is the functional setting of the hamstrings. It is obvious that the system can, and for the most part does, function at lower levels of power and speed than the possible maximum. It seems clear that the hamstrings tend to be powerfully developed in pongids for the size of the animal; in man they are much more slender but appreciably longer. This probably somewhat increases the power differential between the two. Here again there is much variation possible, especially in man where a physically active individual will have well-developed hamstrings compared to a physically inactive individual. But these are minor differences compared to those stemming from differences of skeletal proportion between man and pongid.

Moment arm length is not the same in all stages of the stance phase. The maximum length of the moment arm occurs early in the stance phase and is at its smallest in the closing phase. This means that the hamstrings can exert maximum power early in the stance phase, and the power potential drops as the stance phase progresses. This is true of pongids as well as of man, except that in pongids the moment arm length is a little below maximum if the thigh is very strongly flexed at the hip, as when climbing, so that the mechanical advantage increases slightly as extension begins and then passes the maximum and begins to fall. In normal walking and running, the beginning of the stance phase coincides more or less with the point of maximum mechanical advantage. In pongids the main, extensor, portion of gluteus maximus also has its maximum mechanical advantage early in the stance phase. In man, in contrast, the mechanical advantage of gluteus maximus is least in the early stage of the stance phase and increases as the stance phase progresses. In the pongids the lower limb extensors function in a context that emphasizes power of movement more than speed and the power is concentrated rather more strongly in the earlier part of the arc through which the lower limb can be extended. In man the mechanism is arranged to allow speed of movement at the end of the lever but with proportionately less power, and the application of power is more uniform through the arc of movement of the limb. The normal locomotor activities of man are such that the power aspect is not normally at a premium, but a long lower limb is certainly advantageous for rapid walking. In such activities as lifting the trunk against gravity from a bent forward position, in standing up from a sitting or squatting position, climbing stairs or walking up a slope, and in running and jumping, a reserve power source is available in gluteus maximus. It reinforces the action of the hamstrings in such circumstances; otherwise it is not active or only acts briefly as a source of propulsive power in standing or in locomotion. In fact, as Duchenne (1867) observed, essentially normal locomotion is possible in man even if gluteus maximus has suffered paralysis. Later workers have confirmed that only relatively minor compensatory adjustment is necessary following such paralysis. In normal individuals walking slowly it appears to act systematically only briefly in the early part of the stance phase (Eberhart, Inman, and Bresler 1954; Steindler 1955; Basmajian 1967).

Smith and Savage (1956) discuss the forelimbs of *Dasypus* (armadillo) and *Equus* to illustrate the difference between emphasis on power or on speed and demonstrate that for the main propulsive muscles the moment arm length/lever length ratio is about 1:4 in the former and 1:13 in the latter. *Dasypus* is fossorial and the forelimb emphasizes power because digging is an important function of the forelimb. The horse, on the other hand, is adapted for fleetness. The difference in locomotor habit is very marked in this case, and yet the difference in the two ratios is of the same order of magnitude as in the chimpanzee-modern man example. What might at first sight appear to be a relatively small shift in emphasis in the latter example obviously is important enough in actual fact to make the functional context of the hamstrings and gluteus maximus very different in the two. It should be noted that the variations in the ratio at different positions in the stance phase are not nearly enough to cause the difference between them to be blurred. The variation for the hamstrings is from about 1:4 to 1:5.5 in pongids as compared to 1:10 to 1:13 in modern man.

What conclusions may be drawn about *H. africanus* in this regard? We know, in the case of Sts 14, that the ischium is proportionately shorter than is that of *H. sapiens*. Unfortunately, it is not possible to be certain what the lever length was in this case. We have already seen that the minimum possible length of the femur of this individual was such that it must have been proportionately as long as that of modern man (see chap. 9). The existing piece alone, without any restoration, is proportionately longer compared to the ischium than is usual in pongids. The lower section of the Sts 14 limb is missing; therefore there is no direct means of determining total limb length. In modern man the upper segment of the lower limb is roughly the same length as the segment from the knee to the sole of the foot. If this were true of *H. africanus* Sts 14, then the ratio in this case would have been 1:14—roughly of the order found in modern man. In some higher primates, however, the lower section is shorter than the thigh length; this is true to a small extent in all of the living types of pongid. If it is assumed that this was also true of Sts 14—but in an exaggerated form so that the lower segment was only half the length of the upper segment—the ratio would nevertheless be as small as

1:11.5. This is still of the order found in man rather than in pongids; but clearly this ratio is too large since no higher primate has so great a discrepancy between the lengths of the hind limb segments above and below the knee. There seems little alternative, therefore, to the conclusion that the functional framework for the hamstrings in this individual was specialized in a manner that emphasized speed more than power to an extent comparable to that in modern man and different from that in modern pongids. We have seen that the comparatively small sample of postcranial material of *H. africanus* now known affords no reason for supposing that Sts 14 is aberrant or nonmodal. Furthermore, having regard to the nature of the character here being discussed, it is unlikely that a single population would include individuals with pongidlike short lower limbs as well as ones with manlike long lower limbs. In the circumstances it seems reasonable to suppose that *H. africanus* was characterized by relatively long lower limbs in the manner of man.

In view of the fact that the estimated ratio in this case is rather small even for modern man, and of the conclusion emerging from other evidence that this form was a plains-dwelling hunter with primitive hunting equipment, it is not beyond the realm of possibility that this low ratio means that *H. africanus* was exceptionally fleet of foot. This possibility is perhaps rendered more likely when it is recalled that the individual with this low ratio was a wide-hipped female and that males may have had even more advantageous ratios for speed. In view of the smallness of the sample, however, it is preferable to regard this suggestion as merely a possibility that should not be ignored and consider the ratio to be approximately that of modern man until more evidence is available.

There is another aspect of ischial structure, besides its functional length in relation to the length of the lever moved by the hamstrings, that is important from a functional point of view. This aspect is the orientation of the ischium in the body.

In a quadruped the sacrum more or less forms a roof to the pelvic cavity and the innominates are approximately horizontal. In pongids, because of their relatively long forelimbs, the vertebral column is not horizontal but slopes upward toward the head. Hence the innominates also slope upward toward the head of the animal. The main axis of the ischium slopes in a similar manner. When a pongid adopts erect posture temporarily, the pelvis swings upward so that the innominates are now more or less vertical and the ischial shaft points directly downward. By the time extension of the thigh at the hip joint has brought the femur into a nearly vertical position it is approximately parallel with the shaft of the ischium and the hamstrings have run out of power to extend the femur any further. For this reason the thighs of pongids are seldom extended so far that they parallel the long axis of the trunk. This, and not lack of extensor action on the part of the gluteus maximus as is commonly believed, is a prime reason for the comparatively inefficient shuffling gait of a pongid when in an erect position. This limitation is not felt when the animal is moving in the quadrupedal position since the ischium is then directed backward and slightly downward in a manner that allows the thigh to be extended at the hip joint to a point where it is well beyond the vertical (see Fig. 48).

In the case of man, adoption of the fully erect position as the usual condition has not involved rotation of the pelvis in the above manner. The portion of the innominate that includes the acetabulum and the ischio-pubic region is still in very nearly the same position in relation to the ground as the equivalent part of the pongid innominate is when it is in the normal quadrupedal position. The sacrum is still orientated in a similar fashion, with the pelvic surface of the first two or three sacral elements almost horizontal, and the ischial shaft is directed backward and downward at an angle of roughly 60° from the vertical. The well-developed lumbar curvature of modern man is a consequence of the need to have a vertical spinal column connected to a sacrum that is more nearly horizontal than vertical. Because of the angle of orientation of the ischial shaft, man does not have the problem with extension of the thigh at the hip that is experienced by pongids when walking erect. The femoral shaft of man can be extended well past the vertical before the hamstrings begin to lose capacity to extend the thigh any further. This ability to extend the thigh past the vertical, coupled with the spring action of the relatively inflexible foot under the control of the calf muscles, gives man the ability to stride and to complete a stride with power.

Furthermore, because of the fact that the center of gravity of the pongid body is well above, and forward of, a line joining the two acetabula, coupled with an inability as a rule to completely straighten the lower limb, erect walking in a pongid requires a great deal of energy. In man the reduction of the height of the ilium and the considerable increase in size and weight of the lower limbs have reduced the center of gravity to about the level of a line joining the acetabula. Also, the bending back of the ilium and the nature of the curvature of the spine have moved the center of gravity backward so that it is roughly on a line between the acetabula. As Joseph (1960) has shown, this combination of factors has resulted in a condition in which very little expenditure of muscular energy is needed to maintain erect posture.

It would thus seem that the much greater efficiency of erect-walking in man, as compared to the pongids, is attributable in great measure to the spatial relationship of the fulcrum of the lower limb, the origin and insertion of the hamstring muscles, and the height of the fulcrum above the ground, as well as the orientation of the ischial shaft in relation to the horizontal. No significant changes in muscle function are involved: gluteus maximus is primarily an extensor of the thigh at the hip in both pongids and man, and the action of the hamstrings (including the hamstring portion of adductor magnus) is in principle the same in pongids and in man. A crucial element in the mechanism of erect-walking in man is the fact that erect posture did not involve upward rotation of the pelvis so that the spinal column came to be vertical as occurs in an erect-walking pongid. Instead, there is a different relationship of the trunk to the pelvis so that the spinal column is vertical but much of the pelvis remains in very nearly the same position that it occupies in a pongid in the quadrupedal position. This leaves the ischial shaft orientated in a fashion that allows the hamstrings to extend the thigh at the hip in a manner that allows for striding. The shift of the center of gravity downward and backward, and the increase of the lever length in relation to the moment arm length, improved the efficiency of this mechanism as did the modified structure of the foot.

The structure of the pelvis of *H. africanus* and the presence of well-developed lumbar curvature—the evidence for which is discussed in a later section—clearly indicate that the pelvis as a whole, and the ischium in particular, had an orientation similar to that in *H. sapiens*. This refers not only to the orientation of the sacrum in relation to the ischial shaft, but also the orientation of the latter to the horizontal. These crucial elements of the anatomical basis for erectly bipedal locomotion of the sort found in modern man were thus present in *H. africanus* in a form that is not significantly different from that in modern man but is very obviously different from the condition seen in the living pongids.

Washburn (e.g., 1963) has argued that early hominids have a long and apelike ischium and a manlike ilium, and concludes that this could mean that they could run bipedally but were not capable of long-distance bipedal walking. The assumption basic to this reasoning, that the ischium is apelike while the rest of the pelvis is manlike, is incorrect for *H. africanus*, as we have already seen. It has also been noted that Washburn's erroneous view about the ischium of this form is a consequence of the measure of ischial length that he employs: the distance from the acetabular margin to the nearest edge of the ischial tuberosity. The more suitable measure of functional ischial length—the distance from the center of the acetabulum to the end of the ischial shaft—demonstrates that ischial length in this form is actually less apelike than is that of modern man.

The fact that the acetabular margin is relatively far from the near edge of the tuberosity—that is to say, that the tubero-glenoidal sulcus is relatively wide—is of interest and merits examination to discover if it has functional significance.

The acetabulum is relatively slightly smaller in Sts 14 than it is in man, but the ischium is relatively smaller still in the former than in the latter. As a consequence, the acetabular portion of the functional length of the ischium is relatively greater in Sts 14. For example, in the latter the distance from the center of the acetabulum to the margin is 43.6 percent of the ischial length, but the corresponding figure for man, using mean dimensions, is only 33.5 percent. Since the tubero-glenoidal sulcus is relatively very wide in Sts 14, it follows that the area covered by the tuberosity is proportionately very small in Sts 14 compared to the condition in man. The shank between acetabular margin and tuberosity does not have any specific function of its own other than being a part of the total functional length of the ischium. The question, then, relates not so much to why the shank is relatively long but why the tuberosity is so small.

The tuberosity serves two important functions: it serves for the attachment of (a) the hamstring muscles, including the hamstring portion of adductor magnus, and (b) the sacro-tuberous ligament. There is a very intimate relation between the sacro-tuberous ligament and the tendinous fibers of the hamstrings. In modern man some of the latter continue as part of the sacro-tuberous ligament. Careful dissection allows one to distinguish an attachment area for the hamstring portion of adductor magnus, one for semitendinosus plus biceps femoris, and another for semimembranosus. But the tendons are all intimately tied into a single tendinous mass. For example, tendinous fibers of semitendinosus plus biceps femoris are continuous with the sacro-tuberous ligament. Investing tendinous fibers tie together intimately the tendons of semitendinosus plus biceps femoris and that of semimembranosus. Some tendinous fibers of semimembranosus are continuous with the hamstring portion of adductor magnus, sometimes forming a strong commissure very close to the tuberosity.

Weidenreich (1913) has suggested that the tuberosity in modern man has expanded upward toward the acetabular margin at least partly under the influence of the sacro-tuberous ligament. The latter serves the important function, in erect bipeds, of holding the caudal end of the sacrum firmly to prevent it from rising as a result of the pressure exerted downward on the opposite end of the sacrum by the weight of the trunk resisted primarily through the vertebral column. It seems to me improbable that this is the overriding factor in tuberosity size since the hamstring attachments seem to be at least as important. Furthermore, as already noted, the ligament and the hamstring tendons are very intimately associated into a virtually single mass of attachment.

In *H. africanus* (Sts 14) the tuberosity surface is missing for the most part, and therefore it is not possible to see areas of attachment for the different components of the tendinous mass of attachment. But I can see no reasonable alternative to the conclusion that the tuberosity surface area was small because the hamstring muscle tendons and the sacro-tuberous ligament were relatively small. It has already been noted that much evidence points to the conclusion that the Sts 14 individual had a small and very lightly built body with slender and long lower limbs but with a proportionately large pelvis. It is possible, therefore, that the hamstrings, and hence their tendons, were relatively rather small for the proportionately large pelvis since the rest of the body was slender and lightly muscled. If this is so, then the larger tuberosity in modern man could at least partly be due to larger hamstring muscles in an animal that is heavier and larger, even though the pelvis is not much larger. This situation necessarily also would involve a larger sacro-tuberous ligament.

It is possible that another factor has contributed to the increase in surface area of the tuberosity. Spreading out of the attachment of the hamstrings over a larger proportion of the functional length of the ischium would mean that the length of the moment arm of the muscles could be varied depending upon which muscle fibers were acting. This would allow a more precise adjustment of the effect achieved by the hamstrings. There would be no reduction in the maximum length of the moment arm, hence of the maximum mechanical advantage of the muscle group, since what is involved is not a migration toward the acetabulum of the area of attachment as a whole. All that is involved is the migration up the ischium of the nearer edge of the tuberosity without a compensating migration of the further edge. Clearly the fibers of semimembranosus alone would be operating from an appreciably shorter moment arm than would the fibers of adductor magnus alone, with the thigh in the same position in each case. This situation apparently allows more precise control of extension of the thigh at the hip through finer gradations of action.

These two factors together, it seems to me, could explain the relatively larger tuberosity of modern man compared to that of *H. africanus.* The first one would be a simple consequence of larger body size and the other would be a subtle refinement of the basic mechanism under the influence of selection.

As a single adult individual only (Sts 14) is known in which ischial tuberosity size may be determined, it is not possible to know how typical this individual was of the deme from which it came. The only other evidence available is the adolescent ischium of MLD 7 from Makapansgat. In this specimen, in which the epiphyses had not yet fused, the distance between the acetabular margin and the nearest edge of the tuberosity is 8 mm only as compared to 16.1 mm in Sts 14. The specimen is not yet fully mature, but already the ilium size almost matches that of the fully adult Sts 14. It would be most remarkable if the distance between acetabulum and tuberosity could have doubled by adulthood in this individual when already the ilium has almost reached adult size. This suggests that normal variation in *H. africanus* included individuals with a smaller, perhaps a considerably smaller, gap between acetabulum and tuberosity than occurs in Sts 14. This would be consistent with the situation occurring in some other aspects of *H. africanus* anatomy—such as dental—where a considerable degree of variability occurs. It would thus be well to be cautious of regarding the Sts 14 ischia as being typical of *H. africanus* in this respect. On the other hand, it is improbable that the proportionate length of the ischium would vary greatly so that apelike ischia occurred along with very manlike ischia such as those of Sts 14, all in the same deme. As we have seen, even comparatively small changes in relative length of ischium are associated with very considerable changes in locomotor pattern, and it is very unlikely that different individuals in the same deme would be adapted to very different locomotor patterns. Therefore it might be expected that relative ischial length would be fairly stable in one deme. On the other hand, because relative size of the ischial tuberosity reflects a relatively minor adjustment within the major locomotor pattern in this case, this character may not be as stable as ischial length, especially if the adjustment was in the process of being made at the time represented by the known fossil specimens.

There remains to be discussed the relatively long pubis in Sts 14. As already noted, in relative length the pubis falls at about the upper limit of the standard population range for *H. sapiens* and very close to the mean values for the living pongids. Thus it may provide a clue to at least one part of the reason for the relatively large size of the pelvis in an otherwise lightly built and small animal.

Much of the special character of the hominid pelvis compared to the sort of pelvis seen in the pongids is associated with the maintenance or achievement of a large true pelvis or birth canal while the ilia were undergoing reduction in height. The size of the birth canal is not especially critical in the case of the pongids since the elongation of the ilia causes the birth canal to be large. Schultz (1949) has shown that the size of the newborn, as reflected in head size and shoulder width, generally is closely correlated with birth canal dimensions in monkeys and in man, but in the living pongids this is true only of the gibbon. In the three great apes the newborn is small compared to the size of the true pelvis. Since the gibbon also has elongated ilia, hence a relatively large birth canal, the closer correlation between size of the newborn and the birth canal may reflect greater maturity of the infant at birth since it is proportionately larger. In man the size of the newborn is large compared to the size of the birth canal, and here, for the first time among primates, the sagittal diameter of the birth canal becomes critical. This is because of the changed relationship between the sacrum and the pubic symphysis, which are now closer together due to the shortenings of the ilium. This shortening necessitated modifications around the margin of the true pelvis in order to maintain a large enough birth canal. As already noted, the sacrum widened and the iliac blade expanded backward so that the sacrum was carried backward away from the long axis of the innominate bone. An obvious further place for expansion round the birth canal is the pubis and indeed, as Schultz (1949) has shown, in higher primates in general females have longer pubes than do males. The relatively long pubis of Sts 14, especially as this individual appears to have been female, is therefore not unexpected. Since the expansion in the other parts of the true pelvis had already occurred to an extent comparable with that in modern man, why should the pubes be relatively longer than are those of modern man? The head size of the modern human infant would presumably be noticeably larger than that of *H. africanus* since the latter had one-half to one-third only of the adult endocranial volume of the former. In view of the evidence that the body of Sts 14 was small and slender, the known adult head size of *H. africanus* evidently was relatively great. One may thus speculate that the relatively large pelvis in Sts 14 was possibly in part a consequence of relatively large head size in a small and very lightly built animal. This could be the reason why the pubis, at least in the female, is relatively slightly long compared to that of modern man, since the sagittal chord had become of considerable importance in a pelvis of this structure. Subsequent increase in the absolute size of the pelvis could have allowed the pubis to reduce to a small extent in relative length. Such reduction would allow greater mechanical efficiency since a relatively long and slender pubis would not be as strong as a shorter one of similar robustness.

It should not be overlooked, however, that one is not here concerned with *major* changes of relative pubic length. The Sts 14 condition in this respect is not actually completely outside the standard population range for modern man but falls at about the upper extreme for this range; that is to say, it is not impossible that an occasional specimen of modern man will have pubes relatively as long as those of Sts 14.

Probably either or both of two factors would favor relatively large pelvis size. The first is the more obvious one that has just been referred to: the need for a large birth canal. The second is that in the earlier stages of habitual erect posture a relatively wide pelvis would be advantageous because it would provide a rather wide base from which muscles could operate to assist in balancing the trunk in an erect position. This concerns not only the lateral balancing functions of gluteus medius and minimus at the hip, but also the abdominal muscles. In the case of the lateral balancing action of gluteus medius and minimus, increased lateralward protrusion of the iliac crest compared to the acetabulum increases the length of the moment arm of these muscles, hence also their mechanical advantage. Reynolds (1931) concludes that an expanded iliac crest modifies the relation of the abdominal muscles to the rib cage in a manner that makes lateral balance of the trunk more efficient and is thus advantageous in an erect biped. In *H. africanus* the relatively protuberant anterior region of the ilium contributes to this advantage, as Davis has noted (in Napier 1964). As efficiency of adaptation improved, the relative size of the pelvis could reduce somewhat without reduction of functional efficiency. Perhaps both of these factors were responsible for the relatively large pelvis of *H. africanus.*

This is an appropriate place to refer to the role of gluteus maximus in erect bipeds. Washburn (1950) gave prominence to the view that this muscle functions quite differently in quadrupedal animals, such as pongids, and in erect bipeds. He argued that in pongids it passes from the posterior end of the iliac blade and the neighboring ligaments to the lateral part of the femoral shaft and parts of the deep fascia and that the spatial relationship between the origin, the insertion, and the hip joint are such that the line of function of the muscle passes right over the hip joint in most positions of the femur. The muscle thus acted as an abductor, for the most part, of the thigh at the hip. On the other hand, in man the backward expansion of the ilium carried the origin of gluteus maximus backward so that the line of function moved behind the hip joint; as a consequence the muscle functions primarily as an extensor of the thigh at the hip joint. The muscle itself has also increased in relative size and the powerful extensor action that it exercises in the later phases of a stride constitutes the single most important differences between pongid and hominid bipedalism. The change in function of this muscle was thus seen as being the critical and key point in the development of habitual erect bipedalism.

This argument has been adopted widely, for example by Le Gros Clark (1955), Howell (1965), Campbell (1966), Beuttner-Janusch (1966), and Napier (1967) as well as by the present author in a series of publications in which the fact that this interpretation involved a clear threshold below which habitual erect posture would not develop and above which it almost certainly would, was used in interpreting the emergence of erect bipedality (Robinson 1962, 1963, and 1964).

Since dissecting both pongid and hominid cadavers in 1964 for the purpose of examining this matter, I no longer accept that this interpretation is at all tenable. In the chimpanzee, for example, gluteus maximus has no bony origin from the ilium at all, the nearest being on the lower end of the long sacrum. A thin anterior portion of the muscle originates from the deep fascia overlying gluteus medius. Almost all of the anterior portion of the muscle, which is powerfully developed, originates from the lower end of the sacrum and from the sacro-tuberous ligament. But there is also a very powerful posterior portion of the muscle, and this has a strong bony origin on the ischial tuberosity and passes to the femoral shaft well down from the fulcrum; in fact, the lower fibers reach the femoral shaft only a short distance above the knee. This very powerful posterior portion of the muscle has a line of function extending from the ischial tuberosity to a point well down the femur and is thus an extensor of the thigh at the hip joint. Not merely is it an extensor of the thigh, but owing to the long moment arm and short lower limb coupled with the powerful development of this portion of the muscle, it is a very powerful extensor of the thigh. It is thus a muscle with relatively great mechanical advantage capable of powerful but relatively slow extension of the lower limb at the hip. This function would clearly be of great use in climbing, where extension of the femur is used to lift the trunk against gravity. The anterior portion has both extensor and abductor functions, depending on which fibers are functioning and what the position of the femur is in relation to the trunk. There are also other functions that need not be considered here. It seems to me very clear indeed that by far the most important action of the muscle as a whole in the chimpanzee, as well as in the gorilla where it is similar (Raven 1950), is extension of the thigh at the hip joint. Clearly, the major portion of the muscle, which is that portion originating on the ischial tuberosity and inserting on the femoral shaft, acts just like a hamstring with respect to movement of the lower limb at the hip joint. A more detailed study by Sigmon (Ph.D. thesis, 1969) has confirmed the above conclusions.

Evidently, therefore, the function of gluteus maximus is not basically different in pongids—at least those from Africa—and in *H. sapiens*. Consequently there could have been no threshold involving change of function of this muscle in the evolution of an erectly bipedal posture from an ancestor that resembled pongids in the respect here being discussed. Nor can the difference in efficiency in erect-walking between pongids and man be explained in terms of the action of this muscle. This conclusion follows for two reasons: first, because this muscle functions primarily as an extensor of the thigh at the hip in both groups and, second, because this muscle does not play a dominant role in erect-walking as this theory demands. According to Basmajian (1962), gluteus maximus is electrically silent during normal relaxed bipedal walking on a level plane. Other studies (Eberhart, Inman, and Bresler 1954; Steindler 1955; Radcliffe 1962; Battye and Joseph 1966) indicate that this muscle is regularly used during a brief portion of a stride as the limb reaches its maximum forward position and the foot strikes the ground. It is not here acting as a source of propulsive power but to assist in halting the forward momentum of the limb and then to stabilize it as the limb begins to support the body weight. In these actions it is assisted by the hamstrings. It may also act briefly in the closing phases of a stride, but this is not a consistent feature of normal relaxed walking.

The view that a major difference between human and pongid bipedal walking results from gluteus maximus acting with power as an extensor of the thigh in the closing phases of a stride in man but not in the pongids thus seems unfounded. Confirmation of this conclusion is provided by clinical observations dating back over a hundred years (Duchenne 1867) that show that a person in whom gluteus maximus is paralyzed can walk without serious difficulty, although some compensatory adjustment of the trunk is usually involved.

Dating back at least also to Duchenne is the recognition that gluteus maximus is primarily a reserve power supply for extension of the thigh at the hip. This is to say that its most important postural and locomotor function is to provide additional power when more is needed than can comfortably be supplied by the hamstrings. This is the case when the body has to be raised against gravity, as in walking up a flight of stairs or a steep gradient. In these circumstances, gluteus maximus comes into action along with the hamstrings. This is also the case in the raising of the trunk from a bent over forward position or in standing up from a sitting or squatting position. It seems to me probable, though I know of no experimental confirmation of this, that in taking long rapid strides when walking the hamstrings and gluteus maximus are quite strongly involved in producing maximum extension of the femur—which evidently is only slightly or not at all the case in normal relaxed walking. More strenuous locomotion, such as running, involves considerable effort on the part of these limb extensors (Wheatley and Jahnke 1951).

It would thus appear that gluteus maximus serves an important function in modern man as a reserve supply of power for extending the thigh. Of course it serves other, nonlocomotor functions also, as in copulation, for example. It does not, however, appear to serve any unique locomotor function that is unrepresented in pongids. Since it is not even necessary for normal relaxed walking, it seems most unlikely that it could have been a critical factor in selection for erect-walking, even if it had undergone a change in function.

The relatively inefficient erect-walking of pongids evidently does not relate to the function of gluteus maximus but to a number of other factors, as already noted earlier in this work. One of these is the relatively long ischium in pongids. Another is the relatively short femur and lower limb as a whole. The combination of long moment arm and relatively short lever length means that the hamstrings have relatively great mechanical advantage in pongids and can produce a powerful movement that is, however, relatively slow. The relationship of the spinal column to the innominate is such that in the erect position the ischium points directly to the ground; hence the hamstrings are disadvantageously placed to produce the sort of extensor action of the lower limb at the hip that is required for effective erect-walking and striding. This disadvantage is added to by the manner in which the hamstrings attach below the knee in pongids, which makes it difficult for the knee to be straightened completely. As a result of these two factors—limited ability to extend the femur at the hip in the erect position and to straighten the knee—the trunk has to be supported on limbs that are semiflexed, and this requires a relatively great amount of muscular energy. Furthermore, the center of gravity is both well forward of and above a line joining the femoral heads, the hinge axis of the trunk on the limbs, and the trunk therefore tends to flop forward. This tendency has to be counteracted by muscular effort. In man the center of gravity lies more or less on the hinge axis, and therefore less muscular energy is needed to keep the trunk erect with respect to this particular factor. In both of these respects, therefore, erect-walking in a pongid is much more expensive in terms of muscular energy than is true of erect-walking in man. To this must be added the fact that the muscles of lateral balance are not effectively disposed to serve this function when the animal is using the erect position, and this also means that the feet cannot be kept close to the midsagittal plane.

All of these factors contribute to inefficient walking by pongids in the erect position, and this inefficient locomotion can be performed only at the cost of great expenditure of energy. A good deal of the energy is expended by the frontal thigh muscles, which act as extensors of the leg on the thigh and, among other things, have to prevent the knee joint from collapsing under the strain of supporting the trunk with the knee in a partly flexed state.

In summary, it seems that the change from basically quadrupedal to properly erectly bipedal posture and locomotion did not spring primarily from a clear-cut change in the function of gluteus maximus and did not even involve such a change at all. On the contrary, what seems to have been involved is a series of relatively minor, integrated modifications that have converted a creature that is efficient at quadrupedal terrestrial locomotion and not very efficiently erectly bipedal into one that is efficient at erectly bipedal locomotion and not efficient at quadrupedal locomotion. The set of factors making for efficient erect bipedality includes as crucial elements the ability to control lateral balance efficiently, the ability to extend the femur beyond the vertical reasonably rapidly, to keep the feet close to the midsagittal plane habitually and the possession of a broad foot that is comparatively inflexible and has a strong axis or beam through a fully adducted hallux.

Looking at the *H. africanus* (Sts 14) pelvis and related locomotor structures in terms of these capacities produces a very clear picture. A habitually vertical trunk is indicated by, among other features, lumbar curvature as well developed as it is in modern man. The capacity for efficient lateral balance can be judged in two different ways: (a) by the development of well-defined anatomical features on the gluteal surface of the ilium, which in modern man are associated directly with the functioning of gluteus medius and minimus in the efficient maintenance of lateral balance and are found in hominids alone, and (b) by the fact that the angulation of the femoral shaft in relation to the axis of the knee joint (discussed later in this work) demonstrates that the knees—hence also the feet—habitually moved close to the mid-sagittal plane. This would not be possible had not efficient lateral balance control already been achieved. Furthermore, an extended iliac crest was present and situated suitably to allow the abdominal musculature to control lateral balance of the trunk effectively. Since the orientation of the upper half of the innominate and of the sacrum in relation to the lower half of the former had already been achieved to the same extent as occurs in modern man, it is clear the the ischium occupies the same position in relation to the vertical and horizontal planes as it does in *H. sapiens*. Hence the basic ability to extend the femur past the vertical was present. All of the available evidence indicates that the relative lengths of the hamstring moment arm and the distance from the lower limb fulcrum to the ground (lever length) were similar to those in modern man and quite different from those in pongids. Not only is the ischium short but also the lower limb had already elongated in the manner typical of modern man. This elongated lower limb, coupled with the fact that the ilium is shortened and lumbar curvature is well developed, argues strongly in favor of a center of gravity positioned as in man rather than as in pongids. This implies that balance could be maintained more easily than otherwise it could be. Also, the small ratio between hamstring moment arm and distance from the lower limb fulcrum to the ground indicates that the specialization for speed rather than for power, as found in modern man, was already present. This is added to by the fact that the small area of the ischial tuberosity suggests that the hamstring muscles were not strongly developed. The development of the frontal thigh musculature in the manner of man is suggested by the strong development of the anterior inferior iliac spine in the manner of man. The latter and the well-developed femoral tubercle indicate that the iliofemoral ligament was developed in the specialized manner seen in modern man.

This combination of features demonstrates most effectively that almost all of the essential structural and functional aspects of the locomotor pattern of an efficiently erect biped were already present in *H. africanus*. One key feature remains and that is the nature of the foot.

There is no fossil material of the *H. africanus* foot known from South Africa and therefore there is no direct evidence of what sort of foot was associated with the pelvic remains we have been discussing. If *H.* "*habilis*" from Bed I at Olduvai is, as I believe it to be (Robinson 1965a and b, 1966, and 1967), a member of the species *H. africanus* as used in this work, then indeed there is evidence concerning the foot.

The discovery of a fairly complete foot of an adult individual was announced by Leakey (1960) and was reported on briefly by Day and Napier (1964). The latter authors showed that the foot is remarkably like that of modern man with a relatively small amount of flexibility, with a powerful and strongly adducted hallux, and having "most of the specializations associated with the plantigrade propulsive feet of modern man." The foot is small compared to that of modern man but is robustly constructed. The authors conclude further that a striding gait had probably not been achieved by the owner of the foot. It seems to me that the evidence for the latter conclusion is not strong and that striding may well have been possible; the reasons for this conclusion will become clear further on. This is also suggested as a possibility by the fact that a single terminal phalanx of a hallux has been found a little higher up in Bed I that has features that Day and Napier (Napier 1967) regard as occurring in modern man only and "can with assurance be correlated with a striding gait." While it is not absolutely certain that this bone belongs to the same form as the stratigraphically earlier foot, it is highly probable. Also, since it comes from a later horizon, the latter may be more advanced than the earlier foot. Since all of the phalanges of the latter are missing, we cannot say for certain that the known foot did *not* have a terminal hallucial phalanx of the sort seen in the later specimen (Hominid 10). It is my prediction that, if phalanges are found at the level of the more complete foot, they will support the conclusion that the owner was capable of a striding gait.

As already stated, the totality of the available evidence suggests that *H.* "*habilis*" in Bed I at Olduvai represents to all intents and purposes the same sort of creature (*H. africanus*) as that found at Sterkfontein, Makapansgat, and Taung in South Africa and that both groups date from a roughly comparable time period in the Lower Pleistocene. It seems highly probable, therefore, that the Olduvai foot gives a reasonably sound indication of what the feet of Sts 14 were like. Certainly the evidence of efficient lateral balance, of an elongated lower limb and the manner of functioning of that limb and of feet moving close to the midsagittal plane, imply a foot of the sort seen in the Olduvai specimen. It would be very difficult to accept that these advanced features, in terms of erect bipedalism, could have been associated with a foot significantly less manlike than that from Olduvai. The foot is obviously a crucial part of the adaptation to bipedalism, and if the rest of the pattern had reached an advanced stage of adaptation it is hardly conceivable that the foot could still have been at a primitive stage.

In sum, the evidence seems, therefore, to point rather clearly to the conclusion that *H. africanus* habitually and effectively used erectly bipedal posture and locomotion and that it was so well adapted to this locomotor habit that quadrupedal locomotion would have been about as useless to it as it is to us.

As Dart (1958) has very pertinently remarked, the mechanisms in man that allow the sorts of movement here discussed also allow characteristically human movements of the trunk while the feet are firmly planted in one position. These rotatory and bending movements are of great importance in accurate throwing and many other manipulations relating to tool and weapon use. If *H. africanus* could stand, walk, and run essentially as man does, then it could also have performed these other characteristically human movements while stationary. This implication is of considerable significance for assessing the behavior, ecology, and evolutionary status of this form.

6 The *Paranthropus* Innominate Bone

Material

All of the known material of the *Paranthropus* innominate bone is in the collection of the Transvaal Museum, Pretoria, Republic of South Africa. This includes the following:

SK 50 a right innominate bone of an adult. It consists of about half of the ilium, most of the ischium, and just the acetabular part of the pubis. The specimen is from the older, pink breccia at Swartkrans.

TM 1605 a left ilium of an adult individual. Not all of the ilium is present, and both ischium and pubis are entirely missing. This specimen is from Kromdraai.

The general features of the *Paranthropus* innominate bone are known from a number of publications based on study of the original fossil specimen SK 50 (e.g., Broom and Robinson 1950 and 1952).

Having considered the pelvis of *H. africanus* at some length, we may now turn to the much less complete material of *Paranthropus* listed above. TM 1605 is here described for the first time.

SK 50 from Swartkrans is the worse preserved specimen but the more complete of the two and is the only one so far described in the literature. Much of the iliac blade is missing, including the entire crest. Apparently very little is missing in the region of the anterior superior spine. The posterior portion of the iliac blade is least well preserved. None of the actual surface of the auricular region is preserved, and what there is of the gluteal surface is badly cracked and split. Although the anterior portion of the blade is best preserved, three fairly large cracks run more or less transversely across it. At each a small amount of flexure has occurred that is almost confined to the outer table bone. The worst crack passes through the upper portion of the acetabulum, and this crack has sprung quite badly, being some 9 mm wide at its widest. This break has caused the upper portion of the acetabulum to come to lie in a quite unnatural position in relation to the other portion. The remainder of the acetabulum has been warped so that the pubic margin lies closer to the ischial margin than was originally the case and thereby the acetabular notch is reduced to approximately half of its original width. These cracks clearly are of postmortem origin and appear to have been caused by movement in the fossil deposit, probably slumping. Except for the acetabular portion, the entire pubis is missing. Much of the body of the ischium is present, but most of the tuberosity surface is missing, as is the region of the ischial spine and the lesser sciatic notch. A portion of the ischial ramus is present but is broken and displaced.

The second specimen, TM 1605, is not from Swartkrans but from

Kromdraai. It is an extremely well-preserved specimen consisting of most of the ilium; however, the entire iliac crest is absent and only a small part of the auricular region is present. A small arc of the upper acetabular wall is all that is preserved of the acetabulum, and both the ischium and pubis are entirely missing (see Figs. 49 and 50).

The legitimacy of regarding SK 50 as belonging to *Paranthropus* should be considered since *H. erectus* (= "Telanthropus") also is known from Swartkrans in a manner that indicates that both forms were simultaneously present in this region. What evidence is available suggests that *H. erectus* already had the modern type of locomotor skeleton. SK 50 has the specifically early hominid modifications of the hominid innominate pattern. Furthermore, it is detailedly similar to the Kromdraai fragment that occurred with other *Paranthropus* remains in a site that so far has yielded no remains of any other type of hominid. As will become clear later, the *Paranthropus* innominate is less like that of modern man than is that of *H. africanus*, and this is particularly clear in the case of SK 50. This is the reverse of what would be expected if it belonged to *H. erectus*. Finally, the fact that *Paranthropus* remains vastly outnumber those of *H. erectus* in the Swartkrans site increases the chance that SK 50 belongs to the former taxon. Therefore it seems reasonable to attribute both SK 50 and TM 1605 to *Paranthropus*.

Ilium

The major features of the *Paranthropus* innominate are, for the most part and as judged from this incomplete material, in principle the same as those of *H. africanus*; but the bone is larger. Direct size comparisons are not easy to make since both specimens are broken in ways that remove most of the usual measuring points used in this work. And yet, superimposing TM 1605 on Sts 14 (*H. africanus*) indicates that the distance from the center of the border of the greater sciatic notch to the anterior inferior iliac spine is roughly 1.5 cm greater in the *Paranthropus* specimen. The difference in size between the two is 22 percent of the relevant distance in *H. africanus*. Repeating this procedure with respect to the distance from the base of the inferior anterior iliac spine to the superior anterior spine gives a difference of some 33 percent. On the other hand, the maximum breadth of the acetabulum differs by only 11 percent in the two—though it is less easy to calculate this figure accurately because the only known *Paranthropus* acetabulum is both warped and broken.

These figures give some indication of the general size relationship of the *H. africanus* and *Paranthropus* ilia and also indicate that there are differences of proportion between them (see Figs. 51 and 52).

The *Paranthropus* ilium clearly is broad, with the posterior portion expanded in the same fashion as in both *H. africanus* and *H. sapiens.* The anterior region of the ilium is long and well-developed rather than being small and truncated as in *H. sapiens.* It is better developed, however, than is the case in *H. africanus*; the distance from the acetabular margin at the anterior inferior spine to the anterior superior spine is long. This distance is some 33 percent greater in *Paranthropus* than it is in *H. africanus*, whereas the general size difference between the ilia of the two is less than this. Using the diameter of the acetabulum for comparison, it is found that in *Paranthropus* this diameter is 55 percent of the anterior iliac height whereas in *H. africanus* it is 75 percent; the anterior iliac height is relatively appreciably shorter in the latter form. This conclusion can be checked by using the modified ratio diagram method, employing the acetabular diameter for coincidence as before. This procedure demonstrates that anterior iliac height is proportionately considerably greater in *Paranthropus* (see Fig. 7). Whereas the *H. africanus* value falls close to the mean for *H. sapiens*, that for *Paranthropus* falls well outside of the upper end of the standard population range for *H. sapiens* and fairly close to the pongid mean values. Both of the *Paranthropus* dimensions here involved (acetabular width and anterior iliac height) are estimates. The acetabulum is all present but warped; since it must have been approximately symmetrical in shape, there is no real difficulty in obtaining a close estimate of the original dimension. In the case of anterior iliac height, the task is perhaps even easier because this region is well preserved and apparently not much was broken off in the anterior superior spine region. The present measurement is 72 mm, and the loss has been conservatively estimated at 6 mm. To test what difference errors in these estimates would make, the calculations were repeated using an estimate of the acetabular width 2 mm (nearly 5 percent) greater than that which appears to be most probable, and using the actual anterior height of the ilium as it now is without adding anything for what is patently missing. That is to say, the test was made using an acetabular width estimate that must be about the maximum possible for this individual and an anterior iliac height that is demonstrably too small. The result is to reduce very slightly the resemblance of *Paranthropus* to the pongids and to bring it that much nearer to *H. africanus* and *H. sapiens.* The effect is minimal, however: *Paranthropus* still resembles the pongids far more than it does *H. africanus.*

Clearly, therefore, the marked difference between *Paranthropus* and *H. africanus* in this case is not an artifact of incorrect estimates. Although the orientation of the anterior iliac region is similar in the two, the anterior superior spine is thus appreciably more protuberant in *Paranthropus* than it is in *H. africanus.* Since the iliac crest is missing in both specimens of *Paranthropus*, it is possible that the relatively long anterior iliac height in this form is related to a relatively great total iliac height. The fact that the marked backward expansion of the ilium typical of hominids had already occurred, so that a very well developed greater sciatic notch is present, coupled with the fact that the anterior superior iliac spine does actually protrude well forward, strongly suggests that the total iliac height in *Paranthropus* was of the relatively low hominid type since backward expansion of the ilium is a consequence of reduction of iliac height. Consequently it is highly probable that the long anterior iliac height is due entirely, in this case, to an unusual degree of forward protrusion of the iliac blade.

The fact that *Paranthropus* resembles pongids more than it does hominids with respect to anterior iliac height need not indicate any close affinity with pongids. The gorilla, being the most robust and thick-torsoed of the living pongids, has the iliac blades more protuberant anteriorly than has any of the others. Similarly, there is evidence that indicates that *Paranthropus* was appreciably more robust than *H. africanus*, and this difference is probably the main reason, if not the only reason, for the different degree of protrusion of the anterior part of the iliac blade in the two.

Paranthropus appears to have had a relatively small auricular area for the size of the ilium, such as is found in *H. africanus*. If this is so, then the implication is, as in the latter, that the vertebrae were relatively small compared to the size of the innominate, and presumably the same was also true of trunk size as compared to pelvis size. The evidence for this, however, is much less definite than is the case in *H. africanus*. For example, the auricular surface itself is missing entirely, but the area that it occupied is reasonably well demarcated and is small compared to the width of the rest of the iliac blade. But, as has been established above, the proportionate degree of protrusion of the anterior portion of the iliac blade is considerably greater than is true of *H. africanus*. Consequently a simple comparison of auricular area width with the width of the rest of the iliac blade will be misleading. In this case no sacrum is known, and the only lumbar vertebra known from Swartkrans does not belong to this individual; hence comparing its size to that of the SK 50 innominate is not necessarily illuminating in this respect since it could have belonged to an individual of a different size. Furthermore, since *H. erectus* (="Telanthropus") is known from this site, it is not possible to be sure to which of the two forms the vertebra belonged. So little is known about the vertebrae of early hominids that an assignment in this instance is difficult to make with any certainty. An added difficulty is that the vertebra is from an immature individual. The probability seems to me somewhat greater, however, that it belonged to *H. erectus* rather than to *Paranthropus*. Therefore it is not possible to say any more than that the evidence suggests that perhaps the pelvis in *Paranthropus* was relatively large for the rest of the body but not to the same degree as appears to have been the case in *H. africanus.*

Judging from the shape of the blade, the iliac crest had a clearly developed S-shaped curvature of the *H. africanus* type; that is, with the anterior half of the curve less pronounced than it is in *H. sapiens.*

The gluteal surface in both specimens has the same anatomical features as has that of *H. africanus.* Two planes are defined, and the anterior part of the blade has two buttresses of thickened outer table bone. The acetabulo-spinous buttress is the more pronounced and trends strongly toward the anterior superior spine. There is a slight hollow—more a flattening—between it and the acetabulo-cristal buttress, which trends directly up in the long axis of the innominate in the same position as in *H. sapiens.* In TM 1605 particularly, the broken edge of the iliac blade allows the thickened bone of these buttresses to be seen clearly. The bone is broken in an especially advantageous position to allow measurement of the thicknesses associated with the acetabulo-cristal buttress, which is not as strongly developed as is the other. The compact bone thickness on either side of the buttress, which is easily palpable, is 1.1 mm, and is 1.5 mm at the thickest point. The more strongly developed acetabulo-spinous buttress cannot be measured in the same manner because the specimen is broken at a point where the buttress is rapidly losing thickness near the crest. Since the crest itself is totally absent in both specimens, it is not possible to determine whether an iliac tubercle was present. The presence of an acetabulo-cristal buttress suggests that a tubercle probably was present.

The posterior, acetabulo-sacral buttress, which passes next to the greater sciatic notch, is well developed. A natural break in the case of TM 1605 allows measurement of the thickness of the compact bone, which is 4.7 mm at the point where it arches over the greater sciatic notch.

The anterior inferior iliac spine is very well developed, apparently even more so than it is in *H. africanus*—though this again may simply be a reflection of greater body size. In SK 50 there is very clear differentiation between the area to which the ilio-femoral ligament attaches and that to which rectus femoris attaches. The scar for attachment of the reflected head of the latter muscle is well defined in both specimens.

On the pelvic surface the iliac fossa is well developed, and in both specimens the trend of the bone organization is readily apparent. As in *H. africanus* and *H. sapiens*, the split-line direction is strongly toward the anterior superior iliac spine, with a small subsidiary pattern passing directly onto the anterior inferior iliac spine.

Acetabulum

The acetabulum of SK 50 has been considerably distorted. First, by the broad, sprung crack across the top portion and, second, by warping of the pubic portion toward the ischial portion to partly close the acetabular notch. The acetabulum is like that of hominids but resembles that of *H. sapiens* a little more than that of *H. africanus* in the one feature in which the latter differs from *H. sapiens*. In *H. africanus* the acetabular margin is somewhat ill-defined across the base of the anterior inferior iliac spine, whereas in *H. sapiens* the margin is normally continuous and clearly discernible there. In *Paranthropus* this section of the margin is clearly defined, though perhaps a little less so than is commonly the case in *H. sapiens*. This is probably a matter of no special significance other than as a reflection of body size. *Paranthropus*, being heavier than *H. africanus*, would perhaps need a stronger acetabular wall in this general position for weight-supporting purposes.

The fact that the acetabular wall, especially near the brim, is thicker in *H. sapiens* than in *Paranthropus* even though the former is unlikely to have been heavier than the latter probably has a slightly different explanation. The size of the femoral head is larger in *H. sapiens*; this allows the forces at the hip joint to be spread over a larger surface, thus increasing the mechanical efficiency of the joint as far as the femoral head is concerned. This also increases the depth of the acetabulum; hence the wall has to be thickened to compensate for the greater height.

Ischium

The ischium is perhaps the most noteworthy feature of the *Paranthropus* innominate. The only specimen available is that of SK 50, and it is partly damaged. Clearly the ischium in this specimen was robust and well developed. The gap between the acetabular margin and the nearest edge of the tuberosity, the tubero-glenoidal sulcus, is wide. Although much of the tuberosity surface is entirely gone, a small piece of what appears to be the true edge of the tuberosity is present and the direct distance from it to the acetabular margin is 27 mm. The tubero-glenoidal sulcus is quite strongly curved; the margin is high above the floor of the sulcus and so is the tuberosity surface. The depth of the sulcus is 10.4 mm. In *H. africanus* (Sts 14) the comparable depth is 4.9 mm. The tuberosity was thus apparently rather strongly protuberant as well as being far from the acetabulum.

Few of the dimensions of the innominate used for comparison in this work can be measured on SK 50. As has already appeared, acetabular width and anterior iliac height can be estimated to within close limits. The latter is not suitable for comparison with the ischium length since it was proportionately very long in this form. Acetabular diameter affords a better standard, for reasons previously discussed. In *Paranthropus* the tuberosity-margin distance is 62.8 percent of the acetabulum width, whereas in *H. africanus* the comparable figure is 43.6 percent. The distance from the proximal edge of the tuberosity to the distal extremity of the ischium, where it curves round into the ramus, is also relatively long—27 mm. The tuberosity length is thus as long as the ischial shank (tuberosity-margin) length. On the other hand, in *H. africanus* the tuberosity length is only 74.5 per cent of the shank length. Calculating tuberosity length as a percentage of total ischium length shows that the ratio is 36.0 percent in *Paranthropus* and 26.7 percent in *H. africanus*. The tuberosity is thus proportionately longer in *Paranthropus*.

Relative length of the ischium in these two forms can be compared in various ways. Using the acetabulum diameter as a guide shows that in *Paranthropus* the acetabular diameter is 62 percent of the ischium length, while in *H. africanus* it is 83 percent, using width rather than height of the acetabulum in both cases. This indicates that *Paranthropus* has proportionately a considerably longer ischium than has *H. africanus*. The living pongids have an acetabulum: ischium ratio that usually falls between about 40 and 65 while that of *H. sapiens* usually falls between 55 and 85. From this it is clear that *Paranthropus* has a relatively low value for this index that places it (SK 50) in the zone of overlap between the pongids and *H. sapiens*, while *H. africanus* (Sts 14) has a relatively high value that falls above the *H. sapiens* range. This suggests that there is a significant difference between the two early hominids in proportionate length of the ischium.

Using the ratio and modified ratio diagram method demonstrates the same point in a slightly different manner. All of the living pongids have relatively long ischia compared to that of *H. sapiens*, while *H. africanus* has a relatively very short one. The *Paranthropus* SK 50 ischium falls in a group with the pongids with a proportionate length considerably greater than that of *H. africanus* (see Fig. 7).

Unfortunately, lack of appropriate specimens precludes the making of the other comparisons of ischial length that were made in the case of *H. africanus*, such as comparison with femur length and the transverse diameter of the articular facet of the sacrum with the last lumbar vertebra.

The only two available *Paranthropus* innominate bones have either all or almost all of the pubis missing; consequently nothing is known of this portion of the pelvis.

Nothing meaningful can be said about the sex of the individuals to which these two specimens belonged. The pubis is useful in assisting diagnosis of sex, but in this case no information at all is available from this source.

TM 1605 has much of the greater sciatic notch missing; hence another valuable source of information is absent. In SK 50 the greater sciatic notch is present but damaged enough for the exact contour to be uncertain. If the present contour is little altered from its original state, then it is possible that the specimen belonged to a male individual. Enough ambiguity is introduced by the cracking that has occurred in the back half of the blade of the ilium to make this conclusion very uncertain. TM 1605 is less robust than SK 50 and therefore could have been female. These speculations are so insecure, however, that they do not merit serious attention.

7 Functional Analysis of the *Paranthropus* Innominate

No sacrum of *Paranthropus* is known at present, and therefore we do not know whether it was broad and hominine in its features. Having regard to the close functional relationship between reduced iliac height, posterior elongation of the iliac blade, and increased width of the sacrum, the probability is high that the sacrum of this form had already undergone hominid widening. What is present of the auricular surface on TM 1605 suggests that the sacrum was oriented in the manner found in *H. africanus* and *H. sapiens*. This is certainly consistent with the general similarity in innominate anatomy between *H. africanus* and *Paranthropus* as well as with the fact that iliac shortening and posterior expansion had already occurred.

The nature of the iliac anatomy suggests that the balancing mechanism of *Paranthropus* was as well developed as that of *H. africanus*, and the discussion of the features of the latter apply also to the former as far as the available evidence goes. The anterior superior iliac spine region, however, is appreciably more protuberant in *Paranthropus*; hence the conclusion that *H. africanus* had a chunky trunk with protuberant abdomen appears to have been even more true of *Paranthropus*. The evidence certainly indicates that *Paranthropus* was more robust than *H. africanus*, and therefore a stockier build and more thickset trunk and abdominal region are quite probable, especially since the protuberant anterior iliac region had only moderate medialward flexion.

The prominence of the anterior inferior iliac spine indicates that the specialized development of the ilio-femoral ligament characteristic of an erect biped was present and suggests that rectus femoris was well developed and probably also the associated muscles. The fact that the spine is a little larger than that of *H. africanus*, is probably not significant in terms of level of function; it probably reflects the greater robustness of *Paranthropus*.

None of the evidence seems to suggest any differences of function between the two forms of early hominid with respect to the ilium.

This may not be true, however, of the ischium. From the description of the *Paranthropus* ischium it is clear that it is appreciably longer than is that of *H. africanus* and is therefore more apelike. And yet it is not necessarily immediately clear what is the functional meaning of the relatively long ischium.

As in the case of *H. africanus*, the ilium had already undergone the alteration of orientation in relation to the lower half of the innominate. In *H. africanus* and in *H. sapiens* this reorientation places the ischium in a suitable position to allow the hamstrings to extend the femur at the hip in a manner suitable for efficient walking in the upright position. But whether the hamstrings in *Paranthropus* could function as effectively is not so clear since the longer the ischium the more the effect of the reorientation is offset. Hence one might conclude that *Paranthropus* was less efficient at extending the femur while in the upright position than was *H. africanus*. This would imply a poorer level of adaptation to the erect bipedal position.

The question arises why the ischial tuberosity in *Paranthropus* is so protuberant in being set up high above the long axis of the ischium. The direction in which it is protuberant is away from the pubis end of the innominate. The effect of this raising of the tuberosity, hence of the origin of the hamstrings, is to offset some of the disadvantage of the long ischium length in producing full extension of the femur at the hip. This is so because the protuberant tuberosity in this case has the same effect as would be achieved by rotating the long axis of the ischium a little toward the horizontal, thereby increasing the capacity to extend the femur past the vertical. One might thus argue that the long ischium plus offset tuberosity is functionally very similar to a shorter ischium in which the tuberosity is not offset. Perhaps, therefore, the functional disadvantage for erect bipedality of the relatively long ischium was not as great in this instance as the length of the ischium alone might suggest. Whether *Paranthropus* actually did approach *H. africanus* in functional efficiency in the erect position depends also on the answers to some other questions, the answers for which are not available. Of obvious importance is the length of lower limb associated with this ischium, and at present there is no evidence that will allow an estimate to be made that would be anything other than guesswork.

If *Paranthropus* had indeed achieved functional capacity in the erect position basically similar to that of *H. africanus*—and we do not know for certain that this was so—why should the ischium have stayed long with an offset tuberosity when the same effect could more simply have been achieved by shortening the ischium as happened in *H. africanus* and in *H. sapiens*? Various reasons are possible. For example, since *Paranthropus* was evidently a fairly heavily built animal compared to *H. africanus* and yet had a relatively small auricular area, therefore the articulation of the trunk with the pelvis was relatively very small in a robust animal. Perhaps as a consequence the ligamentous connections between the sacrum and the pelvis were relatively more important, especially the sacro-tuberous ligament that assists in counteracting the turning moment of the spinal column and sacrum around the sacroiliac articulations. A long and substantial ischium would assist in giving the sacrotuberous ligament both firm anchorage and reasonable leverage.

It seems more likely, however, that the long ischium is more concerned with improved mechanical advantage for the hamstrings. It might be argued that if the lower limb is relatively long then the ischium length would be in proportion and therefore in fact not relatively long. It is not known what the length of the lower limb was in *Paranthropus*. But in order for the ischium to have had similar proportions to the lower limb to those occurring in *H. africanus* and in modern man, the lower limb would have had to be long compared to those in the latter two forms. In the absence of direct information about the limb length in *Paranthropus* this possibility cannot, of course, be ruled out. And yet it is a most unlikely possibility. Consequently the probability is high that the ischium was long in proportion to the limb length, as the modified ratio diagram suggests.

If this were so, then the ratio of the moment arm length of the hamstrings to the distance from the acetabulum to the ground would be larger than in either modern man or *H. africanus* and would be approaching pongid levels. Since we have seen that reduction of the moment arm of the hamstrings is advantageous in erect bipeds because of the improved speed of movement, it follows that the long moment arm in *Paranthropus* must have entailed some loss of efficiency in respect to this point. It could be argued that this is a primitive condition in an early hominid and that later evolution would have reduced the moment arm length with consequent improvement of efficiency for erect bipedality with regard to femur extension. This is not an appealing explanation. First, the available evidence suggests that *Paranthropus* had been in existence for a long time by the time that the SK 50 individual was alive and that by that time *H. africanus* had already achieved a short ischium. This follows from the evidence, geological and faunal, that indicates that *H. africanus* (Sts 14) from Sterkfontein was living in the Sterkfontein Valley before SK 50 lived in that same valley in geologically later Swartkrans time. Second, what evidence is available suggests that *Paranthropus* was a stably adapted form that was in principle no more manlike when it became extinct in the Middle Pleistocene than it had been when first it appeared in the known record in the Lower Pleistocene or late Pliocene. The evidence of the protuberant ischial tuberosity suggests that the long ischium was of advantage and that its disadvantages for erect bipedalism were to some extent offset by modifications to the tuberosity. That is to say that the nature of the ischium suggests a compromise arrangement between two sets of factors that were to some extent antagonistic.

It seems then that one should look for a reason why good mechanical advantage (long ischium) should be combined with reasonable ability to extend the femur at the hip joint past the vertical in the erect position (offset ischial tuberosity). An obvious solution would be the reason why pongids have hamstrings with good mechanical advantage—they are very useful for climbing purposes. With the femur partly flexed on the trunk, the long moment arm gives excellent leverage to lift the trunk against gravity.

Quite other considerations have led me to the conclusion that *Paranthropus* was a herbivore, though not a grazer, and climatic evidence indicates that *Paranthropus* occurred in climatic conditions wetter than those in which *H. africanus* could live successfully. Some tree-climbing propensity does not seem out of place. Even a small amount of such activity, concerned probably with both food-gathering and shelter, would be very materially assisted by a long and powerful ischium in a rather bulky animal.

This is perhaps an unusual combination: reasonably effective erect bipedality coupled with a certain amount of arboreal activity. But it does not seem to me totally improbable in a herbivorous primate of this type, and such a way of life is consistent with what I, at least, have been able to make out of the available evidence of the adaptational complex of this animal. I cannot see any obviously more suitable reason why both the specialization for erect bipedality, which is clearly expressed in the pelvis, and specialization for relatively slow but powerful extension of the femur should be coupled together in the same animal.

It would have been very helpful had specimens of the distal end of the femur been available to indicate whether or not the planes of movement of the knees were close to the midsagittal plane. With the degree of development of the pelvic mechanisms for good lateral balance control it seems probable that the knees did so move.

Knowing the nature of the foot would also be very helpful. It has been argued that the fragments of foot from Kromdraai, especially the talus, indicate a more mobile foot in which the hallux is not fully adducted, as it is in man and appears to have been in the Olduvai foot (Broom and Schepers 1946; Le Gros Clark 1955, 1967; Day and Napier 1964). If indeed *Paranthropus* habitually did some climbing, a mobile foot with incompletely adducted hallux is more likely to have been present than the type of foot found in *H. sapiens*. On the other hand, the *Paranthropus* talus does imply a foot that was more manlike than is that of a chimpanzee or a gorilla. This suggests that the foot also reflected a compromise situation: it was more suited to erect walking than is that of pongids but more suited to climbing than is that of modern man.

It might be suggested that if *Paranthropus* was a more bulky animal than *H. africanus*, the ischium would be less likely to reduce in length since considerable power is needed to lift the body into the erect position from sitting, squatting, or bending positions and the extra leverage that the hamstrings and gluteus maximus would have with a longer ischium would be advantageous. This does not seem to be a sound line of reasoning. For example, the proportionate length of the ischium of the living pongids, as shown by the modified ratio diagram, is longest in the gorilla and the gibbon and shortest in the orang. There is virtually no difference in relative length of the ischium in the gibbon and chimpanzee, though the latter is much the larger of the two. The size difference between *Paranthropus* and *H. africanus* is unlikely to have been larger than that between gibbons and chimpanzees. Consequently it is improbable that size difference alone would account for a significant difference in relative length of the ischium. The probability is much greater that an adaptive difference was involved.

Obviously more evidence is needed before it is possible to be sure what the reason is for the relatively long ischium in *Paranthropus*. It seems clear that this form had much of the specialized mechanism necessary for regular use of the erect bipedal position—a conclusion that is supported by the fact that the nuchal portion of the occiput of this form was oriented in the almost horizontal position associated with erect bipedalism. Equally clearly, the relatively long ischium is distinctly more primitive than that of *H. africanus* —if *H. sapiens* is taken as the basis of comparison. In view of this feature, and the evidence of the foot, it seems clear that *Paranthropus* was not as well adapted to erect bipedality as was *H. africanus*.

It has been suggested that the displaced upper portion of the acetabulum of SK 50 resulted from dislocation of the hip resulting from incomplete adaptation to erect posture. This conclusion can be discounted entirely because the displacement in question did not occur significantly before the death of the individual. There is no evidence of healing having occurred subsequent to the break. Probably the damage occurred long after death during the process of fossilization.

8 The Vertebral Column

Material

Very little is known of the vertebral column of the early hominids. In the case of *H. africanus*, at least seventeen reasonably complete vertebrae are known, all of them from the Sterkfontein Lower Breccia. It is a matter of great good fortune that no less than fifteen of these belong not merely to the same individual but also to the particular one that yielded the virtually complete pelvis (Sts 14). The vertebral column of this individual is thus known from the incomplete sacrum already described as well as the entire lumbar region and most of the thoracic region (nine vertebrae). The fifteen vertebrae, two of which are represented by the body alone, appear to represent a proper sequence with no gaps. The sequence can be determined without much difficulty. There are also a number of small fragments of vertebrae, apparently of the same individual, which are too incomplete to be identified or to be informative. All of these vertebrae were grouped together, partly in sequence, in the same very small area of breccia that included the pelvis, the incomplete femur, and the rib bones that are now grouped together as Sts 14. Since this material occurred together in a compact, small pile at least five feet below the surface of the very large mass of fully consolidated breccia composing the Type Site, and having regard to the nature of the preservation and coloring, the nature of postmortem weathering, the nonoverlap of parts, and the size and structural harmony between the various items, there can be little doubt that they all belonged to the same individual. As has appeared from the study of the pelvis, there are reasonable grounds for supposing that this individual was a female that was fully adult at the time of death.

Besides this very fine partial spinal column, there are two other specimens. One is a reasonably complete body of an isolated vertebra that had undergone some weathering before fossilization. This has the catalogue number Sts 73 and is from the Lower Breccia at Sterkfontein. An even more fragmentary vertebra was closely associated with the innominate Sts 65 and has that catalogue number. It consists of the neural canal surface of the body and the bases of the pedicles. This specimen is too fragmentary to be instructive except with respect to size. In view of the scantiness of the vertebral material, this information is valuable. The above material represents all that is currently known of *H. africanus.*

Paranthropus presents a more difficult problem. So far four hominid vertebrae have come from Swartkrans. These are not all from the same individual. One specimen (SK 854) is a cervical vertebra, an axis with the odontoid process, the left superior articular facet and the neural spine damaged but otherwise well preserved. Another (SK 853) is a lower lumbar vertebra of an individual that was not fully mature at death as judged by the fluted and crenulated margins of the body. Two more are probably from the same individual (SK 3981a and b). The former is a complete last thoracic vertebra; the latter is part of a last lumbar vertebra in which most of the neural arch and the right transverse process are missing.

There is a real question, however, regarding the taxon to which these specimens belong. There seems little doubt that all are hominid in structure. But since two types of hominid are known from Swartkrans, *Paranthropus* and *H. erectus* (= "Telanthropus"), and since vertebral material is not known that can be referred to either of these taxa with certainty, the taxonomic identity of the two vertebrae is a little obscure. This point will be discussed more fully later. No vertebral material is known from Kromdraai.

Homo africanus

The *H. africanus* material consists essentially of the fifteen consecutive vertebrae of Sts 14, most of which are complete or nearly so.

The last of six lumbar vertebrae has the laminae and neural spine missing as well as the inferior articular processes and a part of the upper, posterior edge of the body. The missing portions have been reconstructed; the position, at least, of the inferior articular facets is indicated quite well by the structure of the first sacral vertebra. The second last lumbar is missing a portion of the lamina and inferior articular facet on the right side as well as the right transverse process. The next, the fourth lumbar, has the top half of the neural spine missing as well as the right transverse process. Both this and the fifth lumbar have a portion of the body weathered away on the right side. The second and first lumbars are virtually complete.

Nine thoracic vertebrae are present. Since it is not known how many thoracics were originally present, it is not possible to assign numbers with certainty to the existing vertebrae. As the modal number of thoracic vertebrae for *H. sapiens* is twelve, with very little departure from this number, it may well be that this was also the modal number for *H. africanus*. This cannot be assumed, however. In cercopithecoid monkeys the modal number is usually twelve if thoracic vertebrae are identified by the presence of ribs (Schultz and Straus 1945), but ten if identified by articular facets (Washburn 1963), hence by the nature of vertebral movement. Judged either way the living pongids have thirteen as a modal number, except for the orang which agrees with man in having twelve (Schultz and Straus 1945). The modal number of *H. africanus* is thus likely to have been either twelve or thirteen. In view of this uncertainty it is preferable not to assign a number to each of the thoracics since it is the first few that are missing.

The last thoracic is intact, but the third and ninth last are represented by the body alone. The other vertebrae are reasonably complete with damage mostly confined to missing tips of processes. The damage is listed more specifically in the section on the thoracic spine.

Lumbar Spine

According to my interpretation, there are six lumbar vertebrae present in Sts 14. According to conventional criteria, what I have regarded as the first lumbar would be regarded as a thoraco-lumbar vertebra since it clearly had a riblike structure articulating on the right side but not on the left. On the left side there is a well-developed transverse process, but a small foramen runs vertically through the base of the process, just lateral to the pedicle. This seems to represent part of the space that exists on the right side between the articular facet for the head of the appropriate rib and that a little farther back for articulation with the tuberculum of the rib. The facet for the head of the rib on the right side, however, is not the expected sort of shallow depression that occurs on vertebrae that have a complete capitular facet; instead it is a distinct blunt process with a convex facet on the top. Also, the facet for the tuberculum of the rib is rudimentary and set close to the facet for the head of the rib. This suggests that there was not, in fact, a proper rib on this side. It seems probable that the transverse process was not firmly attached but articulated with the rest of the vertebra in much the manner of a rib. Since the ribs of thoracic vertebrae and the transverse processes of lumbars are apparently serially homologous, this partial costalization of the transverse process of the first lumbar is not remarkable. The above interpretation is further supported by the fact that the last thoracic vertebra has the typical structures associated with the last rib: a complete articular facet for the rib head, wholly on the vertebra, and not a demifacet, and it has the typical shallow hollowed structure seen in *H. sapiens*, not the aberrant structure seen on the first lumbar in this case. Also the facet for the tuberculum is properly placed and shaped in relation to the capitular facet—which is not the case on the first lumbar. Finally, the superior articular facets of the vertebra that seems to me to be the first lumbar have the unmistakable size, shape, and orientation found in a lumbar vertebra, characteristics quite different from those of thoracic vertebrae. The next vertebra above has the typical thoracic type of superior articular facets. It seems clear that functionally this vertebra is a lumbar vertebra; hence it is designated as the first lumbar.

The remaining five are typical lumbar vertebrae. The transverse processes are quite long and fairly slender, those of the third vertebra being the longest. Those of the latter and of the fourth lumbar curve upward in a cranialward direction. The sixth lumbar has much more robust transverse processes than have the others, as would be expected, and is sufficiently far recessed into the pelvis to allow powerful ligamentous connections onto the ligamentous surface of the auricular region. The neural spines are fairly slender and, in those cases where the relevant part is preserved, are not much expanded at the apex. The articular facets are well-developed and typical of lumbar vertebrae: the superior facets are directed strongly medialward and a little posteriorly, while the inferior ones are directed strongly laterally and a little anteriorly (see Figs. 53–54).

The pedicles are a little more slender, proportionately, than is commonly the case in *H. sapiens*—though not outside the range of variation in the latter—and they are also relatively high. The effect of this is that the intervertebral notches are relatively large. It is not known, of course, whether these features were characteristic of the population as a whole or merely of this individual. The relatively high pedicles cause the vertebral foramen to be relatively high also.

The bodies are well developed, with the characteristic change in shape from the first to the last lumbar seen in modern man. The shape of the ends of the bodies changes from being relatively deep antero-posteriorly and narrow transversely at the top of the lumbar spine, to being relatively wide transversely and shallow antero-posteriorly at the lower end. The bodies are somewhat wedge-shaped toward the lower end of the lumbar spine, being vertically thickest away from the neural arch. This feature is consistent with well-developed lumbar curvature, bearing in mind that most of the curvature in such cases is due to wedge-shaped intervertebral discs. Hence when the bodies are to some extent also wedge-shaped, the implication is that lumbar curvature was well-developed (see Figs. 55–57).

Table 1 compares the lengths in millimeters of the bodies, in the long axis of the spinal column, next to the vertebral foramen (posterior length) and farthest from the vertebral foramen (anterior length).

From Table 1 it is clear that the bodies become wedge-shaped as the promontory is approached. Because most of the neural arch and the inferior articular facets of the sixth lumbar are missing and have been reconstructed, it is difficult to estimate the degree of wedging of the intervertebral disc between the last lumbar and the sacrum. Judging from the degree of wedging of the bodies of the lower lumbar vertebrae and the marked general similarity of the lumbar spine and the pelvis to those of *H. sapiens*, it seems highly probable that the disc between the sacrum and the last lumbar was strongly wedge-shaped, thus adding considerably to the degree of lumbar curvature.

Schultz (1930) and Schultz and Straus (1945) have investigated in some detail the variation in vertebral number in various portions of the

Table 1
Anterior and Posterior Lengths of *Homo africanus* (Sts 14) Lumbar Vertebrae

Lumbar Vertebra	Posterior Length	Anterior Length
1	19.6	—*
2	19.9	19.3
3	19.5	19.7
4	19.1	19.4
5	17.3	18.4
6	16.8	18.8†

* This measurement is not given because the anterior superior margin of the body is weathered. It is clear, however, that the anterior length was less than the posterior—as in the second lumbar vertebra.

† This measurement may not be accurate since a small amount of reconstruction is involved. The reconstruction was deliberately conservative; hence this figure is more likely to err in the direction of underestimation rather than overestimation.

spinal column as well as in the column as a whole in the primates. According to Schultz and Straus, the number of lumbar vertebrae in man is remarkably constant at five. Schultz gives the average ($N = 1074$) percentage occurrence of five lumbar vertebrae in *H. sapiens* (including Negro, Mongoloid, and white) as 91.9. The next most frequent number is six, with a percentage frequency of 4.8, followed by four with a percentage frequency of 3.2. Among the living pongids, only the gibbons have a high percentage frequency of five lumbar vertebrae: five is the modal number, but four and six occur more frequently than in *H. sapiens*, with four about twice as frequent as six. The great apes, however, seldom have as many as five lumbar vertebrae. In the orang the modal number is four (76 percent frequency) and five occurs with 10 percent frequency only. The chimpanzee and gorilla also have four as modal number, but three occurs with relatively high frequency (38 percent and 29 percent respectively) and five very seldom occurs. Schultz and Straus found no instances of five lumbar vertebrae in a combined sample of 151 chimpanzees and gorillas, but Schultz found 5.5 percent frequency in 55 chimpanzees and 1.2 percent frequency in 82 gorillas. In no case did these authors find six lumbar vertebrae in any of the great apes.

There is no way of knowing what the modal number of lumbar vertebrae was in *H. africanus*. The fact that the one available lumbar spine had six vertebrae implies that this number was at least reasonably common in the population; it may even have been the modal number, but this need not necessarily have been so and certainly cannot be assumed to have been so on the basis of a single specimen. In respect of this character also, resemblance is closer to man than to the great apes.

Thoracic Spine

The thoracic spine is represented by nine vertebrae. Of these, seven are fairly complete but two, the third and ninth last, are represented by bodies alone. The left pedicle is missing on the eighth last. The fifth to eighth last have the ends of the right transverse processes missing, while the left transverse process and part of the superior articular facet on that side are missing from the second last one. The fourth and eighth last have the neural spine apexes missing.

As in the case of *H. sapiens*, the last thoracic vertebra is transitional between the lumbar and thoracic types. For example, the neural spine is essentially of the lumbar type, jutting out more or less at right angles to the long axis of the vertebral column, is broad cranio-caudally, and is somewhat flattened at the apex. The neural spine of the second last thoracic is already of the sloping, more slender type that is characteristic of the middle thoracic region of the spinal column of *H. sapiens*. Similarly, the transverse process of the last thoracic has a double structure, with a superior and an inferior part. The inferior portion appears to correspond with the transverse process of the lumbar vertebrae. The superior portion appears to correspond with the nonarticular ridge that runs onto the postero-lateral aspect of the superior articular facet. In the last thoracic these two are closer together than are the corresponding structures on the lumbars. The second last thoracic, however, has a single transverse process without the double structure seen in the last thoracic. Furthermore, as in *H. sapiens*, the inferior articular facets are oriented in the same way as are those of the lumbar vertebrae since they articulate with the typical lumbar type of superior articular facet. The last thoracic, and all higher thoracics, have the typical thoracic vertebra type of superior articular facet; consequently the second last and higher thoracics have the typical thoracic type of inferior articular facet. In this feature also the last thoracic is transitional between the lumbar and thoracic vertebral types.

From the last thoracic upward, the bodies become smaller (see Table 2) and increasingly triangular or heart-shaped in cross section.

Table 2
Superior-Inferior Lengths of The Bodies of the Thoracic Vertebrae of *Homo africanus* (Sts 14)

Last thoracic vertebra	19.1 mm
Second last	16.8
Third last	15.4
Fourth last	14.6
Fifth last	14.0
Sixth last	13.6
Seventh last	13.3
Eighth last	12.8
Ninth last	12.3

The lengths here given were measured immediately adjacent to the neural canal.

The last thoracic has a complete articular facet for the capitulum of the rib, but no true facet for the tuberculum on the transverse process. Unlike the usual condition in *H. sapiens*, however, the second last thoracic has a demifacet only for the capitulum, though it is very nearly a complete facet, and there is a distinct facet for the tuberculum on the transverse process. In *H. sapiens* it is usual for the second last thoracic to have a complete facet for the capitulum and none for the tuberculum. As in *H. sapiens*, the thoracic vertebrae higher in the series have inferior and superior demifacets and distinct facets for the tuberculum. In the *H. africanus* specimen, however, the demifacets appear to be proportionately somewhat smaller than is the case in *H. sapiens*. This may simply be a function of the appreciably smaller size of the *H. africanus* vertebrae and trunk in general.

A number of fragmentary ribs were directly associated with the Sts 14 spinal column and presumably belonged to this individual. Substantial parts of five ribs are available, the most complete being a small specimen that is 30 mm long and 6 mm wide and apparently is much of the last floating rib on the right side. It has a single capitular facet and no tuberculum. The other four ribs are appreciably more robust but are slender and lightly built compared to those of *H. sapiens* or of the orang and chimpanzee. In each of these specimens the capitulum is either missing or damaged, but the evidence suggests that the capitulum was not much, if at all, expanded. A good deal less than half of each of the ribs is preserved, and it is not possible to identify with certainty to which thoracic vertebrae they belong. The four most complete ones appear to belong toward the lower end of the thoracic spine. Except for being smaller and more slender than those of a female Bushman mounted skeleton that has a height of 4′ 6″, these ribs appear to have the same sort of curvature and suggest the same sort of thorax shape as that in *H. sapiens* and the hominoids in general; that is to say, broad and shallow rather than narrow and deep (see Fig. 63).

Spinal Column as a Whole

As will have appeared from the foregoing description and illustrations, the Sts 14 spinal column agrees very closely indeed with the corresponding portions of the *H. sapiens* spinal column—much more than it agrees with those of the living pongids, though there is a good deal of general similarity between the spinal columns of pongids and that of man. There are many detailed differences, however, between the vertebral columns of man and pongid in body shape, pedicle height, shape and disposition of the neural spines, degree of differentiation between lumbar and thoracic neural spines, and so on. Where such differences exist, the *H. africanus* column in each case resembles that of man rather than that of pongids. This is hardly surprising in view of the fact that the pelvis in particular, and the rest of the known skeleton in general, of *H. africanus* bears a much closer resemblance in both structure and function to man than to pongids.

The pelvis and many other parts of the skeleton indicate clearly that *H. africanus* was habitually erectly bipedal and had a sacrum that resembled closely that of modern man in both anatomy and orientation. Therefore the spinal column must have been functioning in the position and in the manner characteristic of man. For example, there would have to be lumbar curvature and also considerable flexibility of the lumbar spine. The thoracic spine would not need as much flexibility as does the lumbar spine. Hence there would be a tendency to greater functional and morphological differentiation between these two regions of the spine. In pongids the lumbar spine tends to be reduced in length (see Schultz 1930) and the innominates are elongate; hence the lower portion of the back tends to be rather less flexible than it is in forms with longer lumbar spines. In pongids the commonly quadrupedal posture removes the necessity for the development of changes in curvature of the vertebral column, such as the lumbar and thoracic curvatures of the hominid spinal column. Similarly the weight-bearing characteristics of the pongid and hominid spinal columns are not identical.

It may be argued that since the pelvis and spinal column of hominids must have gone through a phase, however short, of adjusting to the change from quadrupedal to erectly bipedal posture, it need not follow that the spinal column of early hominids should resemble closely that of modern man. This is a reasonable point of view. It seems clear from the available evidence of the *H. africanus* pelvis that the level of adaptation to erectly bipedal posture was very advanced. This implies two things: at the time from which we know *H. africanus* it was not in an early stage of adaptation to erectly bipedal posture and locomotion; the spinal column must already have had the basic characteristics of that of modern man—for example, the orientation of the sacrum clearly implies lumbar curvature. Fortunately it is not necessary to rely on deduction in this case since the available evidence confirms that the characteristics of the column are those of the hominid type and not those of the pongid or monkey type. This fact, consequently, can serve to support the conclusions reached about the pelvis—if such support is needed, which hardly seems to be the case. For if the vertebral column already had the basic curvatures and other features associated with well-established erect bipedalism, which seems to be the case, then this implies advanced bipedal anatomical and functional characteristics in the pelvis. The latter is clearly true according to my reading of the evidence provided by the fossil pelvic material.

This may sound like circular reasoning; what I am trying to say, however, is that the close functional relationship between the pelvis and spinal column in the case of erect bipedalism is such that, given one, the basic characteristics of the other could be predicted. In this instance prediction is not necessary since both are known from direct fossil evidence. But the fact that the basic characteristics of erect bipedalism are clearly expressed in both parts of the body makes the argument for advanced bipedalism stronger than it would otherwise have been. Had the evidence been misread in the case of the pelvis, then the lack of congruity with the spinal column evidence would have made this obvious. This is not the case; therefore the validity of the interpretation of the basic adaptational pattern is strengthened. As will become clear, other available evidence relating to posture and locomotion in *H. africanus* similarly fits and strengthens the picture that has thus far emerged from pelvic and spinal column evidence.

As has already been mentioned, one difference between the *H. africanus* and *H. sapiens* spinal columns is obvious: the vertebrae of the former are appreciably smaller than their *H. sapiens* equivalents. This is true even if peoples of relatively small stature, such as the Bushman from southern Africa, are used for comparison. For example, the area of the superior face of the body of the third lumbar (fourth from the sacrum) of Sts 14 is approximately 438 square millimeters, while that of the equivalent vertebra of a female Bushman specimen taken at random is 1,080 square millimeters.

This indicates a considerable difference in size. Since the cross-sectional area of the body is a better reflection of weight-bearing capacity than general external dimensions of the vertebra would be, the implication is clear that the weight-bearing capacity of the Sts 14 spinal column must have been considerably less than that usual in the Bushman (see Figs. 58, 60, and 61). Similarly, the Sts 14 vertebrae are considerably smaller than those of the chimpanzee or orang.

Although the *H. africanus* (Sts 14) individual was a mature adult that had a pelvis differing little in size from that of a Bushman, the sizes of the vertebrae of this individual are markedly smaller than those of the Bushman equivalents. This indicates, especially in view of the very small weight-bearing surfaces even in the lumbar vertebrae, that the trunk must have been very slender and small, for the most part. The ribs suggest a thorax shape similar to that of *H. sapiens*, though small. But the pelvis indicates that the lower portion of the trunk was chunky and the abdomen probably protuberant. It would thus appear that this female individual was small and lightly built but was very broad across the hips and had a protuberant abdomen.

A point of some interest that was noted earlier is that the transverse processes of the lumbar vertebrae of *H. africanus* (Sts 14) are relatively very long, especially those of vertebrae three and four. In spite of the smaller absolute size of the lumbar vertebrae of the latter as compared to those of *H. sapiens*, the transverse processes of the fourth lumbar, for example, have an absolute span as great as, or slightly greater than, that of Bushman equivalents.

Among the structures that attach to the transverse processes of at least some lumbar vertebrae are two important muscles: psoas major and quadratus lumborum. Both of these are important postural muscles in an erect biped. The former, acting with iliacus, is a powerful flexor of the thigh on the trunk and vice versa. Iliopsoas is therefore an important muscle in walking as well as for such activities as raising the trunk from the hips to a vertical position when the body is supine. As Basmajian (1962) has shown, it also is "... an active postural or stabilising muscle of the hip joint. ..." Quadratus lumborum attaches to the iliac crest and to the last rib, forming a strong band of muscle between the two, and also attaches to the transverse processes of most of the lumbar vertebrae. It is therefore primarily a lateral flexor of the spinal column when acting unilaterally from a stabilized pelvis. If acting unilaterally from stabilized upper attachments, it tilts the pelvis up on the side on which the muscle is acting. As in the case of iliopsoas, there is a stabilizing postural function also; this occurs with the muscle functioning bilaterally to assist in supporting the trunk above the pelvis in a vertical position.

The relatively long transverse processes in Sts 14 provide about as much surface area for attachment as is available in the appreciably larger vertebrae of man. It is not possible, however, to tell from the specimens what area actually was covered by these muscles. But quadratus lumborum attaches to the terminal region of the transverse process; hence the fact that the transverse process is long means that this muscle has relatively efficient leverage in acting on the vertebral column. In this case the moment arm is as long as that in the considerably larger equivalent vertebrae of *H. sapiens*. It is possible that there is another explanation of the length of these processes. Since the pelvis is relatively large for the trunk size, it might have been necessary to have longer transverse processes so that the discrepancy between the degree of projection of the rib cage and the ilium, on the one hand, and the transverse processes on the other, was not too great. Whatever the reason for the long processes, the effect would have been to give quadratus lumborum effective leverage in acting on the spinal column. It seems that this is yet another of the long series of instances where the anatomy of *H. africanus* was such that it allowed efficient function in the erect position.

This single, incomplete spinal column does not give any indication whether it was modal with respect to size. Because it belonged to a female, the probability is that it may be a little on the small side since the adult male population is likely to have included at least some larger individuals. This conclusion is supported by the very fragmentary, isolated vertebrae from the same site as Sts 14. The most complete is Sts 73, which consists of an almost complete body with parts of the pedicles still present.

Judging by the shape of the body, this specimen would appear to have come from the lower portion of the spine. That it is a lumbar vertebra is suggested by the fact that there are no costal facets or demifacets. There is actually an ill-defined shallow oval hollow situated low down on the right pedicle, and at first glance this may seem to be a costal facet. This would identify the vertebra as a last thoracic. The "facet," however, is very ill defined, whereas those on the last thoracic vertebra of Sts 14 are raised up and extremely well-defined, with freestanding edges all round, and are situated a little lower down. Then also, the surface of the "facet" in Sts 73 is not actually composed of outer surface bone but is cellular; hence the surface bone is missing. The slight hollow is therefore probably an artifact resulting from a missing piece of cortical bone in this region.

The vertebra appears to have come from high up in the lumbar spine because (a) the body has a heart-shaped cross-section, rather than the more nearly oval shape found farther down the lumbar spine and (b) the body is not wedge-shaped longitudinally. Usually the lower lumbar vertebrae have the posterior height of the body less than the anterior height in man, and this is also the case in *H. africanus* (Sts 14). Although one cannot be certain of the exact measurements in Sts 73 because of a small amount of damage, the anterior and posterior heights appear to be the same. It seems probable that this was a first, possibly a second, lumbar vertebra.

The chief point of interest about this specimen is that it is somewhat larger than the Sts 14 equivalent and clearly was a little larger than any of the Sts 14 vertebrae. Table 3 compares some measurements taken on Sts 73 with the corresponding values from the first lumbar of Sts 14. From these measurements the interesting point emerges that while Sts 73 is noticeably larger than the first lumbar of Sts 14, there is very little difference in body length between the two. Evidently Sts 73 belonged to a somewhat more robust individual, but one that apparently was not significantly taller with respect to trunk height; had trunk length been significantly greater, presumably vertebral length would also have been longer. One might speculate from this very slender evidence that perhaps size variation in this form occurred more in the form of variation in robustness than in trunk height. Total height of the individual involves other variables also that do not necessarily vary in direct proportion to trunk height. It would therefore be hazardous to conclude that robustness varied more than total height.

Table 3
Comparison of the Vertebral Body Sts 73
with First Lumbar Vertebra of *Homo africanus* (Sts 14)

	Sts 14	Sts 73
Posterior vertebral body length	19.6	20.1
Anterior-posterior vertebral body thickness	18.2	22.2
Maximum superior vertebral body width	22.2	31.7
Superior width of neural canal	12.2	15.6

An obvious suggestion is that the greater robustness of Sts 73 was a result of its belonging to a male individual. Such a conclusion would be consistent with the evidence of sexual dimorphism in the dentition, where canine size seems to vary more than does that of the jaws and dentition as a whole. This suggests that differences in robustness and strength of the individual were greater than differences in over-all size as reflected by a dimension such as height. These are simply speculations concerning the possible implications of this evidence; the evidence itself is much too scanty to be a safe foundation for them.

The much more fragmentary vertebra that was associated with the innominate Sts 65 consists of the neural canal surface of the body and the bases of the pedicles—there is practically nothing preserved of the rest of the body. This specimen is so fragmentary that there is no certainty that it actually belongs to *H. africanus*, though I think it likely that this is so. If it is indeed such, then it provides evidence of an individual that was probably a little larger than Sts 73. The posterior body length was 24.5 mm in Sts 65 as compared to 20.1 mm in Sts 73. Unfortunately, not enough of the vertebra is present to indicate from which portion of the spine it came. In view of its size it probably came from low down in the spine. If this is indeed an *H. africanus* specimen, then it and Sts 73 suggest the probability that larger individuals than Sts 14 existed in the population. As the latter was probably a female, such a conclusion is not unreasonable. This evidence, in conjunction with the more significant evidence in this respect from the cranial material, suggests the possibility that sexual dimorphism was no more than moderate. Obviously much more postcranial material is needed to clarify this point.

These very incomplete specimens (Sts 65 and 73) add no anatomical information to that obtained from the much more complete Sts 14 material, other than evidence relating to size.

Vertebrae from Swartkrans

As has already been indicated, two vertebrae only that appear to be hominid in structure have come from Swartkrans, in the material excavated by the late Dr. R. Broom and myself. Because two hominids are known from this site and vertebrae are not known elsewhere that can be attributed with reasonable certainty to either, there seems at present to be no certain means of assigning these specimens to one rather than to another of *Paranthropus* and *H. erectus* (= "Telanthropus") (see Figs. 64 and 65).

SK 854 is an axis vertebra, but it does not resemble at all closely that of *H. sapiens*. The neural spine is broken off and so is the odontoid process, almost all of the left superior articular facet, part of the left inferior articular facet, and the left transverse process.

The body is much thicker at the lower end than it is at the base of the odontoid process and has a strong keel in the anterior midline. The neural arch is robust and rounded with very little flattening. The superior articular facet is large but has an appreciable degree of concavity, rather than being nearly flat as is usually the case in *H. sapiens*. The inferior articular facet is not set out on a distinct protuberance but looks more like an undercutting of the robust neural arch. The vertebrarterial canal is directed quite sharply medialward and downward immediately behind the inferior flange of the superior articular facet. The transverse process is small, especially the very slender posterior limb.

While this vertebra does seem to be more nearly hominid than those of other known primates, it differs very considerably from that of *H. sapiens*—much more so than do any of the known vertebrae of *H. africanus*, although no axis is known of the latter. On these grounds it would seem much more likely that SK 854 belonged to *Paranthropus* than to *H. erectus* (= "Telanthropus"). *H. erectus*, being more advanced in the direction of modern man than is *H. africanus*, should be more like *H. sapiens* in this respect than is *Paranthropus* since the latter is known to be relatively primitive in a number of important respects.

SK 853 is a lumbar vertebra that is well preserved and almost complete, missing only the tip of the right transverse process and a part of the right superior articular facet. Unfortunately, it belonged to an immature individual.

This specimen has the characteristics of a first lumbar vertebra, although the incomplete development makes it a little difficult to be sure of this. In a number of features it is more nearly like that of *H. sapiens* than is the first lumbar, or the second either, of *H. africanus* (Sts 14). For example, the superior-inferior length of the body is proportionately short, as in modern man, rather than relatively long as in Sts 14. Also, the lipping around the edge of the body face, above and below, is a little more marked in SK 853 and in *H. sapiens* than is the case in Sts 14. In the latter the neural spine is narrower in an inferior-superior direction than is the case in either SK 853 or in *H. sapiens*, in both of which it tends to be broad.

Because no ranges of variation are known for either the Swartkrans specimens or for Sts 14 of *H. africanus*, one is clearly in a poor position to know how stable or otherwise these characters may have been. SK 853 gives the impression, although immature, of being somewhat more like the *H. sapiens* equivalent than is that of Sts 14. This is not true of size since SK 853 is a little smaller than the Sts 14 equivalent. Because it is immature, however, this difference is not necessarily significant. But one would expect a lumbar vertebra of *Paranthropus* to be more robust than is SK 853.

According to the rather unsatisfactory evidence available, it would seem that SK 854 is an axis vertebra belonging to *Paranthropus*, while SK 853 is a first lumbar of an immature individual of *H. erectus*.

After the preceding section had been written, two hominid vertebrae were found at Swartkrans by Dr. C. K. Brain, Director of the Transvaal Museum, Pretoria, in the course of investigations at that site that he initiated in 1965. Dr. Brain has afforded me the opportunity of studying this material, and I hereby gratefully acknowledge his kindness in so doing. An account of them appeared recently (Robinson 1970).

The block of breccia containing the two vertebrae was not actually in situ when found but in a large pile of breccia blocks that had been excavated by lime miners during the thirties. Reliable evidence is available (Brain 1967) that suggests that this material came from the front part of the Swartkrans site, in front of which the breccia pile was situated. If this conclusion is correct, then the two vertebrae came from the same breccia mass that yielded most of the *Paranthropus* as well as all of the *H. erectus* (= "Telanthropus") material excavated from this site by Dr. Broom and myself. Among the fossils removed from the breccia dump by Dr. Brain are a number of hominid cranial pieces that, in my opinion, belong to *Paranthropus*.

The two vertebrae, SK 3981a and SK 3981b, came out of the same block of breccia. Both belonged to adult individuals, probably the same individual, of a higher primate. The only higher primate remains so far found among the many thousands of fossils recovered from the Pleistocene early hominid-bearing sites of South Africa are either cercopithecoid or hominid. Both of these vertebrae can easily be distinguished in size and morphology from those of known cercopithecoids. On the other hand, they bear very considerable resemblance to modern hominid equivalents and especially to those of *H. africanus*. Although it would be most unlikely that pongids would occur in the South African sites, this possibility must be considered. The possibility of the vertebrae being pongid can safely be rejected, however, since they resemble hominid equivalents more than they do those of known pongids—with respect, for example, to relative size of the neural canal and the nature of the neural spine and the transverse process.

Having concluded that the vertebrae are hominid, however, is not the end of the matter since again we have to take into account the presence of both *Paranthropus* and *H. erectus* at Swartkrans. The two Swartkrans specimens are the last thoracic (SK 3981a) and last lumbar (SK 3981b) vertebrae, and to the best of my knowledge these are not known for *H. erectus*. The postcranial skeleton of *H. erectus* seems to have been much like that of modern man, and it is therefore probable that the vertebrae were more like those of modern man than like those of *H. africanus* (Sts 14). This conclusion is suggested also by the specimen SK 853, discussed above, tentatively attributed to *H. erectus* because it resembles the equivalent vertebra of modern man more than it does that of *H. africanus* (Sts 14). This last point must be treated with reserve since (a) the assignment to *H. erectus* is tentative and not certain and (b) the vertebra is immature.

The axis vertebra SK 854 is distinctly hominid in general appearance but nevertheless does not closely resemble the modern human equivalent; it was therefore assigned to *Paranthropus* and not *H. erectus.* The lumbar specimen, SK 853, resembles the modern human equivalent more than it does that of *H. africanus* (Sts 14); it was therefore assigned to *H. erectus.* In the cases of both SK 3981a and b the resemblance is closer to the *H. africanus* (Sts 14) equivalents, and for this reason they are here tentatively assigned to *Paranthropus* rather than to *H. erectus.* Reasons for this conclusion are discussed more fully below.

SK 3981a: Last Thoracic Vertebra (Figs. 66 and 67).

This specimen is believed to be a last thoracic vertebra because:

(1) the superior articular facets are typical of thoracic vertebrae but the inferior facets are of the lumbar pattern;

(2) a single complete costal facet is present bilaterally for the capitulum of a rib, and this is situated partly on the pedicle and partly on the body.

This vertebra is thus both the last rib-bearing vertebra and the one that connected a series of vertebrae with only thoracic type articular facets with a series having only lumbar type articular facets.

This vertebra can be distinguished from the corresponding modern human one by a number of features:

(1) It has a proportionately long neural spine that projects approximately at right angles to the superior-inferior axis of the vertebra. In modern man the neural spine is proportionately shorter, having a total length (measured from the neural canal to the apex) that is no longer than the anterior-posterior thickness of the body.

(2) The neural canal is proportionately large. The fossil vertebra is appreciably smaller in over-all size than the modern human homologue, but the neural canal is of the same absolute size in the two. In lateral view this causes the absolute size of the inferior notch of the fossil to be the same as that in the appreciably larger corresponding vertebra of modern man.

(3) The mammillary-transverse process complex is different in the Swartkrans specimen. In the latter the mammillary process is relatively well-developed but slender, resembling a small lumbar transverse process. The transverse process homologue is well-separated from the mammillary process and is no more than a small tubercle well below the latter.

The only other last thoracic vertebra of an early hominid known is that from Sterkfontein (Sts 14), and there is appreciably more resemblance between it and the Swartkrans specimen than there is between the latter and the modern human homologue. For these reasons it seems probable that the Swartkrans specimen is more properly assigned to *Paranthropus* than to *H. erectus*.

Although the resemblance of the Swartkrans specimen to the Sts 14 equivalent is greater than that to the modern human homologue, they are by no means identical. For example, the Swartkrans specimen is larger; the maximum width of the superior face of the body is 123 percent of that of the *H. africanus* specimen. The maximum superior-inferior length of the body, nearest to the neural canal, in the Swartkrans specimen is only 105 percent of that of the Sterkfontein specimen. The neural spine is proportionately larger in the Swartkrans specimen; its length is 107 percent of the maximum antero-posterior diameter of the superior face of the body as compared to 89 percent in the Sterkfontein specimen. The spine is also different in being orientated at right angles to the superior-inferior axis of the body, while that of modern man and of *H. africanus* is directed downward at an appreciable angle. The neural canal is relatively slightly larger in the Swartkrans specimen: its width is 55 percent of the maximum width of the superior face of the body and its antero-posterior height is 79 percent of that dimension, while the corresponding values in the Sterkfontein specimen are 50 percent and 66 percent respectively.

In the Swartkrans specimen the costal facet, situated on the pedicle and the adjacent part of the body, is very prominent, standing well out from the rest of the vertebra and facing quite strongly caudalward. In the *H. africanus* specimen the costal facets are well-defined but are not nearly so protuberant as in the Swartkrans specimen, and they face laterally. In both of these features the Sterkfontein specimen is considerably more like modern man than is the Swartkrans specimen.

In the *H. africanus* (Sts 14) specimen the transverse processes are proportionately appreciably better developed than are those of the Swartkrans specimen. In this respect also the Sterkfontein specimen is more like modern man than is the one from Swartkrans.

The Swartkrans specimen has rather more heart-shaped superior and inferior faces of the body than has the Sterkfontein specimen. In this feature the former is closer to modern man. The difference is small, however, and there is some variability in the modern human condition; larger fossil samples might well exhibit much overlap with each other and with modern man in this respect.

SK 3981b: Last Lumbar Vertebra (Figs. 68 and 69).

This specimen consists of virtually the entire body, the base of the right pedicle as well as much of the left pedicle, and almost all of the transverse process.

As already discussed, the affinities of this specimen are hominid. It does not resemble the modern human homologue very closely, but is considerably more like the corresponding vertebra of *H. africanus* (Sts 14). Unfortunately, since all of the articular facets are missing as well as most of the neural arch and spine, there is not much detail available.

The body is quite robust, and the superior face has a cross-sectional area that is appreciably greater than that of the inferior face. This indicates the body of the first sacral vertebral element was relatively small, as it is in *H. africanus* (Sts 14). This feature is not characteristic of modern man. The preserved transverse process, on the left side, is very large compared to that of *H. africanus* and is considerably larger even than that of modern man, in whom the absolute size of the vertebra as a whole is appreciably greater. As in the latter two forms, the transverse process projects superiorly as well as laterally. In the Swartkrans specimen there is a tubercle on the inferior aspect of the transverse process that covers quite a large area (14.5 × 8.5 mm) but is not prominent. In *H. africanus* (Sts 14) this process is quite different and is very prominent and forms a substantial part of the transverse process mass. Neither is like the usual condition in modern man, but that of the Swartkrans specimen is very slightly more like it than is that of Sts 14.

The neural canal is wide (21.4 mm) but evidently was low judging from what remains of it and from the condition in Sts 14. While much of the neural arch is also missing in Sts 14, the presence of neural arches above the last lumbar and the completeness of this region in the first two sacral elements below it made it possible to reconstruct the last lumbar neural arch with some confidence. The superior-inferior length of the Swartkrans body is greater anteriorly, i.e., farthest from the neural canal, than it is posteriorly—21.3 and 19.0 mm respectively. This "wedgedness" of the body, found also in *H. africanus* and in modern man, argues in favor of well-developed promontory and lumbar curvature.

As in the case of the last thoracic vertebra, the last lumbar from Swartkrans has a superior-inferior body length that is only slightly greater than that of the Sterkfontein homologue. The cross-sectional area is appreciably greater in the former, however; in the case of the superior face, the width is 128 percent of that of the Sts 14 specimen.

Discussion

Of the four known Swartkrans vertebrae, three (axis, last thoracic, and last lumbar) apparently belong to *Paranthropus* and one (immature first lumbar) apparently belongs to *H. erectus*. Because no axis vertebra is known of *H.*

africanus, the only comparable vertebrae of *H. africanus* and *Paranthropus* are the last thoracic and last lumbar. These are more like each other than either are like the modern human equivalent, especially in the case of the last lumbar. The probability is high that SK 3981a and b came from the same individual. This conclusion is based on two points: (1) because hominid vertebral remains are so very rare in the South African cave sites, two occurring together are likely to have come from a single individual; (2) the size relationship between the two specimens closely matches that between the corresponding vertebrae of Sts 14.

The *Paranthropus* vertebrae are larger, more robust, and had longer projections than the *H. africanus* (Sts 14) equivalents and, at least in the last thoracic, had a proportionately larger neural canal. It seems reasonable to conclude from this that the individual from which they came was somewhat more heavily built and more muscular than the *H. africanus* (Sts 14) individual. On the other hand, the great similarity in absolute superior-inferior length of the vertebrae in the two forms provides no evidence of difference in stature. This interpretation depends, of course, upon the premise that the proportions between the different segments of the body were essentially the same in the two forms, and we actually do not know whether this was the case. As will emerge in a later chapter, there are slender grounds for suspecting that the *Paranthropus* lower limb was proportionately shorter than that of *H. africanus*. If this was indeed the case, then the more slender *H. africanus* (Sts 14) individual with a proportionately longer lower limb may have been a little taller than the more robust SK 3981 individual.

The single *H. africanus* (Sts 73) body (first or second lumbar) is similar in superior-inferior length to the *Paranthropus* SK 3981a vertebra (last thoracic), but the cross-sectional area is slightly greater in Sts 73. Allowing for the fact that the latter is from slightly farther down the vertebral column, this suggests that the Sts 73 and SK 3981 individuals were roughly similar with respect to both stature and robustness. From other evidence presented in this work, such as the femur size and robustness and larger innominate and proportionately greater protuberance of the anterior part of the iliac blade in *Paranthropus*, it seems that the body of *Paranthropus* in South Africa was appreciably more robust than that of *H. africanus*. If this conclusion is correct, then one might speculate that the SK 3981 individual was female. The Sts 14 individual of *H. africanus* is fairly clearly female in terms of pelvic features, and the larger Sts 73 is therefore more likely to have been male. It might thus be argued that there was probably overlap in size between males of *H. africanus* and females of *Paranthropus*. In view of the slenderness of the actual evidence now available, however, these conclusions can be no more than speculation at present.

The available evidence indicates that not only are the early hominid vertebrae distinguishable from those of modern man, but also they are distinguishable from each other. Owing to the scantiness of the evidence at present, it is not possible to judge with any degree of reliability whether increasing the samples to adequate size would reveal the present differences as genuine taxonomic differences or merely as individual differences of no taxonomic significance. Because other evidence not related to vertebrae suggests that appreciable and significant differences existed between the two taxa, it is only to be expected that differences also existed in the case of the vertebrae. The differential in robustness between the known vertebrae of the two types might be a genuine difference since this is supported by other evidence. It must be stressed again, however, that these conclusions based on vertebrae are tentative and much more evidence is needed before they can be adequately tested.

9 The *Homo africanus* Femur

Material

The *H. africanus* femur is known from a very small number of specimens, all of which are in the collection of the Transvaal Museum, Pretoria. These are as follows:

Sts 14 — the proximal end of a femur from the left side of an adult individual of which the pelvis and much of the spinal column are also preserved.

Sts 34 — the distal end of the femur from the right side of an adult individual.

TM 1513 — the distal end of a femur from the left side of an adult individual.

All of these specimens are from the Lower Breccia at Sterkfontein.

The proximal piece of femur of Sts 14 was described briefly by Broom and Robinson (in Broom, Robinson, and Schepers 1950); TM 1513 has been discussed at some length by various authors, for example Broom (in Broom and Schepers 1946); Le Gros Clark (1947a and b, 1955, and 1957); Kern and Straus (1949); while Sts 34 was discussed briefly by Broom and Robinson (1949).

Proximal End

The single proximal end, of Sts 14, of a femur known at present is poorly preserved but instructive in some very important respects. It had suffered a considerable degree of weathering before being incorporated into the breccia deposit of the cavern. This has caused the bone to develop a large number of small cracks that follow the pattern of organization of the bone. There has also been a more serious cracking due to mechanical damage. The latter has also warped the shaft. The head itself is gone, along with a small portion of the neck. The whole of the distal end of the bone is missing, as well as much of the lesser trochanter and superficial parts of the greater trochanter (see figs.70–73).

The specimen would thus seem to be a most unpromising one; however, much of real value can in fact be learned from it. The size of the head can be determined with reasonable accuracy because the actual acetabulum into which it fitted is present, though somewhat damaged, on the more complete pelvis (Sts 14) and the acetabulum from the other side of the pelvis is complete and undisturbed. Furthermore, the head was fossilized but lost in the recovery process, and a cast of much of it was present in the breccia surrounding the femur. One portion of the natural cast included part of the neck as well as the adjacent portion of the head, thus fixing the position of the head in relation to the neck and hence also the length of the neck. Another portion of the natural cast represents a roughly circular area of the head surface 22 mm in diameter, and the relation between this and the other patch of natural cast is fixed because they are both in a single piece of solid breccia, or were at the time the specimen was recovered. The head was reconstructed by Broom in plaster while the bone was still in the breccia in order to make use of the natural casts.

The head is thus in the correct position as reconstructed and is also of approximately the correct size. Careful comparison with the perfectly preserved acetabulum of the right innominate of the same individual indicates that the head has been reconstructed slightly too small, with a maximum diameter of 31 mm instead of the more probable 33 mm. The exact limits of the articular surface of the head, however, are not known with certainty except for the one small area where a natural cast included portions of the neck and head at their junction.

The head had a diameter of roughly 33 mm. The distance from a point on the surface of the head farthest from the shaft to the farthest point on the greater trochanter in the axis of the neck is 68 mm. The neck length varies depending on where it is measured. The distance from the estimated position of the edge of the head to the nearest point on the greater trochanter immediately posterior to the deep trochanteric fossa is approximately 27 mm, the approximation being due to the slight uncertainty concerning the exact position of the edge of the head. The femoral tubercle is well-developed, and the shortest distance from it to the head margin (known fairly accurately in this region) is 14 mm.

The anterior-posterior thickness of the neck is 13 mm, and the superior-inferior thickness is 22 mm approximately midway along the neck. These figures may be slightly inaccurate because of cracks in the bone, but the order of magnitude is correct because the specimen is least damaged in this area; hence the error is not likely to be large.

The distance from the most medial point on the femoral tubercle to the farthest point on the greater trochanter is approximately 29 mm.

The shaft has suffered considerable damage; hence its exact shape and diameter at any point are not exactly known. Quite reasonable estimates are possible, however. The best estimate for the diameter well away from the proximal end seems to be 20 mm. It is difficult to see whether the shaft varied a little in thickness or shape along its length, but if so, the variation was not very great.

It would be of especial interest to know the total length of this femur as it originally was because it is the most complete long bone so far known of South African early hominids. An attempt has therefore been made to estimate this from the existing piece, which is 198 mm in length. The length is given by

Broom (in Broom, Robinson, and Schepers 1950) as 215 mm. Apparently this was measured along the curve of the bone, though my attempts to do this give a figure of 201 mm.

Broom and Robinson (in Broom, Robinson, and Schepers 1950) estimate the original length of the femur as "about 310 mm." The estimate was actually made by Broom, and I do not now remember how he arrived at this figure and no indication is given in the publication concerned. Femur length, as used here, refers to the bicondylar length; the shortest distance between two parallel lines, one of which is tangential to both distal articular condyles and the other tangential to the head. This is thus maximum length as measured on a measuring board.

The best guide to the probable original length of this femur would seem to be the evidence for a moderately developed linea aspera in the region of the broken end of this specimen. This becomes sufficiently definite to be called a linea aspera some 50 mm from the broken end of the shaft. In femora of modern *H. sapiens* the linea aspera is variable in its length and degree of development but is well defined and of roughly uniform width for a distance that is very approximately one-third of the distance from the deep trochanteric fossa to the distal extremity at the knee. The linea aspera occupies approximately the middle third, though the distal "third" is commonly slightly shorter than the others. There is nothing exact about these proportions, but they do provide a rough guide for guessing at the possible length of the *H. africanus* bone. Naturally one does not know that similar proportions characterized the latter, but in view of the close similarity of pelvic and spinal column features in *H. africanus* and *H. sapiens*, as well as the evidence that the former was an habitual erect biped, it seems that such an assumption is not totally unreasonable. As will appear later, there are some means of checking on this assumption.

On this basis the femur length of Sts 14 would have been roughly 350 mm. This estimate obviously involves the assumption that the *H. africanus* femur was already as advanced as is that of *H. sapiens*. In this particular context "advanced" means relatively long since man has a femur—hence lower limb as a whole—that is proportionately very long as compared to those of the great apes. We have seen that the postcranial skeleton of *H. africanus* has many advanced features in which it resembles man much more closely than it does other higher primates. This might suggest that it is also advanced with respect to elongation of the lower limb. This is not an unreasonable view to take since some of the functional features already discussed, such as the proportionately very short ischium, make the most sense in combination with a lower limb that was already at least reasonably elongate in the fashion of modern man. But to assume the elongation is really begging the question;

what is needed is independent evidence that indicates that the elongation had already occurred.

Such evidence is available. In Chapter 2 we noted that if the percentage ratio of femur length to ischium length is calculated for *H. africanus* (Sts 14) using the preserved length of the femur as the total length, a ratio is obtained (440) that is considerably above the mean values for the great apes (308–66) and a little below that for the gibbon (504). The implication here is clear that some elongation of the femur had already occurred for an obviously incomplete bone, having the entire distal articular end missing, to have so high a ratio value. Furthermore, the smaller of the two specimens of the distal end of the femur (TM 1513) is also from the left side and appears to be of about the correct size to fit Sts 14. There is no overlap between the two specimens since the linea aspera is still compact at the point of breakage in Sts 14, whereas the supracondylar ridges of TM 1513 are well-separated still at the point where the shaft is broken; the latter break is therefore much farther down the shaft than is the break in Sts 14. Furthermore, since the shaft increases markedly in diameter toward the distal end, there is an obvious width discrepancy between the two at their respective breaks. If one assumes that the distal end did fit directly onto Sts 14 with nothing missing in between, then the femur length would be 271 mm and the percentage ratio of femur to ischium in this individual would be 602. The latter value is appreciably higher than the mean value for the Australian Aborigine (about 560), a race that tends to have rather long lower limbs.

It is very difficult to avoid the conclusion that even 271 mm in this case is underestimating femur length. Even if it is assumed that the linea aspera was proportionately shorter in *H. africanus* than is usual in *H. sapiens*, and that TM 1513 is too large for Sts 14 (which certainly does not appear to be the case), the fact that at the break on Sts 14 the linea aspera is compact and single while at the break on TM 1513 the diverging supracondylar ridges are at least 1.5 cm apart and that the two shafts are appreciably different in thickness where they are broken, indicates that a substantial piece of shaft would be necessary in order to join the two specimens in an anatomically acceptable manner. A length of 271 mm therefore appears rather clearly to be an underestimate of the original length.

From this it may be concluded that the femur of this particular *H. africanus* individual was proportionately about as long as that of man. And yet the previous estimate of 350 mm, based on a femur divided in thirds with the linea aspera occupying the middle third, is almost certainly too long as the distal "third" is in fact usually less than a third. On the other hand, the Sts 14 plus TM 1513 estimate of 271 mm is almost certainly an underestimate. The average of the two is 310.5, which, interestingly enough, is virtually identical with Broom's estimate of 310 mm. Taking all the evidence into account, it seems to me unlikely that this femur could have been less than 300 mm in length or more than 320 mm; Broom's estimate of 310 mm seems a very probable figure.

Since this chapter was written an interesting paper on reconstructing the length of the Sts 14 femur has appeared (Lovejoy and Heiple 1970). Their technique of reconstruction is entirely different from that used by me. It depends upon using the interacetabular width of the pelvis, the distance from the acetabular floor to the femoral shaft axis along the neck axis, the femoral bicondylar angle, and the assumption that the medial face of the distal end of the femur moved in a plane close to the mid-sagittal plane. From this information an elongate triangle is contructed having as its sides part of the mid-sagittal plane, the neck axis produced to meet the mid-sagittal plane, and the shaft axis produced downward to meet the mid-sagittal plane. From this the bicondylar length of the femur is deduced. This method is very sensitive, as the authors point out, to variation in the degree of shaft obliquity (their bicondylar angle). The value they obtained for shaft obliquity is considerably higher for TM 1513 (14°) than that found by Le Gros Clark (7°) or my value of 9°. The latter is close to the mean for modern man (10°) while the figure found by Lovejoy and Heiple is near the upper extreme for modern man (17°—Parsons 1913–14; fide Le Gros Clark 1947). The femur length that Lovejoy and Heiple prefer from their investigation is 276 mm, with a possible range for the true value of 210–330 mm. Using Le Gros Clark's shaft obliquity value of 7°, however, yields them a length of 603 mm, and my value of 9°—of which they were not aware—would give a value in the region of 400 mm. These lengths are manifestly much too great. Their preferred length of 276 mm would allow 5 mm only between Sts 14 and TM 1513 (which they in error call Sts 1513), which is obviously much too little for the transition from a parallel-sided shaft with a linea aspera to a steadily expanding shaft width with supracondylar lines already separated by 1.5 cm. Even assuming that TM 1513 came from a slightly larger individual than Sts 14, which does not appear to be the case, 276 mm would be too short to allow completion of the Sts 14 shaft with the anatomical features known from the same site from TM 1513 and Sts 34. The range of 300–320 mm with a preferred value of 310 mm therefore still seems to be preferable.

The value of 310 mm may now be compared with the mean lengths of pongid femora and that of *H. sapiens* by means of the ratio and modified ratio diagrams based on the twelve pelvic dimensions as discussed in Chapter 2. Doing this leads to the conclusion that, proportionately, the orang has the shortest femur, that of the gorilla is very slightly longer, and that of the chimpanzee is appreciably longer, but all of these are shorter than that of modern man. The gibbon, however, has a femur that is proportionately longer than that of modern man. The estimated value for Sts 14 falls almost exactly on the zero line representing *H. sapiens*. In this case very little difference is made if an estimate of 300 mm is used; Sts 14 falls immediately below the zero line, and in the other case it falls immediately above. In this instance, mean dimensions for females were used (taken from Schultz [1937] for pongids and from Davivongs [1963] for the Australian Aborigine). The gibbon value may be slightly too high since in this case the sample included males also. The relative lengths of the pongid femora obtained in this manner agree well with those obtained by Erikson (1963) in comparing femur length to trunk length

Table 4
Maximum Femur Length

		N	Mean	Observed Range
Medieval English	♂	174	458.3	409–543
	♀	102	418.9	358–477
Japanese	♂	44	407.7	—
	♀	20	379.8	—
Australian Aborigine	♂	150	447.7	405–502
	♀	110	423.6	378–470
American Negro	♂	122	449	379–480
	♀	111	416	355–473
Homo africanus (Sts 14)		1	310	—
Gorilla	♂	84	376	341–423
	♀	54	311	285–352
Chimpanzee	♂	38	300	256–333
	♀	58	286	249–318
Orang	♂	32	269	229–95
	♀	29	234	216–61
Gibbon	♂♀	44	197	172–217

The figures for the gorilla, chimpanzee, orang, gibbon, and Negro were taken from Schultz (1937); those for Medieval English, Japanese, and Australian Aborigine were taken from Davivongs (1963).

in various primates. According to Erikson's data also, the gibbon has a relatively very long femur. The apparent anomaly that the gibbon femur length–ischium length comparison indicates that this form has a shorter femur than either *H. sapiens* or *H. africanus* (Sts 14) results from the fact that the gibbon has a proportionately long ischium, which is exceeded in relative length only by that of the gorilla among the pongids.

This comparison shows, then, that at an estimated length of 310 mm, or even 300 mm, *H. africanus* (Sts 14) had a femur of proportionately the same length as is usual in the Australian Aborigine female. Other comparisons confirm that the *H. africanus* femur is very similar to that of modern man, as is shown in tables 4 and 5.

Table 5
Some Hominoid Femoral and Sacral Lengths and the Ratio Between Them

	N	Femur Length	*N*	Sacrum *f**	$\frac{f \times 100}{\text{femur}}$
Chimpanzee	96	291.5	8	47.3	16.2
Gorilla	138	350.6	14	62.5	17.8
Orang	61	252.4	14	46.5	18.4
Modern man	232	417.4	40	55.2	13.2
Sts 14	1	310	1	35.6	11.5

Femur length data from Schultz (1937), except for Sts 14.
*Length in the sagittal plane of the first two sacral vertebral bodies.

Turning now to the anatomy of the proximal portion of femur of Sts 14, an obtrusive feature is strong curvature of the shaft. This clearly is a postmortem artifact: transverse breaks and cracking are responsible for the curvature, which is, furthermore, in the opposite direction to that seen in higher primates in which the shaft is curved. That is to say, the shaft in this specimen is concave forward. Consequently it is not now possible to determine what curvature the shaft may have had originally.

Since the head is missing, little can be said about its structure. The natural casts of parts of its surface already referred to and the structure of the acetabulum in this individual indicate that it probably had features like those of *H. sapiens*. This statement does not imply distinction from other higher primates since there is great similarity among them in this respect with the head forming approximately three-quarters of a sphere with a clearly defined margin that is not entirely symmetrical in its contour.

Table 6
Some Hominoid Acetabular Widths and Femoral Lengths Showing Ratio Between Them

	N	Acetabular Width	*N*	Femur Length	$\frac{\text{A.W.} \times 100}{\text{Femur}}$
Chimpanzee	20	40.7	96	291.5	14.0
Gorilla	20	55.5	138	350.6	15.8
Orang	20	41.8	61	252.4	16.7
Gibbon	26	20.5	44	197	10.4
Modern Man	50	53.9	232	417.4	12.9
Sts 14	1	39.2	1	310	12.6

Femur length data from Schultz (1937), except for Sts 14.

The neck is well defined and set at an angle of 118° to the shaft axis. Owing to the distortion suffered by the shaft, this angle may be in error by a small amount. The proximal portion has suffered less damage than the more distal parts of the specimen; hence this angle value seems to be reasonably reliable. This value is within the range of variation for modern man: Davivongs (1963) found a mean value of 127.3° and an observed range of 114–39° for female Australian Aborigines ($N = 110$). The same author quotes Parsons (1914) as having found a mean value of 125.5° for female femora in a sample of Medieval English and Schofield (1959) a value of 137.6° for a Maori sample.

As is usual in higher primates, the neck is deeper (in a superior-inferior direction) where it joins the shaft, narrowing as it passes from the shaft to the head. As the measurements already given indicate, the neck is considerably flattened in an antero-posterior direction and the long axis of a cross-section does not coincide with the long axis of the shaft; as in *H. sapiens*, it is at a small angle to the latter. This angle is not easy to measure and may have been altered by postmortem distortion. As near as I can determine, it is 18°.

The neck is relatively long: since it is of quite different lengths depending on where it is measured, such a statement is not easy to document even though it is an obvious impression from visual observation. It can be demonstrated by comparing the head diameter to neck length in obvious places, such as the minimum distance between the lesser trochanter and the adjacent edge of the articular surface of the head, or the shortest distance between the greater trochanter near the deep trochanteric fossa and the head margin. In Sts 14 the head diameter, generously estimated according to acetabulum size as well as to the other criteria already discussed, is approximately equal to the two distances referred to, the former being slightly greater than the latter of these two. The usual condition in *H. sapiens* and the great apes involves a head diameter that is proportionately significantly greater; superimposed upon the neck, it would include much or all of the lesser trochanter in the one case and almost all of the greater trochanter in the other.

In Sts 14 much of the surface of the greater trochanter and of the femoral tubercle is missing and almost the entire lesser trochanter also. There had been a well-developed femoral tubercle, but since the surface of the bone is gone it is not now possible to see whether there was any clear surface distinction between the tubercle and the adjacent portion of the greater trochanter anterior to the deep trochanteric fossa. Manifestly there was strong development of this portion of the greater trochanter and of the femoral tubercle into a mass that is some 14 mm wide at least and that extends, from the bottom of the deep trochanteric fossa, for about 13 mm along the superior surface of the neck.

In the great apes there is normally no detectable development of a tubercle equivalent to the femoral tubercle of man and consequently there is no development of the greater trochanter onto the superior portion of the neck. In *H. sapiens* this region of the greater trochanter and adjacent portion of the neck differs from that of the great apes primarily in the clear development of a femoral tubercle. It seems very clear from the strong development of a tuberosity in this particular region of the Sts 14 specimen that there must have been a tubercle of substantial size, evidently relatively larger even than that usual in *H. sapiens*.

The implication is, therefore, that the ilio-femoral ligament in this individual of *H. africanus* was specialized in the form of a Y-shaped ligament of Bigelow. This is a specially strengthened form of the ligament that has developed in response to the more specialized functions in erect bipeds, where prevention of hyperextension of the pelvis on the thighs in the erect position is an important function. There is little or no need for this particular capacity of the ilio-femoral ligament in quadrupeds.

The greater trochanter is well developed and so is the deep trochanteric fossa. No conclusions can be drawn about special features of the greater trochanter since most of the bone surface is missing. Because the nature of the pelvis and other skeletal parts indicate rather clearly that the hip region approximated that of *H. sapiens* very closely in both structure and function, it seems probable that the features of the greater trochanter similarly will have resembled those of *H. sapiens.*

The area of surface damage extends onto the position of the quadrate tubercle; hence nothing can be said about it either. But the area between the quadrate tubercle and the lesser trochanter is relatively well preserved, and the posterior intertrochanteric line is not developed into a well-defined trochanteric crest but consists of a generalized thickening of the region between the two trochanters.

Almost the whole of the lesser trochanter is missing, though its exact position is clearly indicated by the rising bone surface forming its base. Nothing can be said about its shape or the extent to which it protruded. It is in a slightly unusual position compared to that of *H. sapiens.* In the latter the lesser trochanter is so situated that when the femur is viewed directly from the back, the trochanter protrudes past the medial border of the shaft. It is, in fact, situated almost on the medial border.

In the *H. africanus* specimen this is not the case since the base of the lesser trochanter is situated well onto the posterior face of the trochanteric region. That is to say, it is rotated farther round onto the posterior face from the position that it occupies in *H. sapiens.* In this respect it approaches the condition commonly seen in the great apes, in which the position of this trochanter is somewhat variable but tends normally to be more posterior in position than is the case in *H. sapiens.* Having a single specimen only of *H. africanus* makes it impossible to say whether this individual is modal in this respect. As will emerge later, two femora of *Paranthropus* have the relevant region well preserved. In one the lesser trochanter is situated as in the *H. africanus* specimen, while in the other it is more medial in position but not quite as much as is usual in *H. sapiens.* In view of this evidence it seems probable that the more posterior position is a common feature of *H. africanus* and *Paranthropus.*

It is not clear to me what significance these differences in position of the lesser trochanter might have in functional terms. The amount of variation is such that perhaps differences in position of the order here involved may not be significant in terms of function.

Shaft

The rest of the Sts 14 femur is so damaged that comparatively little can be learned reliably from it, although some information regarding the linea aspera is available.

There is slight indication of a spiral line 4–5 mm medialward of the base of the lesser trochanter. Unfortunately, a spiral crack follows the spiral line—perhaps the crack occurred here because of osteone orientation related to the spiral line; other nearby cracks trend parallel to the shaft long axis. There is a gluteal ridge almost on the lateral border opposite the lesser trochanter. This is a combination of a slight hollow bordered by a low ridge. But here again cracks following the direction of the gluteal ridge make it difficult to follow the detail of the ridge for any distance. The cracks following the ridge and those following the spiral line trend toward each other as they pass down the shaft, in just the manner and position of the lower parts of the spiral line and gluteal ridge of *H. sapiens*. Directly below these, about 55 mm farther down the shaft, the bone is well enough preserved to reveal what clearly is a linea aspera. This is a raised ridge with a flat top some 6 mm wide, which has well-defined medial and lateral margins over a distance of roughly 20 mm to the break below which the rest of the femur is missing. On either side the surface of the shaft slopes quite sharply away.

This linea aspera is not as pronounced as is commonly the case in *H. sapiens*, where it often forms a high narrow crest, but is of the *H. sapiens* type in principle and is well defined, especially in view of the fact that this femur is appreciably smaller and more gracile than is that of *H. sapiens*. In fact it is surprising that Broom and Robinson (Broom, Robinson, and Schepers 1950) made no mention of it. I am also in disagreement with these authors with regard to their statement that the bone is too damaged in the appropriate area to show whether there is a gluteal tuberosity. It seems to me quite clear, both to the naked eye and under the microscope, just where the proximal end of the gluteal ridge is and what its nature is; farther down the detail is obscured by cracking.

The position of the gluteal ridge corresponds with that in *H. sapiens* in starting as high up the shaft as the top of the lesser trochanter and also in being positioned on the posterior face of the bone rather than on the lateral surface. In pongids the gluteal tuberosity is situated on the lateral face and starts appreciably lower down the shaft.

Where the shaft is broken the compact bone is roughly 5 mm thick, but in the region of the linea aspera it reaches a maximum of 7 mm thick.

Distal End

Two specimens of the distal end of the femur of *H. africanus* are known: TM 1513 and Sts 34.

TM 1513 consists of the distal 73 mm of a left femur. It is very well preserved, but a flake is missing from the lateral condyle and with it has gone a small part of the trochlear surface. A small flake has been detached from the edge of the medial condylar articular surface at about the point of maximum curvature. For the most part this specimen is so well preserved that fine surface detail can readily be seen.

Sts 34 consists of the distal 70 mm of a right femur. The two specimens are thus very similar in representing almost exactly equivalent portions of femora from opposite sides. Sts 34 is not as well preserved as is the other one. A substantial piece of the medial condylar articular surface is missing posteriorly, and almost half of the lateral condylar articular surface is missing both posteriorly and laterally. Surface detail has been lost over substantial areas of the specimen, especially posteriorly on the shaft.

TM 1513 is slightly smaller than Sts 34. Diagrams (Figs. 74–75) indicate the size and present various dimensions that are difficult to describe.

Both of these specimens of the distal end of the *H. africanus* femur exhibit marked and very detailed similarities to the equivalent portion of the *H. sapiens* femur. They have the very well developed and characteristically asymmetrical condyles that are separated by a very deep intercondylar fossa. The trochlear or patellar articular surface is very well defined, relatively deep, and with the articular surface margins well raised, the lateral one being distinctly higher than the medial one. In TM 1513 the lateral margin of the patellar articular surface has been damaged by the removal of a large flake of bone from the lateral face of the lateral condyle. Enough is preserved, however, to show that the lateral lip was higher than the medial one, and this condition is clearly shown in Sts 34. In the latter almost the whole patellar articular surface is intact.

As in *H. sapiens*, the condylar articular surfaces can be distinguished reasonably well from the patellar articular surface. Also as in *H. sapiens*, this is more easily done in the case of the medial condyle than in that of the lateral condyle because a distinct shallow groove curves over from the medial edge of the condylar articular surface toward the intercondylar fossa floor. This shallow groove is clearly visible in TM 1513. In Sts 34 it is not so obvious because of some damage to the edge of the condylar articular surface in this region; nevertheless, a part of it can be identified with certainty.

In *H. sapiens* there is usually no corresponding groove on the lateral condyle. Instead a shallow incurving of the lateral margin of the articular surface, sometimes with a poorly and partly developed groove on the surface itself, indicates the region where the condylar articular surface becomes the patellar articular surface. This point can be identified on TM 1513 but is less clear on Sts 34.

In Sts 34 the patellar fossa is a little deeper than is the case in TM 1513, due primarily to a more exaggerated development of the lips of the fossa, especially toward the distal end of the patellar surface. This has produced two rather pronounced bosses or tuberosities that produce a fairly sharp angulation of the curve of the condyle, when seen in side view, approximately halfway from the point where the condylar surface changes to the patellar surface to the proximal end of the patellar surface. This is more exaggerated in the case of the lateral than of the medial condyle, unlike the usual condition in *H. sapiens* and also unlike that in TM 1513, which has the usual *H. sapiens* morphology in this respect.

The shapes and curvatures of the condylar surface resemble those of *H. sapiens* very closely. That is to say, the range of variation seen in the latter includes the characteristics of the two *H. africanus* specimens in this respect. This includes the condition seen in TM 1513 where the medial condylar articular surface narrows asymmetrically quite sharply near its posterior extremity and curves over a little farther than does the lateral condylar surface. These details cannot be seen in Sts 34 because the posterior ends of the condylar articular surfaces are missing.

It seems to me that the relative proportions of the condylar surface areas of the two condyles are much as in *H. sapiens*. This is not easy to document, and it may well be that there is little significance to be attached to this feature in any case. There has been some discussion of it, however, in connection with TM 1513 by Broom, Le Gros Clark, and Kern and Straus. The latter two authors (1949) point out that Broom and Le Gros Clark appear not to agree on the relative sizes of the medial and lateral condylar articular areas, indicating that Le Gros Clark stated, as I have above, that the size proportions appear to be much the same as in *H. sapiens*, while Broom appears to have believed that this was not so. They quote both authors verbatim to indicate the discrepancy in opinion. To my mind, however, there does not appear to be a difference of opinion here; in the passages quoted the two authors are not concerned with quite the same thing. Le Gros Clark was referring to the whole of the condylar articular surfaces; Broom was directing attention to a specific part of the condylar articular surface. This is the anterior portion in which TM 1513 has a slight expansion both medialward and lateralward, thus giving the impression of a rather broad surface where it merges with the patellar surface.

The appearance of breadth in this region is partly due to another feature of this specimen that has aroused discussion: the pronounced upward extension of the intercondylar notch. At that time TM 1513 was the only available specimen of the distal end of the femur of *H. africanus*, and it was assumed that this feature was characteristic of the latter form. But Sts 34 demonstrates that not all individuals were like TM 1513. In the latter the forward extension of the notch is in the form of an asymmetrical notch in the anterior margin of the intercondylar notch. This smaller extension notch was interpreted by Broom—and the other authors—as being the true margin of the intercondylar fossa. Broom therefore thought that there was a widening of the medial condylar surface extending a short distance into the intercondylar notch, hence that this anterior part of the surface was relatively broad. But Sts 34 does not have this narrow extension of the notch; the margin of the latter curves over in a smooth wide contour, as it does in *H. sapiens*. Furthermore, the position of the anterior margin of the notch in Sts 34 indicates that the extension notch in TM 1513 is cut into the margin as present in the former. That is to say, the narrow anterior notch in TM 1513 represents a secondary extension of the position of the notch margin as it exists in Sts 34 and in *H. sapiens*. Having only the two specimens of *H. africanus*, it is not possible to say whether the TM 1513 condition is a rare variant or whether indeed it was as common as the other, *H. sapiens*-like, condition.

The apparent difference of opinion between Broom and Le Gros Clark, referred to by Kern and Straus, was in part due to the fact that the former authors were not actually discussing precisely the same point and partly due to the fact that what seems to me to be a nontypical notch condition occurred in the only *H. africanus* specimen then available. The grounds for considering it nontypical, even though it is found in one of only two specimens known, are that the known features of the femur are so remarkably like those of *H. sapiens* in detail and that the other specimen (Sts 34) has the curvature of the notch so like that in *H. sapiens* that it is probable that the latter was the more usual condition.

The intercondylar fossa is deep and relatively narrow; here again TM 1513 is a little less like the typical *H. sapiens* specimen than is Sts 34 since in the former the fossa is relatively narrower. The relatively wider fossa of Sts 34 is like that of the majority of specimens of *H. sapiens*. In TM 1513 the fossa width varies from 13.6 mm anteriorly, to 12.2 mm near the middle, to 13.8 mm posteriorly. It has a maximum depth of 13.1 mm. The average width is roughly the same as the maximum depth: 13.2 and 13.1 mm respectively. In Sts 34 the width near the anterior end is 14.6 mm, and it widens to 19.6 mm near the posterior end. The posterior parts of the condyles are missing; hence the last figure does not represent the widest part of the fossa; it is probable that at its posterior extremity it was at least 20.0 mm in width. The maximum depth is 14.7 mm. Hence, in Sts 34 the fossa is relatively wider. The depth cannot be measured as accurately and easily as can the width; here it has been measured from a line joining the two fossa margins at the point where the depth is being measured.

In *H. sapiens* the morphology of the intercondylar fossa surface is variable in detail; usually there is no difficulty in identifying the areas of attachment of the anterior and posterior cruciate ligaments. In TM 1513 it is also possible to identify these attachment areas, which are situated as they are in *H. sapiens*. At first sight it might seem that four such areas are readily identifiable in this specimen and therefore that no certainty in fact exists as to the identity of the true attachment areas. This is not so; the true attachment areas can readily be identified under a stereoscopic microscope because of their smooth and compact surfaces. The rest of the fossa surface is rougher and pitted with small nutrient foramina. In this specimen the unusual forward extension of the intercondylar notch, on the lateral condyle side, is in just that region not occupied by the attachment area of the posterior cruciate ligament.

In Sts 34 the surface of the intercondylar fossa is much the same as it is in TM 1513, but the surface detail is not sufficiently good to allow identification of areas of ligamentous attachment.

Much of the popliteal surface of TM 1513 is so well preserved that fine detail is detectable and several points of interest emerge. As in *H. sapiens*, there is a shallow but well-defined hollow in the popliteal surface just above the condylar region, the lower portion of which is pitted by small nutrient foramina. On the medial side there is a well-defined supracondylar ridge leading down to a well-defined adductor tubercle for attachment of a strengthened portion of the tendon of adductor magnus. Vastus medialis should run just medial to the supracondylar ridge, but there is no evidence of this on the bone surface. The medial head of gastrocnemius attaches just lateral to this ridge, and the slightly roughened attachment area, resembling that normally present in *H. sapiens*, is identifiable. Immediately below and behind the adductor tubercle is a well-defined area, also identifiable on the femur of *H. sapiens*, that evidently gave rise to the most distal portion of the medial head of gastrocnemius. This area impinges on the cutaway medial edge of the medial condylar articular surface. There is a well-defined facet on the medial epicondyle for attachment of the medial collateral ligament.

On the lateral side of the popliteal surface there is also detectable detail of interest. The lateral supracondylar ridge is, as in *H. sapiens*, poorly defined at the distal end. A little higher up on the shaft of the *H. sapiens* femur is the better defined of the two supracondylar ridges. This may also have been true of TM 1513, but the bone is broken through at just about the point where the ridge begins to be prominent in Sts 34. At this point in TM 1513 there is what appears to be the beginning of the prominent portion of the ridge.

A more interesting point is that immediately below the indications of a well-defined ridge referred to above there is a completely smooth and featureless area of surface some 16 mm long, below which is a low but unmistakable tubercle. There can be little doubt that the tubercle marks the point of attachment of the slightly thickened group of tendinous fibers marking the distal extremity of the intermuscular septum as found in modern man.

There is a gap in the septum just above the latter point, for the passage of blood vessels, and this is represented by the smooth area between the septal tubercle and the beginning of the lateral supracondylar ridge. This situation is well illustrated in fig. 116 in the sixth edition of Frazer's *The Anatomy of the Human Skeleton.* In actual fact this tubercle is often not detectable in the modern human femur though the small bundle of partly separate tendinous fibers forming the end of the intermuscular septum is easy to detect by even moderately careful dissection. It is my belief that these fibers, as well as many others attaching all the way down the linea aspera and this supracondylar ridge, which assist in making up the intermuscular septum, are fibers from gluteus maximus. This is not intended as a statement of a hitherto unrecognized fact but merely a statement that my dissections lead me to agree with those anatomical texts that trace fibers of gluteus maximus far down the shaft as opposed to many that do not mention fibers extending much below the gluteal tuberosity. There are a number of sets of fibers, running in different directions, in the intermuscular septum. One of these sets, including those running to the terminal tubercle, can usually be traced without difficulty directly to the lower portion of gluteus maximus below the point where the heavy tendon, made up of upper gluteus maximus fibers as well as some coming from the upper part of the ilio-tibial tract, inserts onto the gluteal tuberosity.

Evidently this identical pattern of intermuscular septum with a partly isolated terminal bundle of fibers was present in *H. africanus.*

Immediately medial to the septal tubercle is a narrow, elongate depression that may represent the area of attachment of plantaris.

The entire area of the lateral epicondyle is missing; hence nothing can be learned about it from TM 1513.

Unfortunately, the entire popliteal surface is poorly preserved in Sts 34, and no fine detail can be seen on it. It is therefore not possible to compare this specimen with TM 1513 with respect to the wealth of detail in this area on the latter specimen. The surface damage extends onto the adductor tubercle, part of which remains to mark its position. On the medial epicondyle the facet for the medial collateral ligament is well developed and is very similar to that on TM 1513.

The area of the lateral epicondyle is better preserved in Sts 34, but here also some parts are missing, mainly along the lateral margin of the condylar articular surface. The elongate—and in this case partly divided into two, as is often the case in *H. sapiens*—fossa for popliteus is identifiable. The less clearly defined areas of attachment for the lateral collateral ligament and for the lateral head of gastrocnemius are also identifiable. These have the arrangement and appearance of a sort that is commonly seen in *H. sapiens*, which is somewhat variable with respect to the fine detail of this region.

These two specimens are of very great interest since they give information about the whole of the distal end of the *H. africanus* femur and demonstrate that, with quite minor exceptions such as the slightly narrow intercondylar fossa of TM 1513, the anatomy is simply that of *H. sapiens* to an extraordinarily detailed extent.

For the most part I have been comparing the *H. africanus* femoral specimens with the *H. sapiens* equivalents, and the reason for this is clear since the similarities are very detailed indeed. Similarities with the pongids are not nearly so great. This fact has been demonstrated by Le Gros Clark (1947a and b) and in greater detail by Kern and Straus (1949). For example, the latter authors conclude that "Leaving aside the uncertain notch, the *Plesianthropus* femur is definitely not great-ape in morphology. It resembles the femur of man and those of cercopithecoid monkeys in about equal degree."

With regard to the notch: I have not used the metrical approach employed by Kern and Straus. The reasons for this are that it is difficult to orientate the distal end of a femur correctly if the length and true shape of the missing portion are not known, that slight changes in orientation have significant effects upon the measurements, and that the measurements are in any case not easy to take with accuracy. Finally, since the index they used is derived from two measurements, one of which is wholly contained in the other, it does not appear to be an especially good one. All the same, comparisons of these two specimens with the relevant parts of great ape femora leave no doubt that the *H. africanus* specimens approach those of *H. sapiens* significantly more closely than they do those of great apes.

There is one feature that Kern and Straus believe is inadequate for distinguishing between *H. sapiens* and the pongids and that is the obliquity of the femoral shaft in relation to the axis of the knee joint. Le Gros Clark has pointed out, relying on data supplied by Pearson and Bell (1919) and Parsons (1914), that pongids have femora with little or no shaft obliquity, while man has appreciable obliquity. He further pointed out that TM 1513 resembled man and not the great apes in this respect. On the other hand, Kern and Straus have pointed out that their investigations do not support either Le Gros Clark's conclusion or the evidence on pongid shaft obliquity reported by the above authors. My own observations support Kern and Straus in showing that appreciable angles of obliquity are quite common among pongids. Indeed, orang femora usually have some obliquity, though the African pongids usually have little or no obliquity. As a group the pongids have a range of variation in this character that so far overlaps that of *H. sapiens* that it will not serve as a sorting criterion to differentiate the two.

The angle of obliquity given for TM 1513 by Le Gros Clark and used by Kern and Straus is 7°. My own observations yield a value of 9° for this specimen and of 8° for Sts 34. In spite of the above conclusion concerning overlap, these figures make a much better fit with *H. sapiens* than with pongids according to the data supplied by Kern and Straus. Only the gorilla, of the pongids, is likely to have angles of obliquity this high (mean = 6.0; *N* = 28 and observed range 2.4–11.2). As these authors point out, however, only 7 percent of the 28 specimens studied had angles of obliquity between 7 and 11.2°, while 84 percent of the 61 specimens of *H. sapiens* studied had angles greater than 7°. Since both of the specimens of *H. africanus* appear to have had angles greater than 7°, it is probable that relatively large angles were common in this form, thus resembling *H. sapiens* more than any of the pongids.

The point made by Kern and Straus, however, that a femur with an angle of obliquity of the order of 7° does not automatically belong to an erectly bipedal hominid but could belong to a pongid, remains perfectly valid. In this particular instance the point is merely academic since much evidence from the femur, as well as from other parts of the body, demonstrates clearly that *H. africanus* was a hominid and not a pongid. But Kern and Straus have clarified a point that could be of importance in some other context where a question exists about posture based on a femur in a case where no other suitable skeletal information exists.

Similarly, the point that these two authors stressed very heavily—that the TM 1513 specimen resembles cercopithecoids and *H. sapiens* in equal measure—is academic in this case. In the first place, the statement that the "femur" in this case is noncommittal about whether the owner was a cercopithecoid or a man is too sweeping since not the whole femur but only the distal end is involved. In the second place, their conclusion does not appear to be valid even on the restricted basis of the distal end alone. For example, they point out that in one feature TM 1513 "distinctly resembles the cercopithecoid monkeys rather than man," and this is the "great robustness of its shaft." This statement is based on a robustness index of 56.6 calculated from data taken from Le Gros Clark. It is based on a bicondylar breadth of 56.5 mm and a shaft width of 32 mm measured 65 mm above the standard horizontal plane. They calculated that this shaft width was measured 1.15 times the bicondylar width above the standard horizontal plane; accordingly all of their comparative measurements were taken at a similar distance up the shaft. Unfortunately, while the bicondylar width used is reasonably accurate (according to my measurements it is 56.2 mm), the shaft width that they used certainly was not taken 65 mm above the standard horizontal plane as they believed, since at this distance most of the shaft is missing. The shaft has a diameter of 32 mm at about 52.5 mm above the standard horizontal plane; that is rather less than one bicondylar width (0.94) above the standard horizontal. These distances are so close to the distal end of the shaft that shaft diameter is changing rapidly; hence, the difference in width at 0.94 and at 1.15 times the bicondylar width is significant. The shaft of TM 1513 is not complete at 65 mm above the standard horizontal; generously estimated, the width was 28 mm, a figure that would give an index more comparable to those calculated for comparison by Kern and Straus. The index is then 49. This is probably slightly higher than it should be because of estimating the width generously, but even so it is only just above the lower limit of the rather scantily based ranges for cercopithecoids given by these authors (best sample $N = 17$, mean = 53, observed range 47–60). On the other hand, it does not compare badly with a mean of 45 and a range of 35 to 59 for *H. sapiens*. Clearly this is not a sorting criterion, but the fit is better with *H. sapiens* than it is with cercopithecoids. The strongest point of resemblance to cercopithecoids put forward by Kern and Straus is therefore rendered doubtful, at the very least.

Another point made by these authors is that the depth of the intercondylar notch in TM 1513 "is certainly far removed" from that of cercopithecoids. Here, then, is a clear-cut and positive difference between cercopithecoids and TM 1513 in which the latter closely approaches *H. sapiens* and is approached by an occasional pongid only—that is to say, the resemblance is not close but is closer than in the case of cercopithecoids.

A feature not mentioned by Kern and Straus is absolute size. Size is seldom an important taxonomic feature, but in this case the size of the *H. africanus* femoral distal ends clearly ranges them more closely with *H. sapiens* and the great apes than with cercopithecoids. Furthermore, the detailed structure of the popliteal surface demonstrates virtual identity of TM 1513 and *H. sapiens* in this respect, whereas the similarity to pongids or cercopithecoids is not especially close.

Consequently, it seems to me that on the basis of the distal end of the femur alone it can be shown that *H. africanus* is considerably more closely allied to *H. sapiens* than it is to either pongids or cercopithecoids.

When the upper portion of the femur is also included so that one may properly speak of the *H. africanus* femur without qualification, then it becomes obvious that its affinities are much more clearly with *H. sapiens* than with either pongids or cercopithecoids.

In general, the trochanteric region of the pongid femur is appreciably more manlike in morphology than is that of cercopithecoids. In having a well-developed femoral tubercle, the *H. africanus* femur agrees with that of *H. sapiens* but differs from that of pongids. Cercopithecoids differ appreciably in having a relatively long, freely projecting portion of the greater trochanter, a relatively very large deep trochanteric fossa, and a neck that is relatively rather massive. Curiously enough, cercopithecoids have what appears to be a femoral tubercle that is not too different from that of *H. africanus* and of *H.*

sapiens. This is unexpected; investigation of the reasons for this would be valuable since the femoral tubercle of hominids seems rather clearly associated with the specializations of the ilio-femoral ligament in relation to erect bipedality.

Farther down the trochanteric region, however, the cercopithecoid femur resembles that of man—and of *H. africanus*—more closely than does that of pongids. For example, pongids are quite unhominidlike in having the gluteal tuberosity situated altogether on the lateral face of the shaft and well down so that the tuberosity usually does not reach up the shaft as far as the level of the lower margin of the lesser trochanter. In cercopithecoids, the gluteal tuberosity is not laterally situated but is on the posterior face of the shaft in much the same position as are those of *H. africanus* and *H. sapiens*. As in pongids, however, it reaches up only to about the level of the lower edge of the lesser trochanter. In *H. africanus* and *H. sapiens* the gluteal tuberosity reaches to about the level of the top edge of the lesser trochanter.

Still farther down the shaft, cercopithecoids show greater resemblance to *H. africanus* and *H. sapiens* than do pongids. In the latter there is normally no trace of a structure that can be called a linea aspera; the posterior surface of the shaft is usually smooth and featureless in the relevant region. In *H. africanus*, the only available specimen has a true linea aspera, but it is not pulled out into the powerful crest commonly seen in *H. sapiens*. In cercopithecoids, there is a well-defined structure that can be called a linea aspera that occupies much the same position as does the hominid linea aspera. It usually becomes diffuse relatively higher up the shaft than is the case in hominids.

As we have already seen, the popliteal surface details of cercopithecoids differ appreciably from those of the hominids. On the whole, even though the cercopithecoid femur is more hominidlike than pongidlike in many respects, at its distal end, the femur of *H. sapiens*, of cercopithecoids, and of pongids can be distinguished fairly easily.

In my opinion the available evidence concerning the *H. africanus* femur as a whole is fully consistent with hominid affinities but not with either cercopithecoid or pongid affinities. But it must be conceded that Kern and Straus have demonstrated that some features that thus far have been interpreted as evidence of erect bipedality and hominid status do not necessarily point to such a conclusion. The remarkable number of manlike features in the cercopithecoid femur invite further investigation to determine the adaptational reasons for their existence. On the face of it, the differences in anatomy and locomotor function between cercopithecoids and hominids would seem so great that considerable difference in femoral anatomy would be expected. Elucidation of this problem would be of great interest in its own right and would also be valuable for the interpretation of fossil material.

10 The *Paranthropus* Femur

Material

The *Paranthropus* femur is represented by two specimens only. These are as follows:

SK 82 the proximal end of a femur from the right side of an adult individual from Swartkrans.

SK 97 the proximal end of a femur from the right side of an adult individual from Swartkrans.

Both of these specimens are from the older "pink" breccia from the Swartkrans site and are well preserved. Since they are both from the right side, two different individuals are represented (see figs. 71–73).

SK 82 is a generally well preserved specimen that is 137 mm in length. The head and neck are complete and very well preserved. The greater trochanter has suffered some damage, and portions are missing. The lesser trochanter is almost complete and is well preserved, and much of the trochanteric crest is well preserved. The shaft portion has suffered some cracking, mostly on the posterior, lateral, and medial faces.

SK 97 does not include quite as much as the above specimen and is 118 mm in length. There are a few areas of damage on the greater trochanter, especially along the trochanteric crest from the proximal portion on down onto the quadrate tubercle. The lesser trochanter has suffered a small amount of surface damage, as has the distal portion of what is preserved of the shaft, especially on the lateral face. The quality of preservation of the rest of the specimen is extremely good so that surface detail is very well shown.

Proximal End

Figures 80 and 81 give an impression of the nature of the *Paranthropus* femur specimens and also some dimensions. The proximal portion of the *Paranthropus* femur is of comparable size to that of *H. sapiens*, but there are differences of proportion between the two. Shaft diameter just below the trochanteric region has been chosen as the best indicator of general relative size, and in table 7 this dimension is compared with that for *H. sapiens*.

From table 7 it is evident that the two *Paranthropus* femoral specimens have subtrochanteric shaft dimensions and shape index values that fall easily within the observed ranges for *H. sapiens* males as represented by the local populations cited. The fit is less good, however, when comparison is made with females of *H. sapiens*, for example the Australian Aborigine. Unfortunately, there is no reliable means of knowing to which sex the two fossil specimens belonged.

Table 7
Some Hominid Femoral Shaft Diameters and Platymeric Index Values

	N	Ant.-Post. Diameter	Transverse Diameter	Platymeric Index
Paranthropus SK 97	1	23.3	35.3	66.0
Paranthropus SK 82	1	24.8 (24)	30.4	81.6 (79)
Medieval English (Parsons)	185	28.1	35.6	79.3
Medieval English range*		23–38	30–45	59–100
Maori (Schofield)	43	22.4	34.5	65.2
Ainu (Koganei)	47	23.8	32.3	73.7
Japanese (Koganei)	20	22.7	29.7	76.4
Australian Abo. ♂ (Davivongs)	150	22.9	29.0	79.2
Australian Abo. ♂ range*		18–29	24–34	62–100
Australian Abo. ♀ (Davivongs)	110	20.4	26.0	78.8
Australian Abo. ♀ range*		16–25	23–29	62–96

* Observed range.

In *H. sapiens*, females have smaller femoral heads than do males, as table 8 confirms. This table also shows that the generally more robust specimen of the two fossils has a somewhat larger head than has the other. It is therefore possible that this specimen (SK 97) is from a male individual and the other from a female. In the absence of any idea of what the range of variation is in the head size of the *Paranthropus* femur, this interpretation cannot be substantiated. There are various possibilities: that the two specimens represent smaller and larger females or smaller and larger males, that SK 82 (the smaller specimen) could be from a large female and SK 97 from no more than a moderate-sized male, and so on.

One way of obtaining some sort of check on this point is to compare head size with acetabulum size in the innominate of *Paranthropus*. One innominate only is known (SK 50), and in this the acetabulum has suffered appreciable postmortem distortion. It is clear, however, that the head of SK 97 would fit this acetabulum more successfully than would SK 82—in fact, the former appears to be of exactly the right size for the acetabulum. It is therefore likely to be of the same sex as the innominate. Unfortunately the innominate is sufficiently damaged—most of the pubis is missing, the greater sciatic notch has suffered some damage, for example—to make sexing very

Table 8
Some Hominid Femoral Neck Angles and Head Diameters

	N	Angle of Neck	*N*	Head Diameter
Paranthropus SK 97	1	123°	1	37.1
Paranthropus SK 82	1	118°	1	34.4
Medieval English ♂ (Parsons)	183	126.4°	174	49.0
Medieval English ♂ range*		112–40°		45–55
Maori (Schofield)	43	136.3°	39	46.3
Australian Abo. ♂ (Davivongs)	150	127.8°	150	43.1
Australian Abo. ♂ range*		117–42°		39–50
Australian Abo. ♀ (Davivongs)	110	127.3°	110	38.2
Australian Abo. ♀ range*		114–39°		35–42

* Observed range.

difficult. Moreover, even if the specimen had been intact, we do not know for certain whether the *H. sapiens* sexing criteria would apply equally well to it. Having said all that, I would hazard the very tentative guess that the robustness and rugosity of the SK 50 innominate and the shape of the greater sciatic notch, along with the size of the specimen, are consistent with its having belonged to a male. If this is so—and it is by no means clear that it is—then probably SK 97 also belonged to a male. This is consistent with the shaft diameters fitting the ranges for *H. sapiens* males better than those of females. But this does not prove anything, in fact, since if *Paranthropus* was more robust than *H. sapiens*, female femora might fit the size ranges of male femora of *H. sapiens*.

If SK 97 belonged to a male, then it is not improbable that SK 82 belonged to a female since it has a smaller head, a smaller angle of the neck, a less robust shaft, and a lower platymeric index than has SK 97. These are all features in which the female femur of *H. sapiens* tends to have lower values, though there is much overlap in the ranges for the two sexes. Clearly, insufficient information is available for there to be any certainty about the sex of the two individuals here concerned.

Besides the dimensions of the shaft in the subtrochanteric region, the trochanteric region itself seems also to fit well the ranges for *H. sapiens*. This seems to hold for the general size of the trochanteric region as a whole as well as for the size of the greater and lesser trochanters individually and for their distance apart.

Thus, as far as the robustness of the shaft and of the trochanteric region are concerned, both *Paranthropus* femoral specimens fit very well into the respective ranges for *H. sapiens* males. This is not true, however, of their head diameters. As table 8 shows, both *Paranthropus* specimens fall below the observed range for Australian Aborigine males. Comparison shows (Davivongs 1963) that in general and except for absolute length the Australian Aborigine femur is relatively small; it falls among the less robust racial groups. It is therefore probable that these *Paranthropus* specimens have femoral heads that are absolutely small for males of *H. sapiens* in general.

The SK 82 head diameter is small also for *H. sapiens* females, falling just below the lower end of the observed range of variation for a sample of 110 Australian Aborigine females. SK 97 falls within the lower end of this range. The standard population range in this case is 34.3–42.0 mm. Both *Paranthropus* specimens fall within this range, the smaller of the two being of virtually identical size with the minimum value. In femoral head diameter, then, the two *Paranthropus* specimens fit more or less with the smallest females of *H. sapiens* (see also fig. 82).

In comparison with *H. sapiens*, it is evident that shaft and trochanteric region size suggest that SK 97 and SK 82 are from male individuals of roughly average size, while the femoral head size indicates that they were from very small females. This confirms what is obvious on visual comparison; *Paranthropus* has a proportionately smaller femoral head than has *H. sapiens*. This is well-defined difference between the two forms.

The femoral neck, however, is also involved in this difference since the neck is proportionately longer in *Paranthropus* than is the case in *H. sapiens*. It seems that in specimens of *H. sapiens* that have similar shaft robustness to that in the two *Paranthropus* specimens, the functional neck length (the total length from the medial surface of the head to the lateral surface of the trochanteric region) is of just about the same absolute length as in the two fossil specimens. Since the trochanteric region is of roughly the same size in the two groups, and the femoral head is significantly smaller in *Paranthropus*, the neck is relatively long in the latter.

The neck shape is much the same in *Paranthropus* and in *H. sapiens* and so also is the size at the trochanteric end. The neck narrows as it passes from the trochanteric region to the head, as seen either from the front or the back. This is also the case with the *H. sapiens* neck. Since the neck is relatively long in *Paranthropus* and the head relatively small, the neck of the latter narrows to a greater extent than is the case in *H. sapiens*.

The angle that the neck axis makes with the shaft is within the range of variation for *H. sapiens*, as could have been predicted.

From these facts it is clear that if the head of the *Paranthropus* femur increased in size to that of *H. sapiens* without altering the distance from the medial aspect of the head to the lateral face of the trochanteric region, *Paranthropus* would closely resemble *H. sapiens* in the size and proportion characteristics of the proximal end of the femur. The point I wish to make is that there appear at first glance to be two differences between the *Paranthropus* and *H. sapiens* femora—smaller head and more tapering, longer neck in the former—but these are really two aspects of one difference. Both types of femur have much the same functional distance between the acetabulum and the shaft axis; since the head in *Paranthropus* occupies a smaller part of that functional distance, the neck occupies more.

The Greater Trochanter

The greater trochanter is well developed and is especially well preserved in SK 97, though part of the unciform end is missing and two small areas of damage are present near the area of attachment of gluteus minimus. Much of the greater trochanter is either entirely missing or damaged in SK 82.

In general, the greater trochanter is very similar to that of *H. sapiens*. For example, the bursal area can readily be discerned, just below the area of attachment for gluteus minimus, which is itself clearly defined. There is a well-developed femoral tubercle. The attachment pit for obturator internus is very well defined and situated as in *H. sapiens*. Below and behind it is a deep trochanteric fossa of much the same sort as in the latter. There is a distinct quadrate tubercle of much the shape, size, and position of that in *H. sapiens*.

Allowing for the fact that there is an appreciable amount of variability in the greater trochanter detailed anatomy in *H. sapiens*, there appear to be some differences of detail in *Paranthropus*—though the two specimens of the latter are not identical in these respects.

The area of the femoral tubercle is different in *Paranthropus*. In *H. sapiens*, the tubercle is normally well defined and quite large but smoothed off and without sharp edges in most cases. It is situated a short distance down the region where the neck and trochanteric region join and is commonly, though not always, moderately distinct from the nearest portion of the greater trochanter.

In *Paranthropus* SK 97 there is distinct development of a rather rugose tubercle in the same area as that in *H. sapiens*, but it is not clearly separate from the greater trochanter. The superior surface of the neck is developed into a broad, rounded crest as it nears the greater trochanter just anterior to the area of attachment of obturator internus and, presumably, the gemelli. This crest rises sharply into a powerful tuberosity, much of which appears to be the femoral tubercle. The area of this tuberosity that corresponds in position to the femoral tubercle of *H. sapiens* is partly distinct from the rest, which is just above it. This upper portion has a slightly different surface texture that is like that of the area of attachment for gluteus minimus and is directly continuous with it.

It seems to me that there are at least two ways of interpreting this situation. Either (a) the femoral tubercle in *Paranthropus* is appreciably more strongly developed and extends higher up the femur than is the case in *H. sapiens* or (b) the area of attachment for gluteus minimus extends a little more medialward than it does in *H. sapiens* and encroaches onto the area for the femoral tubercle so that it becomes continuous with the upper portion of the latter. The latter seems to me the more probable interpretation. It seems likely also, judging from the surface structure, that the anterior tendinous fibers of pyriformis (superior retinaculum) were quite powerfully developed and tended to push up the anterior portion of the attachment for gluteus minimus into a well-defined crest that is also thereby pushed slightly forward, thus fusing more intimately with the femoral tubercle.

Unfortunately, SK 82 is too damaged in this region to allow certainty concerning the anatomical features. The bursal area below the attachment of gluteus minimus is all present, however, and is not quite the same as in SK 97. This area is of roughly the same size and shape as in the latter, but its long axis parallels that of the shaft instead of being at an angle of about 30° as in SK 97. It also extends farther down the shaft, reaching to the level of the top of the lesser trochanter, some 8 mm farther down than is the case in SK 97. This means also that the bursal area has a slightly different relationship to the area of attachment for gluteus minimus. This difference, however, apparently is of minor significance and is the result of slight changes only.

The base of a well-developed femoral tubercle is present, and the features of the deep trochanteric fossa and the pit just antero-dorsal of it, for obturator internus and the gemelli, are of such a nature that it seems highly probable that this entire region between the femoral tubercle base and the deep trochanteric fossa had the same anatomy as in SK 97. A small portion of the area of attachment for gluteus minimus is intact, and this also is consistent with the pattern seen in SK 97.

Almost the whole of the rest of the greater trochanter in SK 82 is either missing or damaged, including the unciform end and part of the quadrate tubercle. The whole of the deep trochanteric fossa is present and has the same deep conical structure as has SK 97. It appears to be slightly larger than that usual in *H. sapiens* femora having greater trochanters of comparable size, but the difference is small.

There does appear to be a slight difference in structure between the two. In *H. sapiens* the lip of the greater trochanter, the edge of the unciform portion, is closer down over the deep trochanteric fossa so that the roof of this fossa is a smoothly curved surface that is continuous from the deepest part of the fossa to the trochanteric crest. This is not true of either of the *Paranthropus* specimens. In these the trochanteric crest does not come down quite as low over the fossa, and as a result the surface just underneath the crest is separated by a low, rounded step from the roof of the fossa itself. Evidently the actual area of attachment of obturator externus was smaller than the total size of the fossa available; in the case of *H. sapiens*, the smaller fossa is wholly occupied by the muscle tendon. The difference is not very great and does not appear to be of significance.

In SK 82 there is a distinct hollow traversing the posterior face of the neck diagonally upward and lateralward into the deep trochanteric fossa. There is some evidence of this also in SK 97, but a small patch of the bone surface is missing in this region so that one cannot now see whether the groove was as well developed in this specimen. This groove evidently was associated with obturator externus. Less well developed ones may be found on femora of *H. sapiens*.

The quadrate tubercle is similar to that in *H. sapiens*. It appears to be a little more prominent than is usual but is apparently within the range of variation in the latter. The lateral face of the greater trochanter also is similar to the usual condition seen in *H. sapiens*; it differs slightly in that while the latter usually has a fairly prominent tuberositylike projection toward the anterior side of the lateral face, this is much less pronounced in *Paranthropus* SK 97. The reason for this seems to be that the distal end of the area of attachment of gluteus minimus is not as prominent as it is in *H. sapiens* and is situated a little more toward the anterior face of the femur. The bursal area associated with it faces entirely anteriorly: in *H. sapiens*, associated with the slight shift toward the middle of the lateral face of the distal end of the area of attachment of gluteus minimus, the lower end of the bursal area encroaches slightly on the lateral face of the greater trochanter. Furthermore, there is no trace in SK 97 of the roughened line made by attachment of vastus lateralis curving over from the area near the gluteal tuberosity and forming a crestlike edge to the lower margin of the greater trochanter toward the anterior side of the lateral face. Evidently vastus lateralis was not especially strongly developed in this region in *Paranthropus* or perhaps did not reach quite as high as in *H. sapiens*. The latter is the case in the gorilla, for example, where the equivalent portion of the attachment of that muscle does not approach the greater trochanter as closely as is true in *H. sapiens*, although according to Raven (1950) there is another slip of the muscle that attaches on the greater trochanter itself immediately lateral to the attachment of gluteus minimus. It is not possible to tell from the fossils whether a similar arrangement was present in *Paranthropus*.

As far as can be determined, the differences between *Paranthropus* and *H. sapiens* in this respect are small, but the slight rearrangement, or slightly different emphasis, gives the lateral face of the greater trochanter a somewhat less protuberant appearance than is the case in *H. sapiens*. Enough is preserved of this area in SK 82 to indicate that it also had this appearance—even slightly more so than has SK 97 because of the greater downward extension of the bursal area on the anterior face.

The top of the greater trochanter in SK 97 is flattened over a moderately large area over the unciform part—slightly more so than is usual in *H. sapiens*. This rather clearly defined flattened area evidently gave rise to pyriformis. If this is so, then the area of origin of this muscle in SK 97 is of about the size of that in an *H. sapiens* femur of comparable trochanteric robustness, but the area is slightly differently shaped and placed. Again, the difference is not great, and it might well be that examination of enough femora of *H. sapiens* would turn up a matching condition. In part the difference is due to the medial extension of the area of attachment of gluteus minimus onto the neck between the femoral tubercle and the area of attachment of obturator internus and the gemelli.

The Lesser Trochanter

The lesser trochanter is of the same structure and size as that of *H. sapiens*. That is to say, the lesser trochanters of the two *Paranthropus* specimens can easily be matched in these respects by pulling out a few femora at random from a collection of *H. sapiens* skeletons. There is thus no need to describe them in detail.

In one respect, however, the *Paranthropus* specimens do tend to differ slightly, and this is with regard to the position of the lesser trochanter on the shaft. In specimens of *H. sapiens*, which have trochanteric regions of comparable size to those of the *Paranthropus* specimens, the absolute distance between the quadrate tubercle and the lesser trochanter is of the same order: the *Paranthropus* values falling easily within the observed range of variation of *H. sapiens* in this respect.

The slight difference of positioning of the lesser trochanter here being considered concerns the relationship to the medial face of the shaft. When the femur of *H. sapiens* is viewed directly from the back—or from the front, for that matter—the lesser trochanter breaks the line of the neck as it passes down into the medial margin of the shaft and protrudes fairly prominently. In SK 82 the degree of protrusion is appreciably less but probably not outside the range of variation of *H. sapiens*. In SK 97 the lesser trochanter reaches only just to the medial margin but does not protrude beyond it.

Closer examination indicates that this difference is actually not quite so much due to the precise position on the shaft occupied by the base of the lesser trochanter, which is variable in both *Paranthropus* and *H. sapiens*, but rather to the direction of protrusion. In *H. sapiens* the lesser trochanter juts out almost directly in a medial direction, while that of *Paranthropus* juts postero-medially. The direction of protrusion and placing of the scar of attachment suggests that the direction of pull of iliopsoas may have been very slightly different in *Paranthropus* compared to the usual condition in *H. sapiens*.

The Shaft

In neither of the *Paranthropus* specimens is much of the shaft preserved, and both specimens are damaged to some extent in this region.

From table 7 it can be seen that the platymeric indexes of these two specimens, measured just below the level of the lesser trochanter, are well within the range of variation for *H. sapiens*. This means that the general features of shaft cross-sectional shape can easily be matched in *H. sapiens*. We have already seen that shaft robustness is easily matched in *H. sapiens*, especially among males.

There appears to be a small difference in shaft shape, in spite of the fact that the proportion between the maximum antero-posterior and transverse shaft diameters falls well within the observed range for *H. sapiens* in this respect.

In *H. sapiens* the shaft in this region is more rounded than it is in the *Paranthropus* specimens, where the shaft cross section is more nearly rectangular, or perhaps triangular. In both specimens the anterior face of the shaft is flattened with fairly definite margins. Whether the posterior face is similarly flattened cannot be determined because of the damage in this region; it seems probable that the margins of the posterior face were less well developed than those of the anterior face, thus giving a somewhat triangular cross section. In these two specimens, therefore, there is quite a large, almost flat, surface in the anterior trochanteric region and the adjacent portion of the neck. This includes most of the relatively large anterior surface of the neck, across the base of the greater trochanter and some distance down onto the shaft. How far this flat surface extends down the shaft is not known. In *H. sapiens* this region is usually not as obviously flattened.

The trochanteric line is just visible in both specimens but is not at all obvious. It is situated a little more medialward than is usual in *H. sapiens* (more so in SK 97 than in SK 82), and this may to some extent be a consequence of the relatively greater neck length in *Paranthropus*. As the trochanteric line passes over onto the medial surface of the neck in SK 97, there is a distinct small tubercle the base of which is slightly rugose and is met by the continuation upward of the spiral line. This evidently is the area of attachment of the thickened medial band of the Y-shaped ligament of Bigelow (see fig. 83).

The spiral line is clearly defined in both specimens (more so in SK 97 than in SK 82) and can be traced from the trochanteric line to the broken end of the shaft. Although clearly visible, it is not pulled out into a crestlike structure as is the case in specimens of *H. sapiens* in which it is particularly well developed (see especially fig. 81).

The spiral line curves round near the base of the lesser trochanter onto the posterior aspect of the shaft, as in *H. sapiens*. Also, as in the latter, below the lesser trochanter it is approached by a curving line consisting of the gluteal tuberosity or ridge, along with rugosities that probably were caused by adductor magnus and vastus lateralis. Unfortunately, the exact details of this second line cannot now be seen since both specimens have suffered some superficial damage in the relevant region.

In the case of SK 97 portions of the gluteal tuberosity can be seen; it extended to very slightly above the level of the upper edge of the lesser trochanter. Above this there is no clear evidence of the line that is often present in *H. sapiens*, caused mainly by vastus lateralis, and that curves forward to the base of the greater trochanter. Just below the level of the lower edge of the lesser trochanter part of the curving line, converging as though onto a linea aspera, can be seen on the edge of an area of damage. This scar was presumably made chiefly by adductor magnus. It lies on a crack that spirals in just the way that the muscular markings in this area do on the femur of *H. sapiens*. Presumably this crack reflects the osteone organization in this region, which in turn probably reflects something of the stresses due to the muscles whose scars lie in this region. There is also such a crack running up the spiral line of this specimen.

In SK 97 the gluteal tuberosity is situated in the midline of the lateral face of the shaft. In *H. sapiens*—and also *H. africanus*, as we have seen—the gluteal tuberosity is situated farther over to the posterior face; that is to say, nearer to the lesser trochanter. In being near the midline of the lateral face, the gluteal tuberosity of this specimen (and evidently of SK 82 also) resembles pongids more closely than it does either *H. sapiens* or *H. africanus*. It differs from that of the pongids, however, in reaching up to slightly above the upper edge of the lesser trochanter, while in the latter it does not reach higher than the lower edge of the lesser trochanter.

Most unfortunately, the shaft of SK 97 is broken just above the point where the spiral line and the other curving line could be expected to meet and form a linea aspera. It seems evident that, as in the *H. africanus* specimen, there was a true linea aspera. This is perhaps also suggested by the fact that the thickness of the bone at this point is 9.2 mm, whereas the thickness on the other side of the shaft, directly across from what appears to be the beginning of a linea aspera, is 7.1 mm. SK 82 is of no help in this connection since just precisely the strip of bone that would have included the linea aspera is totally missing; there is a large crack in its stead.

SK 82 is more damaged than is SK 97 and less can be learned about its shaft. Part of the gluteal tuberosity and lower portion of the line continuous with it are detectable, and these confirm what can be seen on SK 97.

It would seem that with regard to the spiral line and the other line, formed by the gluteal tuberosity and other muscle scars, which joins the spiral line to form the linea aspera, *Paranthropus* fairly closely resembles *H. sapiens*. But it differs in that the gluteal tuberosity is farther lateralward of the lesser trochanter than is true of either *H. sapiens* or of the one relevant specimen of *H. africanus*.

Since the above was written Day (1969) has published a preliminary note on a femoral fragment from Olduvai Gorge (Hominid 20), which seems rather clearly to belong to *Paranthropus*, a conclusion of Day's with which I concur. The specimen is more fragmentary than either of the two from Swartkrans but obviously had the same pattern of characteristics. Hominid 20 had a neck slightly longer even than occurs in the *Paranthropus* specimens. Day remarks that the Olduvai specimen shares the following features with the two Swartkrans femora: ". . . the breadth and relative length of the femoral neck, lack of lateral expansion of the great trochanter, absence of the trochanteric line and femoral tubercle, backward direction of the lesser trochanter, and possession of both a deep trochanteric fossa and an obturator externus groove. These shared characters, with the exception of the obturator externus groove which is common to all, distinguish these three femora from those of modern man, Neanderthal man and *Homo erectus*." From the description of the two Swartkrans specimens given earlier in this chapter, it is clear that Day is in error (he was evidently working from casts) about two points in the above quotation; each of the Swartkrans specimens clearly has a trochanteric line as well as a structure at the point where the trochanteric line reaches the greater trochanter that in my opinion is almost certainly a femoral tubercle.

Day has argued from the supposed lack of femoral tubercle and trochanteric line that in *Paranthropus* the center of gravity was anterior to a line between the femoral heads, hence the specialized thickening of the iliofemoral ligament found in man did not occur since it was not needed to counteract hyperextension of the trunk on the thigh. He supposes that the long ischium of *Paranthropus* may have been necessary to counteract the forward torque of the trunk resulting from the forward position of the center of gravity. I believe this argument to be incorrect, not only because the two Swartkrans specimens do indeed have tubercles and trochanteric lines, but also because the gluteal surface of the *Paranthropus* ilium has well-developed acetabulo-cristal and acetabulo-spinous buttresses and because the anterior inferior iliac spine is strongly developed with a clearly defined area for attachment of one arm of the strengthened Y ligament. The well-developed buttresses indicate regular use of the lateral balancing mechanism of erect bipeds, and the anterior inferior iliac spine confirms the development of a specialized Y ligament that is indicated by the femoral tubercle on each of the femora. Moreover, as discussed elsewhere in this work, I believe that the backward expansion of the posterior part of the iliac blade is part of the mechanism that carried the center of gravity back to or behind the line joining the femoral heads. This backward expansion of the ilium is well developed in *Paranthropus*. Finally, as Day points out, the presence of an obturator externus groove on the femoral neck indicates clearly that hyperextension of the hip joint was occurring. The totality of this evidence seems to me to indicate that the *Paranthropus* center of gravity was not forward of the hip joints, and the long ischium of this form was thus not concerned with coping with such a forward position.

The presence of this *Paranthropus* specimen at the level of late Bed I or early Bed II at Olduvai is a useful additional piece of evidence supporting the conclusion that *Paranthropus* and representatives of the *Homo* lineage were sympatric over a very long period of time in Africa.

11 Comparison of *Homo africanus* and *Paranthropus* Femora

Comparison of the *H. africanus* and *Paranthropus* femora cannot be very satisfactory, unfortunately, because the former is best represented by two good specimens of the distal end, but the distal end of the *Paranthropus* femur is not known. Conversely, the *Paranthropus* femur is known from two good, and one less good, proximal ends, whereas the only proximal end so far known of *H. africanus* is in a rather poor state of preservation. In neither case is the length of the femur known, although a reasonable guess can be made in the case of *H. africanus* (see figs. 84–85).

The proximal end of the *Paranthropus* femur appears to have been significantly more robust than the *H. africanus* equivalent. Table 9 lists some measurements that indicate this. As we saw in a previous chapter, Sts 14 very probably belonged to a female individual; therefore it is likely that this specimen would fall in the lower half of the size range for *H. africanus*. SK 97, on the other hand, possibly belonged to a male individual, and SK 82 with much less certainty belonged to a female individual. It is probable that the size difference suggested by these three specimens is a little exaggerated.

Table 9
Some *Paranthropus* and *Homo africanus* Femoral Dimensions and Platymeric Index Values

	Antero-posterior Shaft Diam.	Transverse Shaft Diam.	Platymeric Index	Head Diam.	Head/Lesser Troch. Distance
Homo africanus (Sts 14)	19	21	90.5	33	70
Paranthropus (SK 82)	24.0	30.4	79.0	34.4	84
(SK 97)	23.3	35.3	66.0	37.1	88

The figures in tables 7 and 9 indicate that a female *H. africanus* femur (proximal end) is rather small compared to that of a female *H. sapiens*, although probably not totally outside the range of variation in all respects. For example, the antero-posterior shaft diameter in the subtrochanteric region is within the observed range for females of the Australian Aborigine, but the transverse diameter in this region is outside the observed range (23–29) but not outside the standard population range of 20.9–31.0. As we have seen, the *Paranthropus* specimens agree in size much better with femora of *H. sapiens* males. *If* SK 97 belonged to a male and SK 82 to a female, then it means that a sample of two, including a member of each sex, fits well with a sample of substantial size consisting of males only of *H. sapiens*.

This would suggest that possibly *H. africanus* could be compared in body size to a very small female of *H. sapiens* and *Paranthropus* to an average size male of *H. sapiens*. Such a conclusion, however, would not necessarily be valid. For example, while Sts 14 of *H. africanus* had a pelvis that is comparable in size to that of a small female of *H. sapiens*, thus agreeing with the conclusion based on the femur, the spinal column evidence is that the vertebrae are very small compared to those of a small specimen of *H. sapiens*, which suggests a considerably smaller and more lightly built trunk than is found in the latter.

The evidence for *Paranthropus* trunk size is not nearly as good as that for *H. africanus*, but one clue, the relative size of the auricular region, may be of some use. Unfortunately, the auricular area is not well preserved in either of the two *Paranthropus* specimens. The width of the auricular region, as compared to the distance from the edge of the latter region to the anterior iliac margin immediately above the anterior inferior iliac spine, appears to have been somewhat greater in *Paranthropus* than in *H. africanus*. This suggests that the sacrum was more robust, relatively, than in the latter; hence perhaps that the trunk was proportionately larger also. The point I wish to make is that all the evidence now available seems to indicate that *Paranthropus* was more muscular and more robustly built than *H. africanus*. For example, the skull gives evidence of powerfully developed masticatory and fairly powerful nuchal musculature; the femur indicates that the animal as a whole was more heavily built than was *H. africanus*. But if indeed the auricular region of the *Paranthropus* ilium was proportionately larger than that of *H. africanus*, then this might mean that, besides being more heavily built than the latter form, it did not have as great a disparity in size between the trunk as a whole and the pelvis, as seems to have been present in *H. africanus*.

The available evidence is in fact simply insufficient to allow conclusions on body size to be drawn with any certainty. Putting together the scraps of evidence available (discussed in the final chapter under the subhead "Body Size and Proportions", pp. 231–35), I would hazard the guess that the weight of females of *H. africanus* was of the order of 40–60 pounds, if Sts 14 was a modal individual, while *Paranthropus* individuals probably weighed three to four times as much. It does not follow from this, however, that there need have been much difference in height. The *H. africanus* evidence suggests a height of approximately 4 to 4½ feet for that form, while perhaps *Paranthropus* was no more than 5 feet in height. Since it is clear that, *H. africanus* already had elongated lower limbs of the modern human sort, perhaps the more primitively hominid *Paranthropus* had relatively shorter limbs and may thus have been no taller than *H. africanus*. According to Schultz (1953), among primates relatively great femoral shaft robustness is due primarily to relative shortness of the femur and only secondarily to greater body weight. Since shaft length is not known for *Paranthropus*, relative shaft robustness is also unknown. The disparity in absolute shaft thickness between *H. africanus* and *Paranthropus*, however, is appreciably greater than the difference in femoral head diameter between the two. If the difference in shaft robustness had been due to difference in body weight alone, greater difference in femoral head size could be expected. One might thus conclude that the relatively great shaft robustness in *Paranthropus* was due to both greater body weight and a proportionately short femur, hence lower limb as a whole, but was primarily due to the latter. If this was so, then the modal height of *Paranthropus* was probably somewhere between 4½ and 5 feet.

Lovejoy and Heiple (1970) estimated the *Paranthropus* femoral bicondylar length at 315 mm, using the technique already discussed in connection with the *H. africanus* (Sts 14) femur. They thus concluded that the *Paranthropus* femur was appreciably longer than that of *H. africanus*. There are two disadvantages to their case: (1) neither the interacetabular distance nor shaft obliquity are known for *Paranthropus*; (2) their estimate involves the assumption that *H. africanus* and *Paranthropus* were anatomically and functionally similar in this region—a supposition that is not borne out by the evidence.

Table 10
Some Femoral Indices of *Homo africanus*, *H. sapiens*, and *Paranthropus*

	A.-P. Diam. × 100 / Head Diam.	Tr. Diam. × 100 / Head Diam.	Head Diam. × 100 / Head + Neck Length
Homo africanus (Sts 14)	57.6	63.6	47.1
Paranthropus (SK 97)	62.8	95.1	42.2
(SK 82)	69.8	88.4	41.0
Homo sapiens			
Australian Abo. ♂ ($N = 150$)	53.2	67.4	—
Australian Abo. ♀ ($N = 110$)	53.5	68.0	—

We have seen that the two *Paranthropus* femur specimens have small heads and long necks compared to the robustness of the shaft in the subtrochanteric region and to the general size of the trochanteric region in comparison with the condition in *H. sapiens.* We have also seen that the one *H. africanus* specimen (Sts 14) has the head missing, but that both head size and neck length can be determined within very close limits. The latter clearly also has the relatively long neck of *Paranthropus.* The head appears to have been relatively larger than that of *Paranthropus,* however, as is shown in Table 10. Head diameter is compared with the antero-posterior diameter of the shaft in the subtrochanteric region, with the transverse diameter in the same position and with the distance from the lower edge of the lesser trochanter to the medialmost extremity of the head surface. In Table 10 this is called head plus neck length for brevity, but obviously more than that is involved.

From Table 10 it appears that:

(1) The head diameter (compared to antero-posterior shaft diameter) is largest in *H. sapiens* (Australian Aborigine) and smallest in *Paranthropus.* The *H. africanus* individual is intermediate in size, though a little closer to *H. sapiens* than to *Paranthropus.*

(2) When head diameter is compared with the transverse shaft diameter, the proportions are very similar in male and female Australian Aborigines and in *H. africanus,* while the head of *Paranthropus* tends to be considerably smaller.

(3) Head diameter is proportionately a little smaller compared to the lesser trochanter—head distance in both *Paranthropus* specimens as compared to that of *H. africanus.*

Some of these comparisons obviously are affected by proportional variations in any one of the dimensions involved from one taxon to the other. In each case, however, *Paranthropus* has the smallest head proportionately, with *H. africanus* either agreeing with *H. sapiens* or falling between the latter and *Paranthropus.* This suggests the probability that *Paranthropus* actually had a relatively small femoral head size, while *H. africanus* came closer to *H. sapiens* in this respect. Using the ratio and modified ratio diagram method makes it clear that the two *Paranthropus* specimens have proportionately appreciably more robust shafts than has Sts 14. Put in another way, *H. africanus* (Sts 14) has a proportionately larger head compared to the shaft diameter than has *Paranthropus.*

These conclusions, however, can be no more than tentative in view of the extremely small sample sizes in this case. Nevertheless, the suggestion seems rather clear that *Paranthropus* has a relatively more robust femoral shaft compared to the weight-bearing surfaces of the head and acetabulum than had an *H. africanus* specimen that probably belonged to a female.

The modified ratio diagram may be a bit misleading in this particular instance since the dimension used to minimize difference of size and thus to throw into relief differences of proportion is acetabulum size. This procedure involves the assumption that there are no differences in proportion in the acetabula of the animals being compared. This assumption is sufficiently close to fact when other body dimensions are being considered; when femoral head size is being considered, the use of this assumption clearly begs the question. This assumption, however, does not affect the conclusion stated in the preceding paragraph.

The *H. africanus* and *Paranthropus* femora are similar in having a relatively large, almost flat area on the upper portion of the anterior face of the shaft that also passes onto the neck. In *H. sapiens* this area is less flat, being more convex from side to side.

Both forms appear also to have been similar with respect to the femoral tubercle, its close apposition to the medial end of the attachment area of gluteus minimus and to the extension of the latter onto the neck. Too much of the proximal portion of the greater trochanter is damaged or missing in Sts 14 to allow detailed comparisons or conclusions concerning this region. It does seem possible, however, that the area of attachment of obturator internus and the gemelli was not as pronouncedly pitlike as it is in *Paranthropus*. The deep trochanteric fossa clearly was considerably smaller than that of *Paranthropus*; it seems to have been significantly smaller proportionately than that of the latter. Since *Paranthropus* seems to have been larger and more robust than *H. africanus*, it might be expected that the muscles would tend to be proportionately larger in the former and hence their areas of attachment may be expected to be relatively more obvious without necessarily anything more significant being involved than greater body size. The deep trochanteric fossa in Sts 14 is damaged superiorly; hence it is not possible to see whether it also had the stepped feature seen in *Paranthropus*.

Both early hominids have the lesser trochanter situated in a very slightly more lateralward position than is usually the case in *H. sapiens*.

With respect to the position of the gluteal tuberosity, the upward extension of the latter—by vastus lateralis presumably—and its curving over into distinct prominence at the base of the greater trochanter, the *H. africanus* specimen seems to bear a much closer resemblance to *H. sapiens* than does *Paranthropus*. In the Sts 14 femur the gluteal tuberosity is easily visible and is relatively appreciably nearer the lesser trochanter than is the case in SK 97. In this respect the former conforms perfectly with the *H. sapiens* pattern, but SK 97 does not. In both *H. sapiens* and Sts 14 there is clear evidence of a ridge continuing upward from the top end of the gluteal tuberosity, but no trace can be seen of such a feature in SK 97 and the relevant area is damaged in SK 82.

Unfortunately, the damage to the greater trochanter in Sts 14 extends sufficiently far down for it no longer to be possible to see whether this ridge, which extends up from the gluteal tuberosity, curves over forward. But there is the base of a prominence at the antero-distal end of the greater trochanter that may have been more like that of *H. sapiens* than is the case in SK 97 and SK 82 where it can be seen clearly. Unfortunately, the damage in Sts 14 is too great for one to be sure about this possibility. It is clear, however, that in this general region the *H. africanus* specimen is appreciably more like *H. sapiens* than are either of the two *Paranthropus* specimens.

On the basis of the present evidence there does not appear to have been any difference between the two early hominid types with respect to the linea aspera and related lines. In the case of both *Paranthropus* specimens, the shaft is missing from a point above that at which the linea aspera proper could be expected to occur. The lines that normally are associated with a linea aspera are present in a manner that implies that such a structure did occur in this form. In Sts 14 part of a linea aspera is actually present, but the specimen is too poorly preserved for fine details to be seen.

Pongids do not have a linea aspera; indeed, the posterior face of the femoral shaft is smooth and singularly free of surface detail. Manifestly, whatever the fine detail of structure may have been in the two forms of early hominid, the available evidence is enough to show that they resembled *H. sapiens* much more than they do pongids in this region.

Although the *H. africanus* femur appears to have been somewhat more advanced in the direction of *H. sapiens* than that of *Paranthropus*, both are distinctly manlike in femoral morphology than they are like pongids. It may be noted in passing that some confidence is lent to the conclusion that *H. africanus* has the more manlike anatomy of the femur, although based on so few specimens, because it is consistent with the fact that in other parts of the skeleton (pelvis, skull, dentition, etc.) it exhibits a distinctly more *H. sapiens*-like anatomy than does *Paranthropus*.

As we have seen, there is much evidence that indicates that the pongid femur is not especially like that of *Homo*, differing in almost all of its detailed anatomy from that of the latter. Cercopithecoids are more *Homo*-like in this respect than are pongids. It. is therefore easy to differentiate pongid-type femora from those of man. The evidence available concerning the early hominid femur is more than sufficient to demonstrate that it is not of the pongid type but is very similar to that of man. This is especially so in *H. africanus* in which almost the whole of the femur is known. It is unfortunate that the femoral material is so fragmentary and consists of so few specimens, which are noncorresponding parts. The limited comparisons possible indicate that *Paranthropus* is not quite as much like *H. sapiens* with respect to femur anatomy as is *H. africanus*, but these differences are slight compared to the differences between *Paranthropus* and pongid femora.

12 The Foot

Material

Unfortunately, the early hominid foot is almost unrepresented in the fossil material from South Africa. Nothing at all is preserved of the foot of *H. africanus*; of *Paranthropus* the following few fragments are known from Kromdraai:

TM 1517: incomplete right talus of an adult individual.

Two phalanges: one is an incomplete proximal phalanx, perhaps of the fifth digit on the left foot; the other is an almost complete distal phalanx, perhaps of the second or third digit.

No foot material is known from Swartkrans.

The case for the talus belonging to *Paranthropus* is sound. It was found in the block of matrix that contained the type skull of *Paranthropus*. This block was less than a cubic foot in size, according to Broom (in Broom and Schepers 1946), and it also contained fragments of humerus and ulna as well as bones of the hand of *Paranthropus*. It is probable that these all belonged to the same individual as did the type skull. The two foot phalanges apparently also came from this same block, though Broom is not absolutely clear on this point.

Close association of specimens in sites such as the ones here concerned is not necessarily proof that they belonged to the same taxon. Indeed, close association of fragments of quite different sorts of animal is very common. In this particular case a piece of frontal bone and a phalanx of a small form of extinct baboon, *Parapapio*, were also present in the block of breccia with the *Paranthropus* material. Even so, the fact that some of the rest of the skeleton was present with the head is of some significance here. The skull commonly becomes disassociated from the rest of the skeleton in the circumstances associated with accumulation of the cave deposits here concerned. It seems clear that the caves were not dwelling sites but merely accumulation sites from time to time for material lying about on the surface. This is suggested, for example, by the fact that many specimens clearly had undergone considerable weathering out in the open before getting into the sites—the Sts 65 innominate from Sterkfontein, with its numerous fine cracks, is a good example. Another fact that supports this view is that usually these caves open to the surface through a narrow, vertical shaftlike opening in the cavern roof, as can readily be seen in the modern caves in the area; the caverns therefore would not have been suitable dwelling places for all or almost all of the time during which accumulation was occurring. Only when the cavern was almost filled up would access become a possibility. But even this is doubtful: for consolidation to continue there would have to have been a substantial thickness of dolomite roof to serve as a source of lime. Usually the volume of cementing lime is greater than the volume of sand and other included material (Brain 1958). Indeed, it is clear that the bone breccias are really no more than heavily contaminated deposits of dripstone. Furthermore, the caverns, which in this region have a ceiling that usually follows the dip of the dolomite beds (about 30°), fill up completely to the ceiling with consolidated material. This indicates that the influx was coming from a higher level. The implication is that even in the final stages of the period of accumulation the ceiling of the cavern was well below surface level. The cavern will therefore have been dark and difficult of access even at that stage and will not have been a likely habitation for man or most other animals. Hence, the specimens found in the sites were presumably ones that were lying about on the surface near the cavern opening and were washed in or knocked in. This explains why one seldom finds substantial portions of the skeleton of any animal in the deposits. The animals normally will have been dismembered on the surface and eaten. Most bones will have been consumed by hyenas or other carnivores (see also Brain 1970), hence the preponderance of jaws and teeth; carnivores normally do not eat these. Occasionally, however, a whole animal will have fallen in; Dr. Brain once witnessed such an incident when he startled a hyena in the immediate vicinity of a cavern opening; in its attempts to make off rapidly, it lost its footing and fell to its death by injury and/or starvation. In such a case more of the individual might be preserved, as in the case of the Sts 14 individual at Sterkfontein. The type individual of *Paranthropus* evidently represents a similar case. Therefore, when more than one part of the skeleton of an individual is found in one place, the probability is increased that other fragments in the immediate neighborhood also belong to the same individual.

The other consideration that renders diagnosis probable is that only a very restricted range of primates to which the specimens could have belonged is present in the Sterkfontein Valley cave deposits. Pongids are unknown. Other than early hominids, the only primates present are cercopithecoids, the commonest of which are species of the rather small baboonlike *Parapapio*, though larger forms are also represented. Specimens that are broadly manlike in anatomy and size would clearly not belong to a cercopithecoid. The only confusion that could come is from the remains of true man being mixed with those of early hominids. Since these deposits are of solid rock, and since consolidation occurs synchronously with accumulation, there usually is no question of specimens of one age being buried into deposits of a quite different age—as in the case of the Galley Hill skull. Experience has shown that in any particular level in the breccia deposit one is dealing with specimens that were coeval. In the rare instances where this is not true, the fact is rather obvious from the breccia itself. Thus far, even though tool-making hominids were present in the Sterkfontein Valley in times earlier than the period represented by Kromdraai, there is no sound evidence of their having been present in the valley during Kromdraai time. Had they been, stone artifacts should have been common in the deposit, as they are in the Middle Breccia at Sterkfontein. As has been indicated elsewhere (Robinson 1962), Sterkfontein does not appear to have been a dwelling site. The fact that artifacts are plentiful in that portion of the site cannot thus be explained in that manner, but must have been the consequence of tool-makers frequenting the general neighborhood but not actually living at that spot. The artifacts presumably got into the deposit in the same manner as did the bones. Sterkfontein and Kromdraai are hardly more than a mile apart. If tool-making man had been moving about in this region in Kromdraai time as he had in Sterkfontein Middle Breccia time, then presumably stone artifacts would also have been common in the Kromdraai deposit, which is not the case. Finally, most of the postcranial material is sufficiently different from that of *H. sapiens* to cast doubt on the possibility that it belonged to a true man rather than to *Paranthropus*.

For all of these reasons it seems to me that Broom was justified in attributing the material listed to *Paranthropus*. All of this material has been discussed in the literature and will not be treated exhaustively here but is included simply for the sake of completeness.

The Talus

This specimen has been discussed at some length in the literature, especially by Broom (in Broom and Schepers 1946); Le Gros Clark (1947, 1955 [revised 1964], 1967); and Lisowski (1966). The most detailed discussion is in Le Gros Clark (1947).

More than half of the bone is present; from above it appears to be almost intact, but a good deal is missing from the lower portion so that virtually nothing can be learned about the lower surface (see Figs. 86–87).

The talus is small compared to that of man and is of about the same size as that of the chimpanzee. Because of the missing portions it is not feasible to obtain reasonably accurate overall dimensions, although some idea of size can be obtained from the anterior width of the superior articular surface or trochlea, almost the whole of which is well preserved. Le Gros Clark (1947) gives this width as 20.5 mm in the *Paranthropus* specimen, as compared to observed ranges of 31–37 mm for modern Europeans and of 28–30 mm in modern Japanese. Two specimens of southern African Bushmen had corresponding values of 23 and 25 mm. This indicates an appreciable range of variation in modern man, and it is probable that adequate investigation would demonstrate that the *Paranthropus* value is within that range.

The major characteristics of the *Paranthropus* talus are that it has a moderately broad superior articular surface, the neck is short and broad, the articular surface for the navicular bone is relatively very wide, and the angle between the axis of the neck and that of the trochlea is relatively great.

A cross-section through the superior articular surface is nearly symmetrical, as it is in man. In the chimpanzee and gorilla such a cross-sectional contour is much less symmetrical, and Morton (1926) has suggested that this is a consequence of a different orientation of the subtalar articular facets. Le Gros Clark (1947) suggests that the load line of the femur and tibia of pongids passes through their medial condyles and hence also more to the medial side of the talus. The more symmetrical superior articular facet in modern man and in the *Paranthropus* specimen, he suggests, indicates a more even distribution of body weight at the talus. It would seem clear that weight distribution through the talus and rest of the foot is more even in man than it is in pongids. The matter is a complex one, however, since weight distribution is different in standing, walking, and running—see, for example, Morton (1952). The axis about which the weight is symmetrically distributed in the foot during standing passes between metatarsals II and III, while when running the axis of symmetry passes between I and II—that is to say, weight is shafted more medialward in running than it is in standing.

Another consideration enters here; the orientation of the superior articular surface or trochlea in relation to the main body of the talus. The trochlea is generally treated as though its orientation has remained static but that the angle of the neck with the trochlea has altered. Elftman and Manter (1935) have made what seems to me to be a sound case for regarding the main change, where trochlea and neck are differently orientated with respect to each other—as is true of pongids as compared to man, to be due to changed orientation of the trochlea. This involves also the increased height of the medial ridge of the trochlea in man and modification of the distal ends of the tibia and fibula. These authors have also concluded that the greatest similarity exists between the pongid foot and that of man when the former is plantar-flexed. The degree of mobility at the lower, sub-talar or transverse tarsal joint of the ankle is considerably greater in pongids than it is in man. As Elftman and Manter have observed, dorsi-flexion at this joint in a foot with load applied and resting on a flat surface does not cause abduction of the front portion of the foot but flattening of the longitudinal arch. The well-developed longitudinal arch present in the normal human foot results from the mobility at the sub-talar joint being very restricted, with the parts of the foot on either side of this joint related to each other as they are in a chimpanzee foot that is plantar-flexed. In the pongids the appreciable movement at the sub-talar joint modifies the orientation of the whole talus to the rest of the foot.

Manter (1946) has measured the compression forces in the many and various joints in the foot in man. He has reminded those who tend to think of the various lines of bone through the foot as channels along which the load applied to the talus is transmitted that the foot acts like a structural arch in which load forces do not have to be transmitted but resisted. Compression stresses are thus set up at each of the articulations between two bones. As in any structural arch for load-bearing, the greatest stresses are at the top of the arch and the stresses become progressively smaller as either end of the arch is approached. The joint in the foot that has the greatest load stress is the talo-navicular joint.

The manlike cross-section of the trochlea of the *Paranthropus* talus, in which the medial ridge is built up compared to that of the chimpanzee, indicates that much of the reorganization found in the modern hominid talo-crural joint had already taken place. The reorganization, however, was by no means complete since the orientation of the trochlea appears to have been like that in the chimpanzee rather than like that in man. That is to say, in the chimpanzee and in *Paranthropus* the talar head deviates medialward from the central longitudinal axis of the trochlea to approximately the same extent, which is about twice as much as is the case in man. The situation is more complex than appears at first because the head is large in the *Paranthropus* specimen; i.e., the articular surface is relatively very wide. Indeed, it extends medially approximately as far, relatively, as does that of the chimpanzee but extends laterally to relatively the same position as does that of man. Hence, simply measuring the angular deviation of the central axis of the neck is misleading in this case. Clearly, the great relative width of the *Paranthropus* head articular facet includes both the chimpanzee and the human type of talo-navicular articulation with respect to spatial relationship of the trochlea.

A reasonable explanation of this unusual feature is that it is a compromise. When the load was being resisted primarily through the lateral two-thirds of the talo-navicular joint, the situation would be essentially that found in man. The talo-crural joint, including the built-up lateral margin of the trochlea, would be as in man and orientated in relation to the prime stress-bearing portion of the talo-navicular joint as it is in man. On the other hand, when the load was being resisted through the medial two-thirds of the joint, then the direction of the weight stress in the foot, and the relation of the stressed part of the joint to the trochlea, would be similar to that in a chimpanzee foot. In this case, however, the talo-crural joint would not be quite like that of the chimpanzee because of the manlike medial edge of the trochlea and the related changes in the tibia and fibula.

The broad and fairly shallowly curved articular surface of the *Paranthropus* talar head implies at least a moderately heavy body in this animal, more so than in the chimpanzee in which the articular surface is part of a sphere of appreciably shorter radius. This conclusion is consistent with other evidence already discussed that indicates that *Paranthropus* was not a small animal.

If the above conclusion is correct, that weight could be coped with in the foot either in the manner in which it is in the chimpanzee or as it is in man, then the implication is that the foot of *Paranthropus* was relatively mobile, though perhaps not as mobile as in the chimpanzee. This would almost certainly mean that the hallux was not as immobilely adducted as is the case in man. These are essentially the conclusions to which Le Gros Clark (1947) also came.

It would thus seem that the *Paranthropus* foot was considerably more efficient for erectly bipedal posture and locomotion than are pongid feet. This is to be expected in view of the obvious evidence of frequent use of this locomotor habit to be found in the *Paranthropus* innominate, especially the ilium. As noted in a previous chapter, however, there is evidence in the pelvis of compromise with respect to locomotor function. The ilium indicates that features associated with erect bipedality were well developed—e.g., short ilium that is expanded strongly backward, sigmoid curvature of the iliac crest, and mechanical strengthening on the anterior gluteal surface. But the ischium is proportionately very long; this implies that the ischium-hamstring-femur complex was better adapted for power use, as in pongids, than it was for speed use, as in modern man and in *H. africanus*. The foot seems to have had the same combination of rather definitely apelike functional characteristics associated with much more distinctly manlike ones. Obviously more information is necessary before these tentative conclusions about the foot can be regarded as being well founded. For example, the calcaneus would indicate whether the talus was as free to move as it is in the chimpanzee or whether an antero-lateral process was present on the calcaneus that restricted talus movement, as in man. Similarly, having metatarsals I and II might indicate directly whether the hallux was fully adducted or not.

The Phalanges

Comparatively little can be said about the two phalanges from Kromdraai. The following is Broom's description of them (in Broom and Schepers 1946):

> One is pretty manifestly the first phalanx of a toe, and I think it is the fifth left toe of Paranthropus. It cannot be a finger bone, and it cannot be a toe bone of a baboon. It is, however, a little too imperfect to enable one to determine it with certainty.
>
> We have, however, a distal phalanx which is almost perfect, and is thus more important. This toe bone was found close to the other phalanx. Quite certainly it is not a human distal finger phalanx, and almost certainly not a distal human toe bone. And equally certainly it is not a baboon phalanx. We are thus forced to conclude that it is a distal toe phalanx of *Paranthropus*. It measures 12.5 mm in length and its proximal end is 8.3 mm in width. If I am right in concluding that the proximal phalanx belongs to the fifth toe, it is highly probable that this distal phalanx belongs to the second or third toe.

Broom concluded from these phalanges that the toes of *Paranthropus* were a little longer and more fingerlike than are those of modern man. This conclusion seems reasonable as a tentative conclusion from this not very satisfactory material and it is consistent with the impression of the *Paranthropus* foot that we have gained from the talus.

The *Paranthropus* and Olduvai Feet

The only well-preserved and reasonably complete hominid tarsus and metatarsus known at present is that from FLK NN I at Olduvai (Leakey 1960; Day and Napier 1964; Le Gros Clark 1967; Napier 1967). Preliminary comments are all that have been published so far, and I myself have not seen the original material; hence comparatively little can be said at this stage in comparing this foot with that from Kromdraai.

The Olduvai foot has been officially attributed to *H.* "*habilis*" (Leakey, Tobias, and Napier 1964). Insofar as this indicates that the foot is regarded as belonging to the more gracile of the two forms of hominid at present known from Olduvai Bed I, I am in agreement with this referral. Nevertheless, as has been indicated elsewhere (e.g., Robinson 1965 and 1967), it is my opinion that the taxon *H.* "*habilis*" is neither technically valid nor biologically meaningful. Instead, it would seem that the Bed I specimens that have been referred to this taxon are merely variants of *H. africanus*. This opinion is held also by some other workers, for example, Le Gros Clark (e.g., 1967). It is not suggested that the Bed I material is anatomically identical with the South African material of *H. africanus*, but rather that what differences appear to exist are of so small an order as to be quite inadequate to sustain taxonomic distinction between the two groups. Furthermore, since it seems clear that in Africa and at the time concerned *H. africanus* was in the process of evolving into *H. erectus*, it is to be expected that different demes of that phyletic line, separated either geographically or in time, might differ quite noticeably even though belonging to the same phyletic line. There seems to be no sound and objective basis at present for the view that the material attributed to *H.* "*habilis*" belongs to a different phyletic line than does the *H. africanus* material from South and East Africa. In my opinion it belongs to the same single phyletic line as the latter material does as well as that of *H. erectus* and of *H. sapiens*. If this is so, it is a matter of personal preference how that lineage is broken up taxonomically. On this basis, then, it seems to me that the Olduvai foot from FLK NN I can be regarded as representative of the sort of foot that *H. africanus* at Sterkfontein, Taung, and Makapansgat will have had. What distinctions can be made between that foot and the very much less complete one from Kromdraai in essence will be differences between *H. africanus* and *Paranthropus* feet.

The only direct comparisons that can be made concern the talus, but very little information has been made available about the one from Olduvai at this time. According to Day and Napier (1964), the talus is small "... but its length/breadth proportions approach those of modern man, including Bushman." The trochlea has a well-developed groove, and evidently the lateral and medial margins are well developed since the authors conclude that the axis of rotation of the ankle joint was stationary. This implies that in each of lateral and medial views the trochlea has a profile that forms an arc of a circle and the two circles are of equal radius.

The horizontal angle of the neck is not discussed as such, but mention is made of the fact that this, as well as the angle of inclination, is similar to that of the Kromdraai talus, while the torsion angle of the head is "higher" in the Olduvai specimen. According to Lisowski (1966), the torsion angle is higher (24.5° as compared to 22.0°) and the horizontal angle is lower (32.5° as compared to 36.0°) in the Olduvai specimen. In both respects the latter approaches man more closely than does the *Paranthropus* specimen. According to Lisowski's data, however, neither approaches modern man closely, being very close to the means for the great apes.

These data are puzzling since the Olduvai specimen appears from the photographs to resemble man distinctly more closely than do great ape tali with respect to the horizontal neck angle. I have no means of knowing what the torsion angle is in the Olduvai talus, but in that from Kromdraai the medial margin of the trochlea is built up in the manner of the human trochlea and is not low as in the chimpanzee, for example. As Elftman and Manter (1935) have shown, the apparent change in torsion angle of the talar head is due to the increased height of the medial border of the trochlea, and a line tangential to the two trochlear margins is the base against which torsion is measured. Yet according to Lisowski's data the torsion angle is identical in the Kromdraai talus to the average value for a sample of 85 chimpanzee tali (22.0°) in spite of the fact that in the former the trochlea has the high medial border found in man (torsion angle 41.3°, according to Lisowski) but not in the chimpanzee. Unfortunately, Lisowski's paper is very condensed and his explanation of his measuring technique so brief and general that it is not possible to evaluate his data; hence the reason for this apparent contradiction is not evident.

As already noted, the talar horizontal neck angle is not very illuminating by itself. In the case of the Kromdraai talus it is misleading because it gives no indication of the great width of the articular surface for the navicular bone. Thus it does not bring out the fact that in this specimen both the great ape and the human functional conditions appear to be included. This does not appear to be the case in the Olduvai talus; the relationship of the articular facet of the head to the trochlea, in top view, appears to be essentially of the human type and does not include the extreme medial extension seen in the *Paranthropus* specimen. Day and Napier (1964) conclude that the differences between the Kromdraai and Olduvai tali suggest the possibility that the former belonged to a foot in which the hallux was not fully adducted. The hallux of the Olduvai specimen was fully adducted, as is indicated by a well-developed articulation between the proximal ends of metatarsals I and II.

The conclusion that the *Paranthropus* individual had a more mobile foot than has *H. sapiens*, with a hallux which was not habitually fully adducted, emerged in the discussion of the Kromdraai specimen and was reached a long time ago by Le Gros Clark (1947). Day and Napier's conclusion following their comparisons with the Olduvai specimen therefore is of interest, and it is very fortunate indeed that enough of the latter specimen is preserved to show that the less medially protuberant articular facet of the talar head was associated with a fully adducted hallux.

Day and Napier have noted that the Olduvai foot had a well-developed longitudinal arch as well as a transverse one at the level of the distal tarsal bones. The evidence indicates that this was a compact, strong, and relatively inflexible foot of the human type, "... with most of the specializations associated with the plantigrade propulsive feet of modern man." Further, " ... we must conclude that the individual possessed in his foot the structural requirements of an upright stance and a fully bipedal gait." They point out, however, that the robusticity pattern of the metatarsals ($1>5>3>4>2$ rather than the usual $1>5>4>3>2$ of modern man) and the tilt of the talus suggest to them that the unique striding gait of modern man had not yet been achieved.

The difference in robusticity pattern of the metatarsals is small, involving only a reversal of the positions of III and IV. The robustness and length of these two bones do not appear to be very different; in any case, the robustness index cannot be determined with complete accuracy for either since the distal end of each is missing. It seems to me that this is not an especially compelling reason for doubting the ability of the owner to stride. Naturally, the account by Day and Napier is only a preliminary one, and the case will doubtless be argued more fully at a later date; more detailed discussion of their views must therefore wait.

In view of the detailed evidence presented in earlier chapters concerning the ischium-hamstring-femur complex in *H. africanus* and its great similarity to that of modern man, it seems to me hardly conceivable that such an apparatus was not used for striding. The elongated lower limb and short ischium that is orientated in the human manner indicate a mechanism that is primarily adapted for speed rather than power and possessed the full potential of modern man for extending the femur past the vertical. This apparatus was thus well suited to taking long steps. Furthermore, the gluteal surface of the ilium gives evidence of a well-developed capacity to control lateral balance, and details of hip and knee joints indicate close functional similarity to modern man. The Olduvai foot seems to me fully compatible with the above complex. Since the foot is well arched with a powerful and completely adducted hallux, robust metatarsals, and both the lateral and medial borders of the foot well supported (V is the second most robust metatarsal after I), and upper and lower ankle joints that were essentially of the modern human sort, it seems highly probable that this foot was fully capable of providing a powerful push-off with the heel well off the ground. Napier (1967) reports that Day has studied a distal hallucial phalanx, from higher up in Bed I at Olduvai than the level from which the foot came, and has concluded that it belonged to an individual capable of striding. Most unfortunately, all the phalanges of the older foot are missing; I suspect that if they had been present they would have told the same story as the above phalanx.

The implication of the *Paranthropus* talus evidence—more mobile foot capable of functioning in either the great ape or a more manlike manner and probably having an incompletely adducted hallux—is that striding is unlikely to have been a capacity possessed by its owner. It should not be overlooked that thus far we have extremely incomplete evidence of a single *Paranthropus* foot and rather good evidence of a single *H. africanus* foot, and this is not a great deal to go on. The small amount of evidence in each case, however, appears to fit very well with what can be learned from the other postcranial material available.

Since this chapter was written two interesting papers have been published concerning foot material from Olduvai (Day 1967; Day and Wood 1968). The former reports on a multivariate analysis of Olduvai Hominid 10, the terminal phalanx referred to briefly above. The analysis places this specimen clearly with those of plantigrade men and distinguishes it sharply from those of the chimpanzee and gorilla.

The second paper is a much more detailed analysis of the Olduvai talus than had previously been available. This is a very useful paper that should be consulted since it presents many data that will not be repeated here. A point of great interest brought out in it is that the talar horizontal neck angle is rather large and apelike, which would ordinarily imply that the hallux was not adducted, while the rest of the foot shows that the hallux was in fact fully adducted. The authors point out that this is possible largely because the articular surface of the talar head is not symmetrical about the neck axis; the head axis is turned lateralward to a small extent (about 15°), running at a valgus angle to the neck long axis. In man the horizontal angle of the neck is appreciably smaller than that in the pongids, as we have seen, and in both the talar head articular surface is set symmetrically about the neck axis. The Olduvai talus has a large horizontal angle, like that of the pongids, but has compensated for this by the asymmetric arrangement of the head on the neck to achieve a functional result similar to that in man. This accounts for the seeming contradiction of an apelike talar horizontal angle with a fully adducted and manlike hallux.

The authors point out that while the horizontal angle is much the same in the Olduvai and Kromdraai tali (28.0 and 32.0 respectively) the Kromdraai specimen does not have an asymmetrical head. The implication of this is that the latter probably did not have a fully adducted hallux. As was noted earlier in this chapter, the relatively large articular surface of the Kromdraai talar head implies, it seems to me, the ability to use the hallux strongly adducted or somewhat abducted; evidently it was not habitually fully adducted, as was apparently true of the Olduvai hallux. This evidence from Day and Wood thus tends to confirm the existence of a type of foot in *Paranthropus* that represents a compromise between ape and human forms. The combination of characteristics in the Olduvai talus and foot, which at first sight appear so contradictory, could be explained as a transitional stage between the *Paranthropus* type and that of man—a stage in which some of the apelike characteristics were still present even though the manlike mode of function had already been achieved.

13 The Pectoral Girdle and Limb

Material

Very little is known from South Africa of the pectoral girdle or limb of the early hominids. Nothing is known of the pectoral girdle in *Paranthropus*; in the case of *H. africanus* the following material is known:

Sts 7 the articular end of a poorly preserved scapula from the Lower Breccia at Sterkfontein.

MLD 20 a small fragment of clavicle from Makapansgat.

Pieces of humerus are known from both early hominid types. In the case of *H. africanus* the following material is available:

Sts 7 much of the humerus of the same individual to which the fragment of scapula belonged. The distal end is missing and much of the shaft is crushed flat. The specimen is from the Lower Breccia at Sterkfontein.

MLD 14 a small piece of the shaft of a humerus from Makapansgat.

In the case of *Paranthropus* the following material of the humerus is available:

TM 1517 the distal articular end of a humerus with a very small piece of the shaft, very well preserved. This specimen is from Kromdraai.

Nothing is known for certain of the radius and ulna of *H. africanus*; all that is at present attributable to *Paranthropus* with reasonable certainty is the ulnar fragment from Kromdraai:

TM 1517 the proximal end of an ulna, including the olecranon process and the trochlear notch. This specimen was associated with the type skull of *Paranthropus*, as was the distal end of a humerus, listed above.

As is evident from this list of the known specimens of the pectoral girdle and limb (but not including the hand), there is virtually no duplication of comparable parts in the *H. africanus* and *Paranthropus* material. All of these specimens have already been described in various places in the literature and what follows is primarily a summary of the latter for the sake of completeness rather than a complete re-study of the material.

The Pectoral Girdle

The small piece of poorly preserved clavicle from Makapansgat is not particularly instructive because it is so incomplete. It has been described by

Boné (1955). The only other material known is from Olduvai: two clavicles have been reported from FLK NN I (Leakey [1960] includes a photograph of one) but have not yet been described. Napier (in Tobias et al. 1965) has commented briefly on one clavicle that has been referred to *H.* "*habilis*" (Leakey, Tobias, and Napier 1964) and has provided a drawing of it compared to a modern human equivalent. According to Napier,

> The clavicle . . . is a robust bone distinguishable in its morphology from *H. sapiens* only by the presence of a smooth groove postero-inferiorly towards its medial end and a slight modification of the cross-sectional area of the medial extremity such as might be expected were inlet to the thorax rather steep, the scapulae set relatively higher than in modern man, and the clavicles, consequently, rotated anteriorly around their horizontal axes. The functional and behavioural deductions that can be made from the morphology of the clavicle are limited, but it can reasonably be inferred that the movements of the shoulder girdle were substantially as in modern man, and that therefore there was no morphological barrier to overarm throwing or other activities requiring external rotation of the shoulder joint.

The scapula is perhaps more instructive in terms of functional and taxonomic information than is the clavicle. Unfortunately, the single scapula available is incomplete and poorly preserved. Because the bone of the scapula is mostly thin and because it is embedded in a strongly muscular region, mammalian scapulae are not often preserved as fossils.

The single scapula, Sts 7 from Sterkfontein, consists of a piece of the articular end some 7 cm in length. It has suffered some crushing and is so poorly preserved that it was deemed unwise to attempt to remove it from the matrix. Thus the details of the costal surface are not available since it is the dorsal surface that is exposed. The whole of the glenoid is present and reasonably well preserved but is broken into two pieces that are slightly separated. A small portion of the spine is present, but it is incomplete in length as well as in height. All of the acromion is missing. Broom (in Broom, Robinson, and Schepers 1950, fig. 19) illustrates a substantial piece of acromion; this is not now present on the specimen. That Broom had a factual basis for this portion of his drawing of the specimen is fortunately documented by plate 2, figure 9, of that same work. This photograph, which I took, shows the scapula and associated humerus still in the breccia. A portion of the acromion is clearly visible—not as a natural cast but as actual bone unobscured by matrix. In 1965 I was unable to find this piece of acromion, which has been removed from the specimen for some reason though it was free of matrix. The coracoid process is present and appears to be complete except for a small patch of superficial damage.

The glenoid is almost symmetrically oval in shape and measures 37 mm in length and 18 mm in width. These dimensions do not agree closely with those given by Broom and Robinson (in Broom, Robinson, and Schepers 1950), 33 mm and 20 mm respectively. The glenoid is shaped approximately as in the human ovoid type of Vallois (1928–46) but is relatively narrower than is usual in man. The length/breadth index is either 51 or 61, depending on whether the later or earlier figures above are used. This compares with a mean value of about 77 (observed range 52–92) for *H. sapiens*, according to Vallois. In explanation of the relatively large difference between the 1950 measurements and those given here I can only say that Broom himself made those of 1950 and I am quite unable to duplicate his values in what is an uncomplicated situation. It is possible that some further preparation of the specimen was carried out after 1950, but I can no longer recall if this is so or not (see figs. 88 and 89).

Table 11
Glenoid Cavity Lengths, Breadths, and Length/Breadth Index Values

		Glenoid Cavity Mean		
	N	Length	Breadth	Breadth × 100 / Length
Polynesian	24	36.1	28.4	78.7
French	166	34.9	27.5	78.8
Japanese	87	34.1	26.7	78.3
Australian	14	34.1	24.5	71.8
Fuegian	63	33.6	26.2	78.0
African Negro	100	32.8	25.2	76.8
Negrito	51	28.9	22.0	76.1
Neanderthal (Gorjan.)	6	33.9	22.1	65.2
Homo africanus				
(Broom and Robinson)	1	33.0	20.0	60.6
(Robinson)	1	37.0	19.0	51.4

Unless otherwise credited the above data are from Vallois (1928–46).

The coracoid process is well developed, perhaps slightly more so than is usual in man but not as much as is usual in the orang. It originates almost level with the glenoid plane and the free end curves backward to some extent. In this feature it is a little more like the orang than like man. In having a well-developed area of attachment for biceps brachii, it also resembles the orang slightly more than it does modern man.

It is unfortunate that so little of the scapular blade is preserved because it is not possible to know what the original shape was. Scapula shape is most variable among the higher primates and within any one species of higher primate (Schultz 1930), and the preserved portion of the *H. africanus* scapula is too small to allow even reasonably accurate prediction of the original shape. It does seem highly probable, however, that the blade was relatively narrow compared to that of modern man. Some indication of this can be obtained from the angular relationship of the glenoid plane to the axillary margin.

Oxnard (e.g., 1963 and 1967) and Ashton and Oxnard (e.g., 1964) have discussed this angular relation (their angle *g*) in higher primates as part of a long and interesting series of investigations into the structure and functioning of the primate shoulder and other parts of the pectoral girdle and limb. According to these authors, the mean value of this angle in modern man is 137.4°, while in the brachiators the mean value is 112.7°. Ranges representing 90 percent fiducial limits for these means do not overlap. Their semibrachiator category is exactly intermediate with a mean value of 124.6°; 90 percent fiducial limits for this mean include the means of both the human and brachiator groups. Quadrupedal monkeys are almost exactly intermediate between the semibrachiators and man, with a mean value of 130.6°. The confidence interval for this mean represented by 90 percent fiducial limits includes the means of the human and semibrachiator groups, but not that of the brachiators.

Because so much of the scapular blade of the Sts 7 specimen is missing, and because there is very considerable variation individually and between taxa in the contour of the half of the axillary border nearer the vertebral border, the position of the junction of the vertebral and axillary borders cannot be known accurately for this specimen. Hence the axilloglenoid angle as used in the studies by Ashton and Oxnard cannot be determined accurately. Assuming that the roughly 7 cm of axillary border adjacent to the glenoid accurately reflects the direction of the entire axillary border, the angle in Sts 7 is 106.5°. By assuming that the axillary border deviates downward farther away from the glenoid, which is frequently the case in higher primates, the angle is increased in size. It seems to me that the maximum probable angle on this assumption would be no more than 115°. These two estimated values lie on either side of the mean value for pongids given by Ashton and Oxnard (112.7°) and clearly agree more closely with pongids than with the means of the other groups. According to this particular criterion and the comparative data provided by Ashton and Oxnard, the Sts 7 scapula of *H. africanus* fits very much better with that of brachiators than with that of any other higher primate group.

In the case of Sts 7, a single specimen, not a sample mean, is being compared with the means given by Ashton and Oxnard. It would be valuable to know, therefore, what the ranges of variation were on either side of their means. The authors do not give either observed ranges or estimated population ranges. Their diagram gives only the means and their 90 percent fiducial limits, while their table gives only means with their standard errors. This does allow calculation of standard population ranges (mean ± 3 s.d.). Performing this computation and constructing a diagram of the results show that the standard population range for the brachiators includes the entire range for each of the other forms. Moreover, the semibrachiator range includes the whole of the ranges for both quadrupeds and man. In turn, the quadruped range includes the entire range for man. The respective population ranges are approximately: man, 120–54; quadruped, 90–172; semibrachiator, 71–179; brachiators, 22–203. On this criterion no individual human scapula could be distinguished from scapulae of brachiators, semibrachiators, or quadrupeds. Indeed, except for the extremes of the brachiator range, no individual specimen from any of the groups could be placed in its correct category. In other words, essentially this particular criterion has no diagnostic value. Because of the relative smallness of the range for man it is possible to say that the Sts 7 value, wherever it may fall in the estimated range of 106.5–115°, is least likely to be human. It is nearest to the brachiator mean but falls well within the ranges of the other two groups also and could as easily belong to any one of the three. Even the rather small distance the Sts 7 value falls below the range for man may be misleading. Judging from data provided by Vallois (1928–46), the sample of 66 human scapulae used by Ashton and Oxnard may not be fully representative of *H. sapiens* as a whole. Vallois does not treat his data statistically but gives a series of means and observed ranges for a geographically wide range of samples, much wider than the sample used by Ashton and Oxnard. The data of Vallois suggest a wider range of variability for *H. sapiens*; for example, his tiny sample (2) of Negrillo scapulae has a mean that falls just outside the standard population range estimated from the sample of Ashton and Oxnard. Some means given for particular racial groups by Vallois, which are based on sample sizes larger than the total sample used by Ashton and Oxnard, fall quite far below their mean for man. This suggests rather clearly that had a sample been used by them that was larger and more widely representative of *H. sapiens as a whole*, the estimated standard population range would have been appreciably larger than that which they actually obtained from their smaller and less representative sample. In that event the Sts 7 value would probably have fallen within the lower end of the human range as well as within the ranges of the other three groups. Indeed, the smallest *observed* value given by Vallois is 111°, which is of comparable value to that probable for Sts 7. Vallois defined and measured this angle in the same manner as did Ashton and Oxnard, hence their data are directly comparable. Unfortunately, since we have only a single specimen of *H. africanus*, it is not possible to know whether it was modal or not. It does fall closer to the brachiator mean than to that of any other group here concerned. But this means extremely little in view of the demonstrable fact that this criterion has virtually no diagnostic significance in the higher primates involved in these comparisons.

If a small angle between the glenoid plane and the axillary border is equated with narrowness of the scapular blade, then it is possible to conclude that this *H. africanus* scapula did not allow the range of movement of the forelimb that is found in man. This follows from the studies of Frey (1923, particularly) who showed that narrow scapulae allow a reduced range of forelimb movement as compared to broader scapulae. The above conclusion is supported by the work of Ashton and Oxnard already referred to. In quadrupedal monkeys, in which the scapula tends to be narrow, trapezius and serratus magnus (with deltoideus) tend to produce protraction of the scapula rather than rotation. In semibrachiators and brachiators, as well as in man, these muscles tend to produce rotation rather than protraction. The range of movement of the scapula, hence also of the limb, is thus greater in the latter groups, which tend to have broader scapulae.

As Ashton and Oxnard have shown, the effectiveness of capacity to rotate the scapula can be related to several major factors, among which are the following:

(a) The direction of pull of the two most important muscles (trapezius and serratus magnus) in this connection. If their directions of pull are parallel, the capacity to rotate the scapula will be maximal.

(b) The length of the couple arm connecting the two muscle insertions here involved. The longer this arm, the greater the capacity to rotate the scapula.

(c) The extent to which the insertion of serratus magnus is removed caudalward of the axis of rotation of the scapula. The greater this distance, the more effectively can serratus magnus contribute to rotation of the scapula.

Clearly, the best combination of these factors for scapula rotation, hence for a wide range of movements of the forelimb, will be found in broad scapulae. This is consistent with Frey's conclusion (1923). The most obviously breadth-related of these factors is (c).

In the case of (a), Ashton and Oxnard (1964) use the angle (e) as a reflection of the relation between the lines of function of the two muscles of the couple. This angle can be seen in figure 90. Unfortunately, not one of the four points determining the two lines that form angle (e) is present on the *H. africanus* scapular fragment. Even though this angle appears to be no more than moderately effective as an indicator of the mutual relation of the lines of function of the two muscles involved, it cannot be determined, even approximately, for Sts 7. Nor, indeed, does there appear to be any other way of determining this relation in Sts 7 that would not amount to sheer guesswork on my part.

In the case of (b) and Sts 7 we are in no better position since neither end of the couple arm is present and, in fact, at most only a relatively small portion of the couple arm is represented on the specimen. There appears to be no way in which the couple arm length could be determined between reasonably close limits. The fact that the axillo-glenoid angle is relatively small in Sts 7 does not necessarily mean that the couple arm was short, as can readily be seen by comparing the scapulae of gibbons and orangs. The scapula of the former is relatively long and narrow compared to the short and broad scapula of the latter, but both have relatively long couple arms.

In the case of (c) the actual distance cannot be determined since all of the relevant portion of the Sts 7 specimen is missing. There is, however, obviously a fairly close relationship between the size of the axillo-glenoid angle and the extent to which the insertion of serratus magnus—or, what amounts to the same thing, the inferior angle of the scapula—is extended caudalward of the axis of rotation of the scapula. Since, by definition, the axillary line of the axillo-glenoid angle passes through the inferior angle of the scapula, there would be a very close correlation between the axillo-glenoid angle size and the distance to which the inferior angle is removed caudalward of the axis of rotation, provided that the relation of the other line of this angle, the glenoid plane, had a fixed relation to the rest of the scapula. This it does not have. That this is so can be demonstrated by means of the spino-glenoid angle used by Vallois—that is, the angle between the glenoid plane and the line representing the junction of the spine with the blade of the scapula. Vallois gives figures for man only, and the observed range is 101.5–63.0°. The latter figure is apparently unusually small and occurs in a sample of 113 Finnish scapulae studied by Kajava (1924). A sample of 120 French scapulae gives an observed range of 98–83°. I do not have comparable figures for pongid scapulae, but according to dioptrographic tracings given by Schultz (1930) it would appear likely that there is some variation in the mean value for this angle in the various forms of living pongid. In Sts 7 the value appears to have been about 99°. In any event, as a result of the fact that this angle does vary, the correlation of axillo-glenoid angle size with the degree of caudalward displacement of the inferior angle is obviously less close than otherwise it would have been. Nevertheless, it is clear that scapulae in which the axillo-glenoid angle is large must have a greater caudalward displacement and those with smaller angles less displacement of the inferior angle. Since it is possible to determine with reasonable certainty that the axillo-glenoid angle value for Sts 7 will have been between 106.5° and 115°, it seems very probable that the *H. africanus* specimen had an inferior angle in which the degree of caudalward displacement was relatively small.

Of course, in view of the fact that this is the only australopithecine scapula known and that very little is known of fossil hominoid scapulae in general, it is not impossible that the Sts 7 scapula had an unusual structure that allowed the rotation axis–inferior angle distance to be relatively great as compared to that suggested by the angle between that part of the axillary border that is near the glenoid and the glenoid plane. Such a conclusion cannot be assumed, however, in the absence of any evidence to suggest it. The scapula seems to have been unusual in one respect; the spine appears to have been placed relatively low down on the blade. In this respect it seems to agree more closely with the gorilla than with man or the other pongids. This situation would tend to reduce the degree of caudalward displacement of the inferior angle of the scapula.

The measure used by Ashton and Oxnard to reflect variation in the above feature is the angle, at the inferior scapular angle, between two lines: one runs from the point of junction of the axillary border with the glenoid through the inferior angle of the scapula; the other runs from a point on the vertebral border at the end of the line of junction of spine and blade to the inferior angle. This angle (see fig. 90) will become smaller as the degree of caudalward displacement of the inferior angle increases. Hence, broad scapulae will have small angles and narrow scapulae will have wide angles. According to the data of Ashton and Oxnard, the angle tends to be smallest in the brachiators (mean = 47.6°); man and the semi-brachiators are next (53.6°, 55.6° respectively); and the quadrupedal monkeys have the largest angles (67.0°). Even so, all of the 90 percent fiducial limit ranges overlap extensively. Calculation of standard population ranges again demonstrates extensive overlap of all of the ranges, that of the brachiators again wholly including all of the others.

We have seen that the *H. africanus* Sts 7 specimen appears to have had relatively little caudalward displacement of the inferior angle; therefore the angle here being discussed would have been relatively large. It is not possible to estimate the size in degrees, however, since scapulae vary appreciably with respect to the angular relationship of the vertebral and axillary borders, and this variation would affect the size of the angle with which we are concerned even if the factor being estimated—caudalward displacement of the inferior angle—remained the same. Indeed, it is apparent—as in the case of the axillo-glenoid angle—that the angle being measured is not a good reflector of the factor being studied for the very reason that at least one other factor that is not closely correlated with it contributes to the size of the angle.

Since it is not possible to determine within close limits the angle for Sts 7, its relation to the mean values for higher primates provided by Ashton and Oxnard cannot be determined. The angle is unlikely to have been very small, as has already been noted, nor is it likely to have been extremely large. Therefore it would probably fall in the middle range of values obtained by Ashton and Oxnard—the region that includes man, brachiators, semi-brachiators, and quadrupeds. The data provided by Ashton and Oxnard are thus singularly unhelpful for interpreting the nature of an individual the value of which falls somewhere in this middle region of the ranges they obtained.

This latter point seems to be true in general of the data on the shoulder provided by Ashton and Oxnard (1964) and that by Ashton, Oxnard, and Spence (1965). Trends are shown by these data that are very useful for an understanding of the primate shoulder in relation to locomotor habit. But when unidentified material is being studied to determine to what taxon or locomotor group it belongs, the data are of little help, unfortunately. This point was explicitly recognized in the paper by the latter three authors in relation to scapular characters specifically chosen for their purposes as being only distantly related to locomotor habit. This difficulty of not being able to use data from studies on living forms, however valid in their own right, is frequently encountered in taxonomic work with fossils. It is not always clearly recognized that a great difference exists between comparing known samples by statistical means and determining what, if any, level of significance attaches to the differences between them, and the quite different problem of sorting individual specimens for the first time into samples whose taxonomic status can then be examined. Differences between populations that can only be shown statistically are seldom useful for purposes of initial sorting.

It seems to me, further, that the locomotor classification used by Ashton and Oxnard et al., which was taken from Napier, is somewhat artificial. As the former authors explicitly recognize, *Ateles* can as logically (and to me more logically) be classified as a brachiator than as a semi-brachiator.

Similarly, it seems to me that there are good grounds for placing *Colobus* among the quadrupeds rather than among the semi-brachiators. The two ends of the semi-brachiator category as used appear to be indistinguishable from brachiators at one end and from quadrupeds at the other. The very great amount of overlap in the various features—functional and otherwise—that these authors have studied seems to confirm the already recognized fact that the primate scapula is most variable and also suggests that perhaps the category classification used by them is artificial; a somewhat similar criticism has been made by Ripley (1967). Moreover, it is not possible to assess the meaning of these data effectively because the data for each individual species or genus sample of those that go to make up a particular locomotor category are not provided. Instead, the means and 90 percent fiducial limit ranges refer to the broad group as a whole for each of the major categories used. Therefore it is not possible to see how the data for each species and genus relate to the parameters for the locomotor category to which they have been assigned. These comments refer not so much to these studies as carried out for their original purpose but to the usefulness of the data provided for the purposes of the present study.

The reader is referred to comments on the Sts 7 scapula by Campbell (1966) and to those by Oxnard (1968) who discussed the views of Campbell. These discussions will not be repeated here for two main reasons. The first is that the discussions are based on the work of Ashton, Oxnard, and Spence, already referred to, and we have seen that this has a number of drawbacks for our present purpose, however illuminating it may be in its own right. The second reason is that my studies on the original specimen—and, indeed, also the drawing by Broom (in Broom, Robinson, and Schepers 1950) on which these authors based themselves in the absence of access to the original specimen—do not substantiate the angle values used by them in their discussions. For example, Oxnard uses a best estimate of 17.5° (possible range 15–20°) for the angle between the spine and the axillary border, whereas I obtained a value of about 7°; he uses a value of 103° for the angle between the glenoid and the axillary border, whereas my best estimate from the existing fragment is 106.5° but could be as much as 115° since so much of the axillary border is missing. The specimen is simply too damaged for firm conclusions to be drawn from it.

It is well to recall here Schultz's (1930) demonstration that variation in scapular shape is very great—between-taxon as well as within-taxon—and also Frey's (1923 and 1924) demonstration that variations in the manner of attachment of muscles to the scapula cause variations in its contour. Thus it would seem that this bone *by itself* is suitable neither for taxonomic purposes nor as an indicator of locomotor behavior in fossil forms. In the case of *H.*

africanus the difficulty is compounded because a single, very incomplete specimen is all that is known of the scapula. It seems that very little can be learned of *H. africanus* locomotor behavior from this specimen. It clearly was not a typically hominid scapula as judged by that of *H. sapiens*. Indeed, it appears to have been less hominidlike than any other part of the skeleton of this form known at present.

The Humerus

The best humerus specimen of *H. africanus* is the greater portion of a humerus belonging to the same individual as the scapula fragment, Sts 7 from Sterkfontein. This was described by Broom and Robinson (in Broom, Robinson, and Schepers 1950). The only other fragment known is MLD 14 from Makapansgat. This is a piece of humeral shaft but is too fragmentary to be instructive and is regarded as being "suspicious" (Boné and Dart 1955). It has been discussed by Boné (1955).

The Sts 7 humerus has the upper or proximal end very well preserved and complete. For some distance from the surgical neck down the shaft has been crushed almost completely flat so that it is less than 1 cm thick by about 3 cm wide. Not all of the shaft has been crushed in this manner, and a portion that appears to have constituted about the third quarter seems to be uncrushed. In cross-section the shaft is somewhat triangular in shape in this region and measures 23 mm by 21 mm. Broom and Robinson estimated that approximately three-quarters of the shaft is present:

> Though the distal quarter of the bone is lost we can form a fair estimate of how much is gone. We can be moderately safe in considering that the whole humerus was about 300 mm in length. It cannot have been less in our opinion than 290 mm, nor do we think more than 310 mm.

With a specimen in the condition of this one, and in the absence of any other reasonably complete early hominid humeri, this estimate obviously cannot be of a high order of accuracy. Because of the changing shape of the humeral shaft down its length, however, landmarks are available that should ensure that this estimate is not unrealistic. Reexamination of this specimen has suggested no modification of the earlier estimate. The dimensions of the well-preserved head are given in Figure 91.

The anatomy of the upper end is very similar to that in *H. sapiens*. The head has the typical structure of *H. sapiens* with the anatomical neck clearly developed. The neck is perhaps a little less well defined where it separates the head from the lesser tuberosity than is usually the case in *H. sapiens*. The bicipital or intertubercular groove is large, has well-defined edges, and runs straight down the shaft from the actual edge of the head surface nearest to it. This condition closely resembles that in *H. sapiens* and is quite similar to that in the chimpanzee and gorilla, though less like that in the orang. The edges of the groove, especially that on the lesser tuberosity, are rather more sharply defined than will usually be found in *H. sapiens*. The lesser tuberosity, to which subscapularis attaches, is very well defined and is a little narrower and more elongate than is commonly the case in man. The distal portion of the tuberosity narrows sharply to a well-defined ridge that is a part of the anterior wall of the bicipital groove. Between this ridge and the near edge of the head is a well-defined hollow. In possessing this straight, well-defined, and long ridge, as well as the hollow between the lesser tuberosity and the head, this specimen is less like the great apes. In the latter the hollow is not so obtrusive and the ridge is shorter and usually curves inward, away from the bicipital groove.

Table 12
Humerus and Femur Lengths and Humero-Femoral Index Values

		N	Humerus	Femur	$\frac{\text{Humerus} \times 100}{\text{Femur}}$
White	♂	122	321	434	74.2
White	♀	110	292	399	73.2
Negro	♂	122	329	449	73.3
Negro	♀	111	303	416	72.8
Homo africanus		1	300 (♂)	310 (♀)	96.8
Gorilla	♂	93 (84)	437	376	116.2
Gorilla	♀	60 (54)	368	311	118.3
Chimpanzee	♂	45 (38)	306	300	102.0
Chimpanzee	♀	62 (58)	290	286	101.4
Orang	♂	41 (32)	365	269	135.7
Orang	♀	39 (29)	323	234	138.0
Gibbon	♂ + ♀	50 (44)	224	197	113.7

Data from Schultz (1937) except for *H. africanus*. The sample size in parentheses refers to the femur when sample size is different from that of the humerus.

The greater tuberosity is well developed and is very similar to that of *H. sapiens*. It also resembles that of the orang rather closely, but is a little less like that of the chimpanzee or gorilla. The facets for attachment of supraspinatus, infraspinatus, and teres minor are clearly defined, and the rounded ridge between the attachments for the former two is prominent. This ridge is 2.5 mm high and does not usually occur in *H. sapiens*. Below the greater tuberosity there is a prominent, rounded ridge running down the shaft. This almost certainly is an artifact due to the severe flattening of the shaft in this region; this ridge is one edge of the flattened shaft.

The Sts 7 bone appears to have been fairly robust. Assuming that the estimate of 300 mm for the length is a reasonable approximation, it is of about average length for *H. sapiens* females as represented by American whites and Negroes (see table 12) and a little shorter than the average for males. The head diameter also agrees well with that of modern man. According to Dwight (1904–5), the mean transverse diameter of the humeral head in American whites ($N = 200$) is 44.7 mm in males and 37.0 in females. The corresponding diameter in Sts 7 is 40.0 mm. Shaft thickness at about the estimated position of midshaft tends to be slightly greater than average for *H. sapiens*, though certainly well within the range of variation for the latter (see table 13).

Table 13
Diameter of Humerus at Midshaft

			Max. Diam.			Min. Diam.	
		N	Mean	Range	*N*	Mean	Range
North Chinese (Black)	♂	39	22.2	20–25	39	17.1	15–19
North Chinese (v. Bonin and P'an)	♀	47	19.0	16–21	7	13.4	12–15
North Chinese (v. Bonin and P'an)	♂	133	22.5	—	133	17.4	—
Ainu	♂	46	22.6	—	46	17.3	—
Ainu	♀	29	21.0	—	29	15.7	—
Homo erectus (Pekin)	♂?	1	20.7	—	1	15.4	—
Homo africanus	♂?	1	23.0	—	1	21.0	—

Data from Weidenreich (1941) except for *H. africanus*.

Long bone robustness can be measured in various ways; one index that has been used for this purpose is the girth-length index of Schultz (1953). This consists of girth at midshaft × 100/maximum length. According to Schultz's data, this is a stable index of some interest. Only means and observed ranges are given; hence standard population ranges cannot be derived from the data provided, some of which are set out in Table 14. From this table it appears that quadrupedal climbers tend to have more robust humeri while the true brachiators have more slender humeri. Man falls in between the two groups. From table 14 it is clear that there is much overlap of the

Table 14
Humerus Girth–Length Index

			Girth-Length Index	
		N	Mean	Observed Range
Cebus	♀	5	21.2	—
Cebus	♂	11	22.6	—
Ateles	♀	5	17.0	—
Ateles	♂	4	17.5	—
Macaca	♀	12	23.3	21.6–25.2
Macaca	♂	12	25.9	23.4–28.2
Hylobates	♀	24	13.4	12.3–14.9
Hylobates	♂	24	13.6	11.7–15.5
Pongo	♀	12	19.4	16.0–22.1
Pongo	♂	12	16.9	16.9–24.2
Pan	♀	14	25.0	22.4–28.0
Pan	♂	12	25.7	21.7–28.4
Gorilla	♀	12	22.5	20.9–26.4
Gorilla	♂	12	25.6	22.1–30.2
Homo sapiens	♀	12	21.1	18.2–22.7
Homo sapiens	♂	20	21.6	19.2–24.3
Homo africanus	♂?	1	23.2	—

Data from Schultz (1953) except for *H. africanus*.

observed ranges of variation and that this character is of little use in a strictly taxonomic context. Because the *H. africanus* Sts 7 specimen is still partly embedded in matrix, the girth of the uncrushed piece of shaft could not be determined with as much accuracy as would otherwise have been possible. The value obtained, 69.5 mm, gives an index value of 23.2. This is well above the mean for *H. sapiens*, being more or less intermediate between the latter and those for the chimpanzee and gorilla, the two pongids that are quadrupedal climbers. Midshaft shape is given in table 15.

Clearly there are too many uncertainties with respect to the Sts 7 dimensions for this solitary figure to be a reliable guide to humerus robustness in *H. africanus*. Taking the evidence at face value suggests that *H. africanus* had a more robust humerus than modern man usually has but less robust than is usual in the heavier pongid quadrupedal climbers. Least resemblance is shown with those pongids that are true brachiators.

Table 15
Humerus Midshaft Index

		N	Midshaft Index
North Chinese (v. Bonin and P'an)	♀	47?	70.5
North Chinese (v. Bonin and P'an)	♂	133	77.3
Ainu	♂	46	76.5
Ainu	♀	29	74.8
American White (Hrdliĉka)	♂	2326	81.6
American White (Hrdliĉka)	♀	1000	78.8
American Negro (Hrdliĉka)	♂	131	83.4
American Negro (Hrdliĉka)	♀	66	79.0
Homo erectus (Weidenreich)	♂?	1	73.5
Homo africanus (Robinson)	♂?	1	91.3

The Sts 7 scapula and humerus were very closely associated in the site with the largest mandible of *H. africanus* known at present; this also has the catalogue number Sts 7. This mandible seems clearly to have belonged to a male individual because of the large size of the corpus and ramus and also the large size of its canine teeth as compared to other *H. africanus* specimens. If the length estimate of the humerus is correct, then in length the bone fits well with that of the *H. sapiens* female equivalent but shaft robustness fits much better with that of *H. sapiens* male equivalents. This suggests that a large male individual—of advanced years since the teeth are so worn that little of the crowns remain—had a robust but slightly short humerus compared to that of modern man. The single humerus of *H. erectus* from Choukoutien (Weidenreich 1941) has an estimated length of 324 mm but is appreciably more slender at midshaft (table 13). It is of course entirely possible that 300 mm is an underestimate for the length of Sts 7, but there does not appear to be any reasonable way to check this point at present. Methods such as those used by Le Gros Clark and Thomas (1951) for the humerus of *Pliopithecus* (= "Limnopithecus") *macinnisi* and by Napier and Davis (1959) for that of *Dryopithecus* (*Proconsul*) *africanus* cannot be used here because of the crushing and breaking of the shaft and because a substantial part of the distal end is missing.

Some insight into locomotor function in primates can be gained from the proportional relationship of the upper and lower limbs, and it would be valuable to know this relation for *H. africanus*. The nearest to a reasonable basis for estimating this relation is the Sts 7 humerus and the Sts 14 femur. The estimated length of the former is slightly less than that for the latter, giving a ratio [(humerus × 100)/femur] of 96.8 (table 12). This indicates a proportionately longer humerus than is usual in *H. sapiens* but one that is proportionately shorter than is usual in pongids. According to Schultz (1960), values of this index in cercopithecoid monkeys range between about 77 and 85 (presumably these are mean values). Unfortunately, the author does not make clear what sample sizes are involved in the above case, nor are ranges of any sort given. According to the available figures, then, pongids range from having the humerus at least slightly longer than the femur (chimpanzee) up to its being very much longer (orang). One might expect that the very long-armed gibbon would also be in the proportionately very long humerus class; the upper limb as a whole is indeed appreciably longer than the lower in this form, but the femur is a proportionately long segment of the lower limb and this reduces the ratio in this case to roughly that of the gorilla. On the other hand, man (and cercopithecoids) have a proportionately short humerus. *H. africanus* is intermediate in having, on the above estimates, a proportionately slightly short humerus. This conclusion, however, is almost certainly misleading. It seems clear that the Sts 7 humerus belonged to a large male individual and the Sts 14 femur to a female. In man and in the pongids the mean for humerus length in males is greater than that for females; in modern man the difference is about 8–9 percent of the male mean and in pongids it may be as much as 15–16 percent (gorilla). Using the above percentage difference for man and a length of 300 mm for a male *H. africanus* humerus, the humerus for the Sts 14 female would have been of the order of 270 mm in length. This would give a humero-femoral index of about 88.

Obviously many uncertainties are involved in these estimates. And yet they do suggest that it is reasonable to suppose that *H. africanus* did not have the proportionately long humerus of the pongids but was somewhat more like man in this respect. This conclusion is consistent with that, based on quite other evidence, that the lower limb of *H. africanus* had already increased in proportionate length in the manner of modern man.

The distal end of the humerus of *H. africanus* is not known from South Africa. A small amount of material is known from Olduvai but has not yet been described.

In the case of *Paranthropus*, the scapula is unknown and so is the humerus except for the distal extremity. A single specimen, TM 1517, is known, and this was associated with the *Paranthropus* type skull at Kromdraai. This specimen has been discussed at considerable length in the literature, for example by Broom (in Broom and Schepers 1946), Le Gros Clark (1947), and Straus (1948). In view of the detailed analysis to which this specimen has been subjected, a brief summary only will be given here; the original writings should be consulted for more detail. The major dimensions (as measured by me on the original specimen) are given in figure 92, and the anatomical features are illustrated in figures 92–95.

The specimen is very well preserved with a few slight areas of surface damage, the most obvious being along the proximal part of the anterior articular surface of the capitulum. The specimen is 55 mm in length; hence little of the shaft is present. It is most unfortunate that there is no comparability between the *H. africanus* and *Paranthropus* humeri—what is present in the one form is missing in the other.

In size, TM 1517 agrees closely with both the *H. sapiens* and the chimpanzee equivalents. It also appears to be similar to that of the orang in size and is smaller than that of the gorilla but larger than that of the gibbon. Straus (1948) gives the most complete metrical comparison, although only in the case of modern man and the chimpanzee are the sample sizes large enough to be reasonably reliable. The gorilla and orang samples include females only.

According to these data (table 16), the *Paranthropus* specimen falls within the standard population range, for each of nine dimensions, of both modern man and the chimpanzee. In the case of five of these dimensions (CW, AW, BW, SW, and SD) as well as one angle, the *Paranthropus* specimen agrees more closely with the appropriate mean for man than it does with that for the chimpanzee. In the case of the other four (TW, TD, MW, and FW), agreement is closer with the chimpanzee. There thus appears to be a very slightly greater resemblance in size of the *Paranthropus* to the human equivalent than there is to that of the chimpanzee. In all of these dimensions, however, man and the chimpanzee are very similar indeed.

Broom considered that the *Paranthropus* specimen was nearly human in its anatomical features but in some respects resembled neither man nor any of the pongids. Le Gros Clark also noted some features in which the *Paranthropus* specimen differed from man but seems to have regarded it as being more clearly manlike than even Broom believed. Straus disagreed with

Table 16
Some Dimensions of the Distal End of the Humerus in Man, Chimpanzee, and Paranthropus

	Man (N = 37)		Chimpanzee (N = 28)		*Paran-*
	Mean	Range	Mean	Range	*thropus*
Trochlea width (TW)	22.8	16.1–29.5	22.0	16.5–27.5	20.0
Trochlea diameter (TD)	25.7	18.7–32.7	25.6	19.8–31.4	23.0
Capitulum width (CW)	16.0	11.6–20.4	17.3	13.6–21.0	16.0
Art. surface width (AW)	43.3	33.0–53.6	44.8	36.8–52.8	40.0
Bi-epicondylar width (BW)	58.6	44.7–72.5	61.1	47.2–75.1	54.0
Med. epicondylar width (MW)	24.1	15.5–32.7	19.9	12.9–27.0	19.0
Olecr. fossa width (OW)	27.0	19.7–34.5	25.4	18.6–32.2	21.0
Shaft width (SW)	37.6	24.7–50.5	39.3	28.5–50.1	31.0
Shaft diameter (SD)	16.6	11.6–21.6	16.9	12.9–20.9	13.0
Inclination angle	95.9°	85.8–106.0°	94.8°	83.2–106.4°	96.0°

Data from Straus (1948). Range = Mean ± 3 s.d.

this viewpoint. On the basis of his study of a cast of the *Paranthropus* specimen as well as human and pongid comparison material, he concluded that: (a) the *Paranthropus* specimen actually agrees with pongids (especially the chimanzee) in rather more points than it does with modern man; and (b) the distal end of the humerus is so similar in pongids (especially the chimpanzee) and man "... that it is of extremely limited value in taxonomic and phylogenetic studies. It is no real clue to the structure of the humerus as a whole, not to speak of its inability to predict the structure and usage of the entire forelimb."

My own studies lead me to agree basically with Straus's conclusions. Clearly, there is much similarity between pongid and hominid distal humeral ends in anatomy as well as function. Certainly the differences in structure and function of the cercopithecoid humeral equivalent as compared to that of either man or pongids are of a quite different order—indicating that not all higher primates have the same features in this respect.

Using Straus's basic metrical data to plot ratio and modified ratio diagrams points up rather sharply the general similarity of pongids and modern man in this respect. It also leaves no doubt that the pattern of proportions is more similar in the case of the chimpanzee and the *Paranthropus* specimen than in either of these compared to modern man (figs. 96 and 97). It is interesting that the most obvious deviation of the pongid pattern from that of modern man involves the medial epicondylar width. The two true brachiators (gibbon and orang) depart fairly markedly from man in this dimension, and the chimp and gorilla are intermediate between them and man. The *Paranthropus* specimen falls with the chimpanzee and gorilla. What the significance is of the proportionately very small medial epicondylar protrusion in brachiators is not clear to me.

It is a pity that the samples available to Straus of gibbon, orang, and gorilla were so small and that the latter two included females only. If these three samples do indeed give a reasonable indication of the populations from which they came, then the *Paranthropus* specimen agrees more closely with the quadrupedal climbers, among the pongids, than it does with the true brachiators. Furthermore, the *Paranthropus* specimen appears also to agree more closely with the pongid quadrupedal climbers than it does with modern man.

In view of the generally very close anatomical and functional similarity of this part of the humerus in pongids and in man, the *Paranthropus* humeral fragment by itself is not especially informative with respect to locomotor habit. Straus appears to be justified in concluding that this specimen does not provide evidence that the forelimb was not used for locomotion. On the other hand, it also does not appear to provide evidence that the forelimb was used for locomotion.

The Ulna

The only ulna specimen known at the present time is a single small fragment of *Paranthropus* from Kromdraai, TM 1517. This consists of the proximal end, about 1½″ long, which includes the trochlear notch and the olecranon process. This specimen was found in association with the distal humeral fragment (TM 1517), already discussed. These two fit so well together that there can be little doubt that they belonged together in the same right elbow, presumably of the same individual to which the type skull belonged. This specimen has been discussed by Broom (in Broom and Schepers 1946) and by Le Gros Clark (1947).

What there is of the specimen is moderately well preserved, but both the proximal and distal portions of the trochlear notch are missing. This fact makes it difficult to estimate the depth of the notch and to compare it with that of modern man. Le Gros Clark regards the curvature of the notch as being less marked than is that of the latter; however, this conclusion might to some extent be due to the fact that apparently substantial portions of either end of the notch are missing. The entire coronoid process is missing, and the area from which bone is missing at the proximal end is fairly large. Much of the depth of the notch in modern man is due to a bony prominence in precisely the two areas from which bone is missing in the *Paranthropus* specimen (see figs. 98–100).

The olecranon is not strongly developed, being much as it is in modern man. Le Gros Clark regards it as being not quite as markedly developed as in *H. sapiens*. It seems to me possible that this conclusion also was influenced by the missing proximal end of the trochlear notch. The absence of the usual well-developed prominence in this region tends to make the proximal end of the ulna look somewhat weakly developed. The structure of the olecranon differs appreciably from that of pongids. Broom regarded several minor features of the olecranon as distinguishing it from that of *H. sapiens*. He drew attention especially to a prominent ridge on the medial face. This appears to be the ridge to which the posterior fibers of the ulnar collateral ligament attach, and indeed it is more prominent than is usual in *H. sapiens*. The difference seems slight, however, and does not appear to place it outside the range of variation in modern man.

Because this ulna specimen is so very incomplete, and because the structure and function of the pongid and hominid elbow joints are very similar, this specimen is not very informative. Furthermore, because of the missing portions of the trochlear notch, its present appearance may be somewhat misleading. On the basis of the available information, there appear to be no features that differentiate this specimen in a significant manner from the modern human equivalent—which it seems to resemble more than it does those of modern pongids. It seems wiser, however, to draw no definite conclusions from so incomplete a specimen, especially since the ulna is not known in either *H. africanus* or *H. erectus*.

14 The Hand

Material

Very little is known of the early hominid hand. In the case of *H. africanus* the following material is known:

> A capitate bone, complete and well preserved. A small fragment of what is apparently the proximal end of a proximal phalanx, possibly of a second digit. Both of these specimens are from the Lower Breccia at Sterkfontein.

Of *Paranthropus* the following hand material is known:

TM 1517 (a) most of a metacarpal, probably of the second digit of the left hand. Much of the proximal end is missing.
(b) a proximal phalanx, probably of the same digit as the above metacarpal.
(c) an incomplete proximal phalanx, possibly of the fifth digit. All three of these specimens are from Kromdraai and were associated with the type skull of *Paranthropus.*

SK 84 a complete thumb metacarpal of the left hand. This specimen is from the oldest breccia at Swartkrans.

The Carpus

Almost nothing is known from South Africa of the carpus of the early hominids: of *H. africanus* a single capitate bone is known and of *Paranthropus* nothing at all.

The *H. africanus* capitate is well preserved and complete. It has been discussed at some length by Broom (in Broom and Schepers 1946) and by Le Gros Clark (1947).

The bone is small, having a maximum length of 18.3 mm measured from the proximal end to the middle of the distal surface. It is thus slightly smaller than that of a small Bush female and also that of a large male baboon, for both of which Broom gives lengths of 19 mm. He also gives figures for the equivalent length in a medium-sized gorilla male (26.5 mm), a large male chimpanzee (26.5 mm), and a very large male orang (31.7 mm). Le Gros Clark has pointed out that the bone is shorter and broader than are those of the chimpanzee and gorilla, in which it is elongate in conformity with the elongation of the wrist and hand in general. In this respect the *H. africanus* bone closely resembles that of man.

The distal articular facet is more nearly like that of modern man than like those of the pongids. The facet for articulation with the third metacarpal is gently curved, as in man, but without an obvious deflection to allow for the styloid process, such as that seen in the latter. On the other hand, in the chimpanzee the articulation area for the third metacarpal is fairly complex with a number of well-defined facets corresponding to those on the metacarpal. In the gorilla the articular surface is not as complex, but more so than is true of either the modern human or the *H. africanus* capitate.

In the case of the articulation with the second metacarpal, the resemblance is also close between *H. africanus* and modern man. The facet concerned is elongate and a bit more than half of it is on the distal surface of the bone and the other remaining portion is on the medial surface. In the gorilla and chimpanzee the facet is rectangular and small and is not at all on the distal surface; instead it is wholly on the radial face of the bone.

There does not appear to be a definite facet for articulation with the fourth metacarpal. In this respect it differs from the average human capitate. Le Gros Clark notes, however, that this facet is not present in 10 percent of Europeans.

The surfaces for articulation with the scaphoid and trapezoid bones are appreciably smaller, proportionately, than they are in modern man. In this respect the *H. africanus* bone resembles those of pongids more than that of modern man. Another feature in which the *H. africanus* bone is more nearly like that of the pongids is the relatively deep excavation of the radial face. This probably received a well-developed interosseous ligament. This concavity lies opposite a smaller concavity on the opposite face, thus giving the bone a "waisted" appearance in palmar view. This is not true of the modern human capitate, which is usually slightly convex on the radial face and consequently appears to be appreciably more robust than does the *H. africanus* specimen. In the chimpanzee and gorilla the concavity is even more strongly developed than it is in the *H. africanus* specimen. The degree of excavation is slight in the orang. Finally, in possessing a blunt projection from the distal end of the palmar surface, the fossil bone resembles that of the chimpanzee.

It seems that the *H. africanus* capitate bone is, therefore, very small—which is not surprising in view of the other evidence for small body size in this form—and is generally more like that of man than it is like those of pongids. It is thus somewhat intermediate in its characteristics. Le Gros Clark concluded that the movements between the capitate and adjacent bones were less free in *H. africanus* than they are in modern Europeans—and presumably in modern man in general—but much more free than they are in the hands of the living pongids. Broom concluded that *H. africanus* must have had a well-developed proximal end on the second metacarpal. He believed this to be associated with substantial development of the thumb and suggested the possibility that *H. africanus* may have had a well-developed thumb, or at least a more useful thumb than is possessed by pongids. The more complete hand from Bed I at Olduvai unfortunately has most of the thumb missing. From the nature of the trapezium Napier (1962a and b) has concluded that the thumb was better developed than it is in pongids but not quite as well as in modern man.

Some carpal bones are known from Bed I at Olduvai (Leakey 1960; Napier 1962a and b). These include single specimens each of trapezium, scaphoid, and capitate. The age of the individual or individuals to which these bones belong is not certain; they were associated in the FLK NN I site with seven phalanges of a juvenile and two phalanges of an adult, along with two incomplete phalanges and a fragment of metacarpal whose ontogenetic age also is uncertain.

The juvenile hand material has been attributed to the supposed taxon *H.* "*habilis*" (Leakey, Tobias, and Napier 1964) as part of the type specimen. The latter includes a mandible, an isolated upper molar, and two parietal bones as well as the juvenile hand bones. According to the interpretation followed in this work, *H.* "*habilis*" of Bed I is synonymous with *H. africanus*, and thus it would seem that the above carpal material, if it is regarded as being part of the same taxon as the type material, is representative of *H. africanus*. Unfortunately, no detailed study of this material has yet been published; hence little can be said about it here.

There is a problem with respect to this hand material. Napier believes, on what seem to be reasonable grounds, that the adult and the juvenile phalanges belong to the same taxon. But in view of the fact that *Paranthropus* has also been found in Bed I very close to this site and level, the possibility exists that the hand bones belong to that taxon rather than to *H. africanus*. Because the type mandible and the parietals belong to a juvenile and so do some of the hand bones, it is a natural assumption that all belonged to the same individual. While this is a natural assumption, it, of course, does not follow that it is a correct assumption, especially since it is clear that more than one individual occurred in the site. There is comparatively little to guide one in this case other than the morphological congruity or otherwise of the various specimens from the site, including also a clavicle and much of a foot.

At first glance the hand material appears to be too primitive (compared to modern man) to belong to the same sort of creature to whom the foot belonged. Since the foot would almost necessarily have to be advanced in morphology in an habitually erect form such as *H. africanus* evidently was, the hand might well be much less advanced in the same form if it was still in an early stage of developing facility with culture. Since the stone artifacts associated with the material here concerned are among the least advanced known, it seems not unreasonable to associate the less advanced hand with the more advanced foot. The initial doubts about a single taxon being involved are thus allayed by this process of reasoning.

But further thought raises doubts again. Throughout this work there is much evidence that *H. africanus* was small and slenderly built. Also, the one available capitate from South Africa is smaller than is that of even a small Bush female. The phalanges from Bed I, on the other hand, are remarkably robust and give clear evidence of powerful ligamentous attachments that indicate, as Napier has pointed out, that the hand was capable of greater strength than is that of *H. sapiens*. Also, the phalanges tend to be quite strongly curved, and it seems that the thumb may have been relatively somewhat short compared to that of *H. sapiens*. This evidence of robustness and power is perhaps a little incongruous in relation to a creature that appears to have been appreciably smaller and more lightly built than even a small form of modern man such as Bush people.

In contrast, the evidence from South Africa indicates—as will appear later in this chapter; see also Napier (1959)—that the hand bones of *Paranthropus* seem to have been robust and strongly curved with the thumb relatively short and the hand capable of considerable power. It thus seems a reasonable possibility that the hand bones from Bed I belong to *Paranthropus*. It is possible, of course, that the above characteristics were shared by both *Paranthropus* and *H. africanus* and that the Olduvai material is quite properly to be placed in *H. africanus*. Again, however, in view of the considerable volume of evidence indicating that, as hominids, *Paranthropus* was more primitive than *H. africanus* and was considerably more robust and muscular, it does not seem probable that their hands were of similar size, robustness, power, and degree of similarity to that of modern man.

Tuttle (1967) has also drawn attention to the possibility that the Olduvai hands do not belong to the same taxon as does the cranial material. He has suggested that the anatomy of the hand bones is not inconsistent with some use of the hands in locomotion, e.g., knuckle-walking. If the interpretation of *Paranthropus* locomotion presented in this work is even approximately correct, then such use of the hands might have been involved when the animal was on the ground, especially if the lower limb was still relatively short. This seems to be very much less likely in the more efficiently bipedal *H. africanus* in which the lower limb had already elongated. Furthermore, the strength of the Olduvai hands, the curved phalanges, and apparently shorter thumb are features that are consistent with a certain amount of climbing.

I am not attempting to urge here that the Olduvai hands forthwith be accepted as belonging to *Paranthropus*. This interpretation seems to me to be at least as good as the other and in some respects even better. What I am suggesting is that we should be open-minded about their affinities until sounder and more complete evidence is available. It seems to me clear that the grounds for assuming them to be associated taxonomically with the cranial material from the same site are wholly inadequate. Deductions about the nature of the hand of man in the early tool-using phase, based on the assumption that the Olduvai hands belong to *H. africanus* (= *H.* "*habilis*"), could well turn out to be very misleading.

The Metacarpus

Nothing is known of the metacarpus of *H. africanus* from South Africa.

Two metacarpal specimens are known from Swartkrans; one of these has been referred to *H. erectus* (= "Telanthropus") and the other to *Paranthropus* (Napier 1959). A single metacarpal is known from Kromdraai that has been referred to *Paranthropus*. The Kromdraai specimen has been described and illustrated in detail by Broom (in Broom and Schepers 1946) and the Swartkrans specimens have been discussed in detail by Napier (1959); the specimen that has been attributed to *Paranthropus* has also been discussed in Broom and Robinson (1949) and in Napier (1964) (see Figs. 101 and 102).

The most complete of the above specimens is the thumb or first metacarpal from Swartkrans. This specimen has a fine crack across the shaft slightly distal to midshaft, but this does not seem to have caused any modification of the shape. Napier believes that the crack caused a slight "forward angulation of the distal fragment which has had the effect of exaggerating the longitudinal curvature of the shaft." I cannot detect any such modification

Table 17
Metacarpal I Length

	N	$\bar{X}$	S.D.	Observed Range	Standard Pop. Range
European man	166	44.6	3.1	37.3–53.0	35.3–53.9
Bushman	19	40.0	4.3	34.5–48.9	27.1–52.9
Chimpanzee	13	38.6	4.5	35.0–48.5	25.1–52.1
Gorilla	16	46.0	4.8	39.0–55.5	31.6–60.4
Orang	10	50.8	4.2	45.5–56.5	38.2–63.4
Paranthropus SK 84	1	35.0	—	—	—

Modified from Napier (1959).

and, if indeed such exists, it must be of the most minor character. The bone belonged to an adult individual since epiphyseal fusion had already occurred.

The dimensions of this specimen, as given by Napier, are as follows:

Maximum length	35.0 mm
Antero-posterior width in midshaft	7.5
Transverse width in midshaft	9.5
Maximum transverse width at base	12.5
Maximum transverse width at head	11.0

Compared to the length of metacarpal I in the living great apes and in modern man, the *Paranthropus* specimen is rather short. This can be seen from Table 17, which is modified from Napier, as well as from Fig. 103.

The following is a concise general description of the bone given by Napier (1959):

> The fossil bone is an adult specimen and is short and robust and quite strongly curved in its longitudinal axis; the shaft is triangular in cross-section and bears strong muscular markings on both medial and lateral aspects. . . . At the proximal end there is a well-developed, saddle-shaped articular surface. The distal articular surface is asymmetrical, being flattened on its lateral aspect; the articular surface narrows from front to back, so that its appearance is somewhat heart-shaped. There is a prominent beak projecting forwards from the anterior edge of the articular surface; this beak is flanked on either side by short, shallow grooves presumably related to sesamoid bones; the medial groove is deeper than the lateral one.

The robustness of the specimen is not very different from that of modern man, being slightly higher than the mean for a sample of 166 specimens of modern European man as given by Napier. He used an index of mean midshaft diameter × 100/length, where midshaft diameter is the mean of two diameters measured at right angles to each other. The *Paranthropus* specimen, however, is more robust than is usual in the living great apes (see table 18).

Associated with the moderately high robusticity is evidence of well-developed muscularity in the form of strongly marked areas of muscular attachment. A very obvious feature of the bone is the marked curvature of the shaft. The first metacarpal in modern man is quite straight; those of gorilla and chimpanzee are either straight or slightly curved. Of the great apes the orang alone has marked curvature of metacarpal I. The well-developed central protuberance on the anterior edge of the distal articular surface rather clearly

Table 18
Metacarpal I—Robustness Index

	N	$\bar{X}$	S.D.	Observed Range	Standard Pop. Range
European man	166	23.0	1.9	18.9–27.3	17.3–28.7
Bushman	17	21.6	2.5	15.1–25.5	14.1–29.1
Chimpanzee	16	19.1	2.2	16.5–22.2	12.5–25.7
Gorilla	13	20.6	2.2	18.3–26.1	14.0–27.2
Orang	10	16.2	2.0	13.2–19.0	10.2–22.2
Paranthropus SK 87	1	24.3	—	—	—

Modified from Napier (1959).

indicates that well-developed sesamoid bones were present on either side of it. A very small protuberance of this sort may sometimes be found in modern man but none is known in any of the great apes. The fossil specimen appears to be unique in this respect.

The proximal articular surface is well developed and saddle-shaped. It has depth and length characteristics similar to those in modern man but extends onto the medial side of the shaft in a manner that is more like the condition in the gorilla than like that in man.

The head of the bone appears relatively robust compared to the base; this is true also of modern man and of the gorilla, but not of the chimpanzee and orang. The curvature of the head from side to side is marked, forming an arc of a circle of 6 mm radius—as Napier has pointed out and as I can confirm. This feature resembles the condition in pongids and not that in modern man, where the articular surface of the head is more flatly curved. Napier found that 87 out of 99 first metacarpals of modern man examined by him had the radius of curvature between 10 and 30 mm. There is a moderate amount of obliquity of the distal articular surface to the lateral side. This is a feature common in the great apes but is seen only occasionally in modern man and is then only poorly developed.

Evidently the thumb with which the metacarpal was associated was very mobile. This is indicated, first, by the relatively great side-to-side extent of the proximal articular surface. This indicates appreciable mobility at the base of the metacarpal where it articulates with trapezium. Second, thumb mobility is indicated here by the strong curvature and lateral obliquity of the distal articular surface. This indicates, as Napier has pointed out, that "active abduction was habitually carried out at this joint," as is the case in the great apes.

Napier has argued that the type of thumb mobility indicated by these characters of the *Paranthropus* thumb metacarpal is associated with brachiation. The elongated fingers, which have curved metacarpals and phalanges at least to some extent, act together to form a hook to secure overhead holds in swinging along under branches. For this reason the thumb is short so that it does not get in the way. It seems to me that it does not necessarily follow that a relatively long thumb would be an impediment, especially in view of the mobility of this digit in pongids. It may well be, on the contrary, that the relatively short thumb of active brachiators is a consequence, not of reduction of the thumb, but of relatively great elongation of the finger metacarpals and phalanges.

This is indeed suggested by various comparisons. For example, if the thumb metacarpals of hominoids are compared by the logarithmic ratio diagram method, it appears that the orang has a relatively longer bone than has the chimpanzee, gorilla, or man. The first metacarpal of the gibbon is proportionately even longer than that of the orang (see fig. 104). This becomes even more obvious if the thumb metacarpal in each case is compared with some size-related anatomical structure that is not a part of the pectoral limb. We have already seen that there is comparatively little variation in the size of the acetabulum either within or between hominoid species or genera. Since the acetabulum is known in one individual of *Paranthropus* from the Swartkrans site, the above comparison can be made in this case. Using the modified ratio diagram method, and employing the acetabulum for coincidence, it appears that the thumb metacarpals of man and the gorilla are of relatively exactly the same size; that of *Paranthropus* (SK 84) is very slightly smaller, that of the chimpanzee is relatively long, that of the orang is relatively very long, and that of the gibbon even longer.

Schultz (1930) has shown that the orang has the shortest thumb digit (metacarpal plus phalanges) of the pongids and the gorilla the longest. He has also shown that changes in proportion involve the phalanges and the metacarpals approximately equally; that is to say, the proportion of metacarpal length to phalanges length is much the same in a long digit as it is in a short one of the same ray. It would seem, therefore, that even though the thumb of the orang is relatively very short compared to the rest of the hand, its metacarpal is actually long and has shared in the elongation of the hand but not to the same extent as have the finger bones. It appears that the thumb metacarpals of modern man, the gorilla, and *Paranthropus* (SK 84) are of much the same relative size with respect to acetabular width. It should be borne in mind that the *Paranthropus* ratio is based on a single metacarpal and on a single acetabulum, probably from two different individuals. These forms thus do not share the elongation of the thumb metacarpal seen in those pongids that engage in at least a little brachiation.

Table 19
Some Proportions in the Skeleton of the Hand

	N	$\frac{\text{Ray I} \times 100}{\text{Ray III}}$	$\frac{\text{Metacarpal I} \times 100}{\text{Metacarpal III}}$
Gibbon	7	44.5	58.0
Orang	4	34.6	48.8
Chimpanzee	2	39.5	50.2
Gorilla gorilla	9	44.6	52.9
Gorilla beringei	5	49.8	
Homo (Negro)	18	62.6	66.2
Homo (White)	14	64.4	

Data from Schultz (1930).

It is pertinent to question whether either the elongation of the hand as a whole or the disparity in length between the first digit and the others is due to adaptation to brachiation. The gibbon is by far the most active brachiator of the living pongids and therefore it should be the most extreme among pongids with respect to these features. Schultz (1930) has provided some relevant information that is included in table 19, along with some that is derived from his data. Unfortunately, the samples were rather small. It is clear from the Ray I × 100/Ray III index that the hominoid with the greatest discrepancy between the general hand length and that of the thumb is the orang. Next in order comes the chimpanzee; then only comes the gibbon with an index value much like that of the gorilla. On this evidence the shortest thumb in relation to hand length is found in a form that does not use brachiation much (Schaller 1961). On the other hand, the gorilla, adults of which do not brachiate (Schaller 1963), has much the same thumb length to hand length ratio as has the gibbon, which is a highly accomplished brachiator. From this it would seem clear that brachiation is not a prime factor in accentuating the difference in length between the first digit and the others.

This appears to be true also of the relationship between the first metacarpal length and that of metacarpal III. In this case, again the orang has the greatest discrepancy, followed by the chimpanzee and then the gorilla. The gibbon, in fact, has the least discrepancy between these two metacarpal lengths of all of the living pongids and is exceeded by man alone of the living forms here being considered.

On the other hand, when considering the proportion between either metacarpal I or III lengths and acetabular width, it is clear that the gibbon hand has undergone more elongation than have those of any of the other forms being considered. The most effective brachiator therefore also has the greatest degree of elongation of the hand and the least effective one has the least elongation. This evidence is thus consistent with the suggestion that adaptation to a brachiation locomotor habit involves elongation of the hand. Even so, it is apparent that a considerable amount of elongation of the fingers can occur in an animal that is at best an occasional brachiator, e.g., the orang. It follows that even in this case adaptation to brachiation is not the sole cause; other factors must also be contributing to elongation of the hand.

Similarly, since well-developed mobility of the thumb is not confined to those pongids that brachiate the most but is characteristic of all of the pongids, it seems highly unlikely that adaptation to brachiation is the prime cause of this feature.

It might be argued that all of the pongids used to be brachiators and that the features here being discussed were developed then, but that increased size of the larger animals has decreased the amount of brachiation that they engage in but the features of brachiators have been retained. This might seem to be a likely conclusion in view of the fact that the smallest living pongid is the most active brachiator and that the largest is the least active brachiator. The fossil evidence, however, suggests that true brachiation was developed rather late in the Tertiary and in pongid history. Furthermore, some pongids were evidently already of gorilla size (e.g., the largest of the *Proconsul* species of the Miocene) even before brachiation characteristics were recognizably developed in any of the known pongids. It is consequently a distinct possibility or even probability, that some pongids never have been active brachiators. Indeed, it seems to me that the traditional clichés that pongids are brachiators and that characteristic anatomical features of the living pongids—including those here being considered—are consequences of adaptation to brachiation badly need careful reevaluation. For this reason it seems to me that the classification of nonhominid higher primates into three locomotor groups—brachiator, semi-brachiator, and quadruped—is quite unrealistic if the term "brachiation" is to retain any semblance of meaning.

Napier's (1959) argument—that since the *Paranthropus* metacarpal I gives evidence at both ends of well-developed mobility, and since this is a feature that forms part of the brachiator adaptational complex, and because the latter complex also includes a relatively short thumb digit as compared to the others, therefore we can anticipate that the *Paranthropus* hand was long compared to the thumb—does not seem to me to be well founded in the light of the above discussion. As will emerge later in this chapter, it may well be that the *Paranthropus* hand was in fact long when compared to the thumb, but it does not seem to me that this conclusion is a legitimate deduction from the data presented by Napier.

The other metacarpal specimen from Swartkrans (SK 85) is, according to Napier (1959), a left metacarpal IV that belonged to an adult individual. The specimen is very well preserved and consists of the distal end, representing perhaps half of the original length of the bone (see Figs. 101 and 102).

The following dimensions are taken from Napier:

Over-all length of fragment	32.5 mm
Transverse width at narrower part of shaft	6.5
Antero-posterior width at narrowest part of shaft	8.0
Transverse width of distal articular surface	8.0
Antero-posterior width of distal articular surface	11.0

The articular surface of the head is well curved, both antero-posteriorly and from side to side, but there is a slight flattening on the left side that makes the head slightly asymmetrical. There is only the faintest trace of sesamoid impressions and a central prominence. The shaft is mildly convex dorsally and, according to Napier, is a little more curved on the right side than on the left when seen from the front. The muscular ridges that mark the dorsal margin of the insertion of interosseus muscles are well defined and form an interosseus crest near what was the middle of the bone.

The specimen is regarded by Napier as belonging to digit III or IV because the asymmetry of the head is very slight and in modern man it is much greater in digits II and V but slight in III and IV. The asymmetrical curvature of the shaft when seen from the dorsal aspect indicates that it belonged to digit IV.

According to Napier, the characters of this specimen agree very closely indeed with those of modern human fourth metacarpals. On the other hand, it differs clearly from those of the great apes in a number of respects. For example, the head of this metacarpal in the great apes is set at an angle to the shaft when seen in side view so that it is deflected toward the palmar surface of the hand rather than being a direct continuation of the shaft axis—as it is in modern man and in the fossil specimen. In the chimpanzee and gorilla and occasionally also in the orang, this offset articular surface meets the shaft dorsally in the form of a shelf that serves to prevent hyperextension of the digits during knuckle-walking (see also Tuttle 1967). This shelf is not present either in modern man or in the fossil specimen. In the gorilla and the orang the dorsal interosseus ridges and crest are absent and are but poorly developed in the chimpanzee but are well developed in the fossil specimen. These ridges imply that the interosseus muscles were well developed, hence that the hand was broad with divergent metacarpals.

Since the fourth metacarpal so closely resembles that of modern man but the first metacarpal from this site is very different from that of modern man, and since Swartkrans is known to contain remains of both *Paranthropus* and *H. erectus*, Napier concluded that the fourth metacarpal belongs to the latter and that the unmanlike thumb metacarpal belongs to the former. This seems to be much the most reasonable conclusion to draw.

Apart from the thumb metacarpal from Swartkrans, one other metacarpal of *Paranthropus* is known. This is the incomplete specimen (TM 1517) from Kromdraai that was associated with the *Paranthropus* type skull. It has been described and illustrated by Broom (in Broom and Schepers 1946; see fig. 105).

The specimen belonged to an adult individual and was found in the piece of breccia that enclosed the type skull of *Paranthropus* so that, as Broom describes, the distal end of the metacarpal was almost in contact with the maxilla of the skull. The distal two-thirds of the bone is perfectly preserved and complete and almost all of the palmar surface of the rest is present; Broom estimated that about 3 mm of bone length is all that is missing, at the proximal end. The dorsum—indeed, most of the thickness of the shaft—is missing from the proximal one-third. On the assumption that the bone is now 3 mm shorter than when complete, the original length was 70 mm and the width of the distal end was 12 mm. Broom believed the specimen to be a second metacarpal on the basis of the features of the head.

The bone is long and slender compared to the usual condition in modern man. According to Fick (1926), the mean length of metacarpal II in a sample of 68 specimens of modern man was 63.1 mm.

According to Broom, the Kromdraai specimen resembles metacarpal II of the Bushman very closely, and this appears to be a fair appraisal; Broom had mostly Bush specimens with which to compare the fossil, and since these do not differ in any significant respect from other races of man with regard to metacarpal II, his conclusion does not imply that the fossil is more like Bush equivalents than it is like those of other races of man. Indeed, since Bush metacarpals are smaller than average for modern man and the fossil is rather long, the resemblance is actually a little closer with larger-bodied types of modern man. The resemblance to pongids is less close. For example, neither the offsetting of the head from the long axis of the shaft when seen in side view, nor the dorsal articular shelf where head and shaft meet, which are usual features of the great ape metacarpals but not of modern man, is present in the Kromdraai specimen. In one feature, however, it is not typically human, and that is in possessing grooves for two sesamoid bones on the palmar surface of the head with a slight elevation between them. Neither the grooves nor the elevation are nearly so conspicuous as are those of the thumb metacarpal from Swartkrans. This complex of features is not characteristic of pongids, according to Broom, but a single sesamoid may be found in man in the case of metacarpals II and V and two are usual in the case of I. According to Wood Jones (1942), a sesamoid is usually associated with the metacarpal head in digits II and V and two in the case of digit I. He believed that sesamoids are associated with the presence of intrinsic flexors: the thenar muscles in the case of I, the hypothenar muscles in the case of V, and flexor brevis indicis in the case of II. Broom states that paired sesamoids may sometimes be found with any or all of the digits in modern man, and they do occasionally occur in the great apes.

Since the metacarpal I specimen from Swartkrans is unusually short and fairly robust, the fact that the metacarpal II from Kromdraai is longer than average for man is intriguing. At first glance it suggests that the thumb of *Paranthropus* was short compared to the length of the hand. Unfortunately, the data are not adequate to allow for certainty on this point, especially as the first metacarpal is not merely from a different individual as the second metacarpal but also is from neither the same site nor the same time period. Consequently, there are a number of possible explanations of the facts. For example, *Paranthropus* may have had a short hand with a thumb of human length during Swartkrans time but a long hand with a thumb of the same proportionate length by Kromdraai time. Or, the hand may have been of the same average length in both populations, with a thumb of human proportions in both, but the Swartkrans metacarpal I might be smaller than average for that population while the Kromdraai metacarpal II might be longer than average for its population. Or, the hands may have been of similar length in both populations and in both the thumb could have been relatively short, so that the two known specimens together accurately represent the condition that existed in both populations. It seems unlikely that the disproportion in length, which is rather marked, could be due entirely to accidents of sampling. Hence, it seems to me likely that the *Paranthropus* thumb was proportionately shorter compared to the rest of the hand than is the case in modern man. If this is indeed the case, then in this respect *Paranthropus* resembles the great apes more than it does man. Unfortunately, there seems to be very little usable information of a comparative metrical nature suitable for documenting the condition in the great apes as far as variation in length of metacarpal II is concerned or its proportion to metacarpal I.

One point seems clear in the Kromdraai specimen: there is no trace of the articular shelf where the head joins the dorsum of the shaft. This shelf is regarded as being responsible for preventing hyperextension of the phalanges on the metacarpal during knuckle-walking. Therefore its complete absence in the *Paranthropus* specimen suggests that knuckle-walking was at best a very insignificant part of the locomotor activity of this form.

The Phalanges

The only phalangeal specimen from South Africa that has been attributed to *H. africanus* is a very incomplete fragment from the Lower Breccia at Sterkfontein. This has been illustrated and very briefly described by Broom (in Broom and Schepers 1946).

The specimen consists of a fragmentary portion, less than 2 cm long, of the proximal end of what Broom believed to be a proximal phalanx of probably the second digit. More than half of the articular end is missing. Broom believed that the anatomical features of this specimen closely resemble those of the proximal phalanx of the second digit of the human hand but do not closely resemble those of either hand or foot phalanges of the great apes. He believed that, though the resemblance was much closer to man than to the great apes, the fossil bone was nevertheless not completely human in its features in that the proximal end is slightly narrower and the shaft slightly broader than the corresponding bone in the Bushman. He argued that this indicated that the finger to which it had belonged was a little longer and a little more slender than is usual for modern man.

Broom may have been entirely correct in his identification of this specimen and in the conclusions that he reached about it. In view of the small and fragmentary nature of the specimen, however, it seems to me that there can be no certainty that it belonged to *H. africanus* and that, even if it did, it is so incomplete that nothing of significance about the hand of *H. africanus* can be deduced from it with any acceptable degree of security.

Two phalangeal bones from South Africa have been referred to *Paranthropus*; both are from Kromdraai and were associated with the type specimen of *Paranthropus*. One was identified by Broom as the proximal phalanx of a second digit and the other as a proximal phalanx of a fifth digit. These specimens have been described briefly and illustrated by Broom (in Broom and Schepers 1946).

The proximal phalanx of the second digit, Broom believed, belonged to the same finger from which came metacarpal II, which was lying next to it.

The specimen is complete and well preserved. According to Broom, it was 45 mm in length; evidently this figure is either a recording error or a typographical error since the bone is 35 mm in length. This length is compared with corresponding values in modern man in table 20. The fossil phalanx is clearly within the range of variation of mean values for modern forms of man, though it seems to be slightly shorter than the average for modern man.

Table 20
Length of Proximal Phalanx

	Digit II	Digit V	$\frac{V \times 100}{II}$
Paranthropus	35.0	31	88.6
European ♂	38.8	32.4	83.5
European ♀	37.0	30.6	82.7
Japanese ♂	37.7	31.5	83.6
Japanese ♀	36.0	29.9	83.1
Hottentot	32.5	26.1	80.3

Comparative data are taken from Martin and Saller (1959). Sample sizes were not given.

The second phalanx in this small group of bones associated with the type skull of *Paranthropus* is an incomplete bone that Broom believed to be the proximal phalanx of digit V, presumably from the same hand as the other phalanx and the metacarpal. The proximal end of the bone is entirely gone, though evidently only a comparatively small piece is missing. As preserved the specimen is about 27 mm long; its original length was probably about 31 mm. This estimated length is compared with the corresponding dimension in modern man in table 20. It seems that the fossil bone is proportionately very slightly long compared to the average for modern man as represented by the samples here available. That is to say, the estimated length of the fossil specimen is slightly higher (31) than the mean of means (30.1) of the human samples. In the case of the digit II proximal metacarpal, the length (35) is a little below the mean of means (36.4) of the comparison material. The differences are slight, however, as can be seen by comparing the ratio between the proximal phalanges of digits II and V in Table 20. Although the necessary figures for proof are not available to me, it seems likely that the differences in proportion here involved are less than the within-population variation in each of the populations represented. Furthermore, the value for *Paranthropus* may not be entirely accurate since the second specimen's length is an estimate. This comparison, however, does tend to support Broom's identification of the second specimen as part of the fifth digit since it so closely approximates the size relation between the proximal phalanges of II and V in man. If the second specimen is treated as belonging to any other digit, the size relation with the first specimen would be very unusual in each possible case.

A number of hand bones have been found at Olduvai Gorge, as already discussed in an earlier section of this chapter. These include two proximal phalanges (adult), four middle phalanges (juvenile), and two pieces of middle phalanges of uncertain ontogenetic age as well as three terminal phalanges (juvenile). A preliminary announcement has been made (Napier 1962), with some illustrations, but too little detail has been provided to allow reasonable comparisons to be made. Fuller descriptions will therefore have to be awaited.

Napier expresses the opinion that these hand bones are neither typically human nor typically pongid. The phalanges differ from those of modern man in: (a) being more robust; (b) having greater dorsal curvature of the shaft; (c) having a more distal insertion of flexor digitorum superficialis; and (d) having greater development of the fibro-tendinous markings. The phalanges resemble those of modern man in: (a) the broad and stout nature of the terminal phalanges of all digits and (b) the ellipsoidal form of the metacarpo-phalangeal joint surfaces. Napier concludes that the fossil bones indicate "a short powerful hand with strong, curved digits, surmounted by broad, flat nails and held in marked flexion. The thumb is strong and opposable, though possibly rather short."

These hand bones appear to have the same characteristics; hence Napier has drawn the reasonable conclusion that the adult and the juvenile belong to the same taxon. By definition this taxon is that which Leakey, Tobias, and Napier (1964) named *H. habilis*, since the juvenile hand bones were included in the type material of the taxon. As has been mentioned earlier in this chapter, however, the fact that it appears reasonable to refer juvenile hand bones to the same individual as yielded the juvenile mandible and parietal bones does not prove that they in fact do so belong. This sort of situation is an extremely troublesome one in paleontology, and it seems to me that great caution should be exercised in such cases to be certain that one is not creating a chimera and thereby drawing misleading conclusions. In this instance caution is especially important since we know so little about the evolution of the higher primate hand and since much depends on whether these hand bones are attributed to *H. africanus* (which is what I believe the Bed I *H. "habilis"* cranial bones represent) or to *Paranthropus*, both of which are present at this general level of Bed I. It seems to me to be wise not to associate together as a type specimen different skeletal parts where clear proof is not available—and it seldom is available—that they belonged to the same individual. It is preferable that the most diagnosable and generally useful specimen be designated as the official type to which the name is attached; of the material available in the case of *H.* "*habilis*," the juvenile mandible alone would have been a better choice for the purpose. In the event, it seems to me that there is reason to doubt whether the robust, curved, and powerful hand deducible from these specimens is compatible with the picture that has emerged in this work of *H. africanus* as a small, slender, and lightly built animal. On the other hand, *Paranthropus* is demonstrably a more robust and powerfully muscled animal. Indeed, one of the most robust specimens of *Paranthropus* so far available is from this general level of Bed I at Olduvai and from the same area from which the hand material came. The robust and powerful hand from Olduvai might more reasonably be associated with this robust type of hominid. In view of this conclusion, it seems to me that more substantial evidence is required before it can be accepted that the Olduvai hands belonged to the more manlike of the Bed I hominids and were thus the hands of early tool-making man.

If the Olduvai hands do actually belong to *Paranthropus*, then we have virtually no evidence of the nature of the *H. africanus* hand. The *Paranthropus* hand would seem to have been robust and powerful with a relatively short thumb compared to the rest of the hand. It apparently was capable of gripping with great power but commanded a poor level of manipulative precision.

15 Summary and Conclusions

The concern of this book is to give an account of the postcranial skeleton of the early hominids insofar as it is known and to see what conclusions may reasonably be drawn about their locomotor habit. Before drawing together the main threads of this evidence, it is worth reviewing very briefly the chief conclusions that have been suggested to me by the more complete and plentiful cranial material. This will allow the evidence from the postcranial skeleton to be related to a broader framework and will indicate where it amplifies conclusions drawn from the cranial material or, perhaps, where it is in disagreement with the latter.

Prolegomena

The conceptual framework within which a student works shapes the viewpoint from which he approaches the material being studied and powerfully influences the conclusions that he draws. It is of course not possible to outline here the bases of my biological viewpoint, but it is worth mentioning some specific points that directly affect the interpretation of the early hominid fossil material here being dealt with.

An apparently unavoidable conclusion from current biological evidence and thought is that all natural populations possess adaptedness and, at least to some extent, adaptability. That is, no natural population can exist for any significant length of time in an environment without being shaped to some extent by that environment. Since no population can exist in a vacuum, a natural population may be regarded as having two poles that are joined together by a dynamic bond. One pole is the population itself and the other is the environment—which includes also all other living organisms not belonging to the population being considered. The net result of the dynamic interaction between the two poles is natural selection. Natural selection is an integral and unavoidable aspect of the existence of a natural population. The opinion that was commonly held until relatively recently, that natural selection does not necessarily operate all the time and that only some characters of an organism or population came under its influence, seems not to be correct. Two developments have been largely responsible for the change in viewpoint. One involves the realization that pleiotropy is not a somewhat rare phenomenon but is in fact the normal state of affairs. For example, genes that have detectable phenotypic effects seem usually also to influence viability. A gene that is selectively neutral in all respects must therefore be a rare phenomenon that, if it should exist, is of a temporary nature only. The other line of evidence, stemming from the work of, for example, Ford and of Dobzhansky and their coworkers, has shown that selection pressures in natural populations are commonly enormously greater than the 1 percent level that used to be thought a maximum. Consequently, as has been demonstrated, natural selection is not the necessarily slow-moving process it used to be thought; extremely rapid and fine adjustments can be brought about in a population by its means.

The import of the above in the present context is that differences between early hominids are interpreted in terms of adaptive advantage rather than as vagaries of variation. The significant differences in anatomy of the masticatory apparatus of *H. africanus* and *Paranthropus*, for example, appear thus to require an explanation in terms of adaptive difference rather than to be ignored or shrugged aside as of no importance. Furthermore, where a character appears to be part of a well-defined adaptive complex or pattern, even if only a very few specimens are available to demonstrate it, I assume that the same population is unlikely to contain other individuals in which that character is quite differently developed. For example, if a specimen of *H. africanus* is complete enough to demonstrate clearly that the ischium was short and the femur, hence the whole lower limb, was long—therefore, that the hamstring specialization was for speed more than power—I assume that the chances are against other individuals in that deme having had a long ischium, short lower limb, and hamstrings specialized primarily for power rather than for speed. Manifestly this argument applies only where clear grounds exist for interpreting a character as being associated with an important part of the adaptation of the population. It is not an excuse to base an argument on scanty evidence. The point I wish to make is that where a character is rather obviously part of a particular adaptive pattern and is therefore under the influence of a well-defined selection pattern, it is unlikely that normal variation in that character in that population will be such that variants will be common that are completely inconsistent with that adaptation. The point to be careful of here is that abnormal individuals do occur; for example, a number of dental abnormalities are known in *Paranthropus*. Usually such abnormalities are detectable on careful study, but one does run a risk of being quite mistaken if a single specimen only is known exhibiting the character concerned.

A point that follows directly from the above point of view, and that, indeed, is self-evident and seems to be overlooked sometimes in discussions of the evolution of man, is that adaptation can occur only in terms of the present. The interactions that result in maintenance or modification of gene frequencies in a population can occur only in a living gene pool—that is, at a time when the organisms bearing the genes are alive. The evolutionary process occurs at the interface between the future and the past in what might be called the nascent zone of a lineage, the growing tip. The past has gone and the future exercises no magnetic creative influence so far as present evidence goes. Hence it is not possible to interpret the emergence of one character as a step taken in order that a much later event can occur. For example, erect posture cannot be explained as something that occurred in order that man could become culture-bearing at a later point in time. Instead it should be interpreted in terms of survival value at the time it was occurring, regardless of whether its existence subsequently was the basis for an adaptive change of importance.

It might be argued that the above viewpoint, that natural populations are very closely controlled by natural selection and that observable anatomical features reflect the basic adaptational pattern of the population, does not apply to man because he possesses culture. It is clear that culture is an enormously important and powerful adaptive mechanism that buffers the gene pool of man to a considerable extent from the influence of the environment. Hence natural selection is no longer the overriding influence in the case of man that it is in other animals. Culture has altered the evolutionary process in man from the quasi-finalistic process of other animals to a genuinely and increasingly finalistic process. Aspects of this matter have been explored at some length by a few biologists, notably Waddington (e.g., 1942 and 1960) and Julian Huxley (e.g., 1947 and 1955), but also by Simpson (e.g., 1949) and Dobzhansky (e.g., 1956), among others, and in a somewhat different context by Teilhard de Chardin (1959). So far this thinking seems to have had little general impact on students of man, though it seems to me crucial to an understanding of man (past, present, and future) and of the evolution of the evolutionary process itself. The viewpoint that observable features must be analyzed in terms of adaptation and survival value is thus at best only partly applicable to fully culture-bearing man. For this reason it is probably misleading to analyze modern human populations strictly in terms of the concepts that have been developed by studying other animals. The unique character of human evolution, however, will hardly have been in existence in the transition period when hominids were coming into existence and when culture was beginning to emerge as an adaptive mechanism. I have proceeded on the assumption, therefore, that the adaptive and evolutionary mechanisms involved in the case of the early hominids were very much more nearly like those in other animals than they were like those of fully culture-bearing man.

For this reason it seems to me incorrect to approach the analysis of the early hominids in terms of a conceptual framework derived from a study of fully culture-bearing man. An example is the view that since man is a hominid and depends on culture for survival, and since *H. africanus* and *Paranthropus* are classified as hominids, therefore they also were continuously dependent on culture for survival. From this basis conclusions are drawn, such as that the reduced canine teeth must have been due to their function having been taken over by tools and weapons, and so on. Surely this is begging the question. The change from a broadly ape level of organization to the fully human level is a major one, and culture is an essential element in it. But presumably culture did not come into existence instantaneously and fully fashioned. Once it was strongly operative and powerfully controlling adaptation, it is unlikely that more than one major lineage could have existed that exhibited it, but this need not have been so in the transition stages. There may well have been several lineages in which it was developing. There may have been some lineages that were part of the transition but which did not develop culture to any significant extent. We need to approach the fossil material with caution and allow *it* to demonstrate to us what actually was happening in this transition phase. If we approach the material with preconceived views such as that, being hominids, they must have been culture-bearing and hence there could be only one lineage at a time, then, if the fossils are not all alike, we conclude that it must have been a variable lineage. We thus successfully close the door on the possibility of seeing more than one lineage simply because of the belief that there must have been one only. What is needed is to let the *fossils* demonstrate whether there was a single lineage or not. It is usual when a new grade of organization is achieved that adaptive radiation occurs. There seems to be no a priori reason why this could not have been true also in this case. That any such radiation is unlikely to have persisted seems clear; once the cultural adaptive mechanism was well established, the most advanced lineage in due course would wipe out any opposition and from then on cladogenesis is unlikely to occur. But this does not mean that there could have been no other lineages in the transition phase before culture was fully developed.

As part of the process of determining how many lineages were involved, it seems to me important to work from the basis of well-defined local samples. For this reason the Swartkrans and Sterkfontein samples are extremely good: they allow a remarkably good picture to be built up of the nature and variation of populations of *Paranthropus* and *H. africanus* respectively in the same geographical region, on the basis of the largest known early hominid samples from one site each and with reasonable assurance that the samples are not mixed with respect to the two forms concerned. Once one moves from these samples to other, smaller ones in other areas, the problem immediately becomes more complex. If one attributes such less secure samples to one or other of the clearly defined taxa in South Africa and does so mistakenly, the taxonomic picture is immedately ruined and a correct interpretation becomes impossible. For example, if one assumes that the Bed I hands and foot belong to *H. africanus* (= *H.* "*habilis*") and the tibia and fibula to *Paranthropus* when in fact, say, the tibia, fibula, and the foot belong to *H. africanus* and the hands to *Paranthropus*, then clearly the conclusions one draws about the two taxa will be invalid and misleading. Therefore, it seems to me faulty procedure to allocate all the known material, wherever it comes from, to definite taxa and then to assess the validity of those taxa on this basis. The taxonomic picture, it seems to me, should be assessed on the basis of those specimens or samples that can be identified with reasonable security. Others must remain in the suspense account until they can be related to known taxa with reasonable certainty.

This raises another point to ponder. There is a tendency at present to stress that any samples used for making statistical comparisons for taxonomic purposes must include all known specimens of that taxon. For example, the *H. erectus* sample for mandibular dentition would include the Javanese, Chinese, Mauer, North African, and sub-Saharan specimens. At first sight this appears to be not merely proper procedure but self-evidently so, and I have been criticized because I quite frequently do not follow it in making certain comparisons for taxonomic purposes. Further thought suggests, however, that the pooling of samples is not wise in all circumstances. The usual relatively simple parametric statistical procedures used in these comparisons, which depend upon determining the mean and the distribution of variation about that mean in a sample to estimate similar parameters of the population from which the sample came, depend rather heavily upon the sample not being biased. They are for use on samples drawn unbiasedly from local Mendelian populations, at least in a taxonomic context. Serious bias can be introduced even here by lumping males with females if there is substantial sexual dimorphism involved, or by including specimens of different growth stages, and so forth, even though these are all members of the same species simultaneously present in a localized area. When moving from a sample taken at a single locality to samples from different localities, much greater complexity is involved. As is abundantly clear from the taxonomic literature and from genetic studies, local samples of the same population taken from different localities will have different characteristics. It is frequently possible to demonstrate statistically significant differences between such local samples even though they are genetically continuous and clearly are members of the same species. Similarly, samples taken from the same local population at substantially different times can be shown to be statistically significantly different. It follows from these very well known facts that if one lumps together samples from localities scattered over much of the Old World and ranging in time over hundreds of thousands or even millions of years, one has gone a very long way from the unbiased sample requirements of statistics based on the normal curve. The estimated ranges of variation will certainly be too large and will thus reduce the possibility of distinguishing between populations that are in fact distinct.

An example concerns the validity of my argument that *H. erectus* is present at Swartkrans as well as *Paranthropus*. It has been argued (Wolpoff 1968) that it is improper to use only the Swartkrans sample of *Paranthropus* in the comparison between the two forms, that all known specimens should be used to make the comparison with the two Swartkrans specimens that are believed to belong to *H. erectus*. As Gutgesell (1970) has rightly pointed out, however, the *Paranthropus* and supposed *H. erectus* specimens at Swartkrans come from the same bed at a single site, hence only two possibilities exist: either the supposed *H. erectus* specimens are variants of the local population of *Paranthropus* or else they belong to a genetically isolated, distinct taxon. Therefore, the only *Paranthropus* specimens relevant to making the decision whether the suspect specimens belong to *Paranthropus* or to another taxon are those from the local Swartkrans population. Including specimens from East Africa and Java in the sample simply introduces bias into the estimation of the parameters of the *Paranthropus* population that lived in the Swartkrans region at the time the suspect specimens lived there.

This is a more specialized application of the general principle because it concerns a decision relating to a purely local situation. The more general application can also be seen from the example of *Paranthropus*. This genus appears now to be known over a period of three or four million years, and apparently body size reduced during this time. This reduction apparently also affected tooth size. To lump all of the known specimens together as a basis for estimating population parameters concerning dental dimensions would produce exaggerated ranges of variation that were never characteristic of any living population of this form at any time in its known history. It is therefore fallacious to use such supposed parameters as a basis for taxonomic comparisons with other samples.

The soundest basis from which to work taxonomically is surely a good local sample. In such a case—whether concerning extinct or extant animals—different species are distinguishable with far greater ease and certainty than in any other situation because specimens either belong to the same local breeding group or are genetically isolated and hence belong at least to different species. Complicating factors sometimes obtrude in such cases, but usually the position is quite clear; even with complications it is still more clear-cut than cases involving more than one locality or time horizon. It is then possible to work away from the best local sample to analyze the relationship to it of other samples or individual specimens. For this reason it seems to me a very fortunate circumstance that such relatively good local samples exist as those from Sterkfontein (*H. africanus*), Swartkrans (*Paranthropus*), and Choukoutien (*H. erectus*), and especially that the two South African sites lie side by side in the same small valley, thus eliminating geographical differences. Moreover, the specimens occurred in solid cave limestone that consolidated as it accumulated, thus the chances are greatly reduced of finding specimens of quite different ages occurring together. For determining population ranges of variation and other characteristics, these samples are extraordinarily useful. Conclusions concerning relative variableness of *H. africanus* and *Paranthropus* are much more meaningful when based on these local samples than when pooled samples are used. Pooled samples will bring in geographic variation and variableness through time as well. These are different characteristics and are valuable to know. They are, however, poorly measured by pooling samples; treating the successive time or geographic samples individually would be more effective for this purpose. These local samples do involve some time depth the extent of which cannot be determined with certainty, but it is less than is the case if samples are pooled.

To avoid misunderstanding, let it be clear that I believe that all known specimens that are attributed to a taxon on reasonable grounds must be taken into account in taxonomic comparisons. The point I wish to make is that lumping specimens from widely different geographic regions and quite different time horizons and treating them statistically as if they came from a single locality and a single time horizon is not the way to do it. Such a course will lead to conclusions that at best are misleading and at worst are biologically nonsensical.

Conclusions from Cranial Evidence

One may turn now to a summary of the conclusions derived from cranial evidence as a background against which to view the conclusions from the post-cranial material. The general features of the early hominid skull are well known. The endocranial volume is within the range of variation of that of the great apes; the face is less prognathous than is usual in the great apes but more so than in modern man; the dentition is more nearly like that of modern man than it is like that of pongids, with canines that do not project significantly beyond the adjacent teeth when unworn and there are no diastemata; there is no mental eminence in the chin region; the brow ridges are not powerfully developed; the nasal area is almost or completely flat; and the foramen magnum in the adult is not situated quite as far aborally as it is in the great apes.

Three major influences on the architecture of the higher primate skull are: (1) locomotor habit, (2) degree of development of the brain, and (3) the use to which the masticatory apparatus is put.

1. Locomotor Habit

Quadrupedal forms have an occiput that consists almost entirely of nuchal plane, with little or no occipital plane. Furthermore, the superior margin of the nuchal plane is situated relatively high up the braincase, usually at roughly the same level above the Frankfurt Plane as the brow ridges. In erectly bipedal man the occiput has a well-developed occipital plane and the nuchal plane usually rises to about the level of the Frankfurt Plane; that is to say, its superior margin is at about the same level as the inferior margin of the orbit. The nuchal plane thus lies in a more nearly horizontal position in erect bipeds than it does in quadrupeds.

In this respect there appears to be uniformity in the early hominid material: so far, only the low, more nearly horizontal type of occiput has been found. In this respect all of the early hominid occipital material is indistinguishable from that of erect bipeds and is easily distinguished from that of quadrupeds. There is no suggestion of division into two or more groups.

2. Degree of Development of the Brain

The change from the ape level of organization to that of modern man involved an approximately threefold increase in endocranial volume. This change had a marked effect upon the appearance of the skull.

Because the cranial roof in the higher primates is relatively thin and closely follows the contours of the brain with its investing membranes, the position of the vault reflects approximately the position of the upper surface of the brain. The early hominid material falls into two clearly defined groups on the basis of this character.

In *Paranthropus* the vault is relatively low while in *H. africanus* it is relatively high. This can be demonstrated rather well by means of Le Gros Clark's supra-orbital height index (Le Gros Clark 1950; Ashton and Zuckerman 1951; Robinson 1962, 1963), which reflects the extent to which the vertex rises above the upper orbital margin. In *Paranthropus* the value of this index falls among the mean values for the living pongids, while the value for *H. africanus* is appreciably higher, outside the range of variation for the pongids, and approximates to that for *H. erectus* and *H. sapiens*.

It has been argued (Tobias 1967) that this difference in index value does not relate to expansion of the brain but is merely a consequence of the braincase being "hafted" onto the face higher in *H. africanus* than in *Paranthropus*. It seems to me that this is not so. In the first place, the Frankfurt Plane is used as the zero line for measurement; by definition, therefore, strictly comparable parts of the face and braincase are being used as a base line. In the second place, the only part of the face that is involved is the actual orbital height; that is, the distance between the upper and lower margins of the orbital cavity—the thickness of the brow ridges is not involved, nor is the rest of the face below the orbits. Orbital height is not significantly different in *H. africanus* and *Paranthropus*. Therefore, Tobias's statement can be correct only if relatively more of the *Paranthropus* braincase is below the Frankfurt Plane than is true of *H. africanus*; that is, if the external auditory aperture is relatively higher on the braincase of the former than of the latter. This evidently is not so, as can readily be demonstrated in the case of *H. africanus* Sts 5 and *Paranthropus* SK 48, respectively the most complete skull from Sterkfontein and the most complete one from Swartkrans, some three-quarters of a mile away. If natural size drawings in true side view of these skulls are superimposed (see Fig. 106) using the Frankfurt Plane and porion for orientation, it will be found that glabella coincides and inion very nearly coincides in the two. Furthermore, the occipital condyles overlap and are virtually the same distance from the Frankfurt Plane. Also, the maxillary tooth rows partly overlap, those of *Paranthropus* being slightly farther below the Frankfurt Plane than are those of *H. africanus.* It is clear, therefore, that these two skulls have a remarkably similar appearance in side view with respect to the major dimensions: braincase length is virtually identical; orbital height is virtually identical; vertical facial height is very closely similar; and the base of the braincase is in virtually identical position in relation to the Frankfurt Plane in the two. Evidently, therefore, the braincase is not hafted onto the face in a significantly different manner in the two cases. What is not similar in the two is the position of the skull vault; it is appreciably higher in the *H. africanus* skull, as the supra-orbital height index clearly indicates.

The latter index, however, by no means indicates the full extent of the difference since the frontal region, in side view, of the *Paranthropus* skull is flat or even concave, whereas in *H. africanus* this region is strongly vaulted. This difference is well demonstrated by the fact that though the braincase length is so similar in these two skulls, basi-bregmatic height in Sts 5 is 105 mm while the best estimate for SK 48 is 86 mm only. *Paranthropus* has no "forehead," whereas *H. africanus* has one that is rather well developed. It should also be taken into account that the frontal sinuses in *Paranthropus* are appreciably larger than are those in *H. africanus.* It is difficult to avoid the conclusion that the *H. africanus* brain is expanded upward in the parietal and especially the frontal region as compared to that of *Paranthropus.* It surely is not without significance that, even though the supra-orbital height index does not adequately characterize the difference in braincase height, it nevertheless has a high value in those forms in which brain expansion of the hominid type has occurred, *H. erectus* and *H. sapiens*, but has a relatively low value in all nonhominids, and that *Paranthropus* falls squarely in the low index while *H. africanus* falls in the high index group. Although few specimens are complete enough to measure in order to calculate this index for the early hominids, the very different frontal anatomy in the two forms is readily apparent to naked-eye inspection in an appreciable number of specimens.

The endocranial volumes of *H. africanus* and *Paranthropus* were either the same or that of *Paranthropus* was slightly larger. Estimates made by me in 1962 on six specimens of *H. africanus* gave a mean of 430 cm^3 and an estimated standard population range ($\bar{X} \pm 3$ s.d.) of 275–580 cm^3. No *Paranthropus* specimens are known that appear to fall outside of that range. The best available value is for Olduvai Hominid 5 and that is 530 cm^3. This specimen was from a large male, hence this value probably lies above the mean. The known South African specimens appear to have had volumes smaller than this. Even if the mean value for *Paranthropus* was slightly higher than that for *H. africanus*, account must be taken of the fact that the latter was a smaller and more gracile form. Probably the *H. africanus* endocranial volume therefore was proportionately significantly larger than that of *Paranthropus*. This conclusion, added to the demonstrable facts that the *H. africanus* skull had a much more vaulted frontal region and a braincase shape much more nearly like that of *H. erectus* than had *Paranthropus*, suggests that the former had already embarked upon the hominid brain expansion but that *Paranthropus* had not. Since this chapter was written a more detailed study of early hominid endocranial volume has appeared (Holloway 1970), the conclusions of which are in essential agreement with the above.

3. Masticatory Apparatus Function

In mammals the masticatory apparatus is closely adapted to the dietary habit of each species. This adaptation involves at least the teeth (we may ignore for present purposes those forms that have lost their teeth or have had them modified unrecognizably), the jaws in which they are set, and the muscles that operate the mandible. The nature of the specializations in these features usually has a profound effect upon the architecture of the entire skull. In order to understand the structure and functioning of a mammalian skull, it is necessary to have some understanding of the basic characteristics of the diet. Eating is always the primary function of the masticatory apparatus in mammals. Being homoiotherms they need a plentiful supply of food, and much of the ecological diversification seen in mammals directly involves dietary specialization.

There are also, however, secondary functions of the masticatory apparatus. The most common of these, probably, relates to defense and offense. Other functions include grooming, carrying, and communication. In man, communication is an extremely important activity that involves the masticatory apparatus. In this case, however, there is little or no modification of the hard parts that can be identified as being related directly to this function. The use of the masticatory apparatus for defense and attack that is not related to eating may involve observable anatomical specializations that are not related to the eating function. This is true, for example, in the Chacma baboon, the males of which are formidable fighters and have maxillary canine teeth that are very elongate with sharp posterior cutting edges that are constantly sharpened by wear against the third lower premolars. The chief food of this animal is, ordinarily, grass rhizomes that are collected by hand. Where such nondietary fighting specialization exists, it is usually readily detectable because of being more obvious in the males than in the females.

The early hominid material falls into two groups with regard to the anatomy and function of the masticatory apparatus. This applies to teeth, jaws, and muscularity as well as to other anatomical aspects that will be referred to.

In general, the teeth of *Paranthropus* are larger than those of *H. africanus*, but a considerable amount of overlap in size occurs between the two in all teeth along the tooth row. While the ante-premolar teeth, but especially the canines, are smaller on the average in *Paranthropus*, the postcanine teeth are larger in this form. Therefore, to say simply that there is overlap in size of each tooth type in the two forms obscures a very interesting fact: the proportions of teeth along the tooth row are not the same in the two forms. *H. africanus* has proportions along the tooth row that are similar to those in man, especially *H. erectus*. As we have seen in earlier chapters of this book, at least the females of this form appear to have been appreciably less robust and more gracile than individuals of *Paranthropus*. The latter, however, has very large postcanine teeth (exceeded only by *Gigantopithecus* among the Hominidae) but has canine teeth that are among the smallest known among hominids. This marked difference in the relationship between the canine tooth crown size and that of the postcanine teeth is obvious in *Paranthropus* as compared to other hominids and is best seen in a single tooth row, upper or lower. It also shows up very well when average size values for each tooth of the tooth row are compared with those of other hominids. Moreover, this characteristic is just as clear in two specimens from the Olduvai region that probably differ from each other in age by about a million years as it is in the South African material. The only exceptional specimen of which I am aware is a juvenile individual from Swartkrans (SK 27) from which I removed the completely unerupted maxillary canine that was still wholly within its crypt. In this specimen the canine is relatively large. Anatomically, it is quite aberrant, however, agreeing with neither the *H. africanus* nor other *Paranthropus* canines or, for that matter, with any other hominid canine of which I am aware. Other such examples of gross developmental abnormality are known in the Swartkrans tooth sample, for example the right P_4 of SK 6. Mandibular dentitions of the two forms are shown in figure 108.

In order to recognize this proportionate difference in canine size, it is necessary to compare the ante-premolar teeth, but especially the canine, with the postcanine teeth *in the same individual or within the same species.* Tobias (1967), for example, has argued that there are no significant dental differences in this respect between *Paranthropus* and *H. africanus*. In saying this he has based himself primarily on a tooth by tooth comparison of the two forms; that is, comparing the central incisor of one form with that of the other, and so on down the tooth row. He found what has been clear for many years (Robinson 1956) that there is overlap in size between the two. Curiously enough, he makes no mention of the fact that while, for example, the canines of *H. africanus* tend to be larger than those of *Paranthropus*, the reverse is true of the molars. But this tooth by tooth comparison is not very meaningful in this context since it does not efficiently demonstrate proportionate differences. The anterior teeth of *Paranthropus* are not functionally related to those of *H. africanus*; they *are* functionally related to the rest of the teeth in the same mouth.

An excellent method of comparing the proportional relationships along the tooth row of these and other forms is to use the logarithmic ratio diagram of the modified sort employed on postcranial material earlier in this work. In this case the modification consists of relating all of the "curves"—they are not true curves but a series of points—to a standard size of M1 equal to the mean of that tooth for the form being used as the standard of comparison, the zero line of the diagram. M1 was chosen because it changed least in size in the course of hominid evolution and because it usually has a relatively low coefficient of variation compared to the other teeth. The indicator of size used in this case is the robustness index, the product of length and breadth of the crown. This approximates to the cross-sectional area of the tooth crown, though this is not so for incisors because of the shape of the crown. For this reason, and because the known samples of them are so small, incisors have not been included in the comparison. In those cases where incisors are present in a jaw with other teeth, the incisors tend to follow the same pattern of size as the canine. The cross-sectional area approximates to the occlusal surface area in the case of the molars and only a little less closely in premolars. In the case of canines the robustness index does measure size, but not as effectively as it does for the postcanine teeth; it tends to underestimate differences in size. For example, the canines of *H. africanus* not merely have greater girth than those of *Paranthropus* but also are significantly taller, and this is meaningful in terms of canine function. Crown volume, in this case, would be a better criterion for comparing size. Volume is much more difficult to measure, however, because of the nature of the material. Moreover, since most available specimens have suffered varying degrees of loss of volume through wear, volume could not be determined accurately in any case. Cross-sectional area is easier to measure and does not begin to be reduced by attrition until a very late stage in crown wear. The product of length and breadth, of course, does not measure surface area exactly, but since corresponding teeth in the forms being compared are so similarly shaped it is accurate enough for the purpose. Accurate measurement of the area would be considerably more time-consuming to carry out and the improvement in accuracy would not justify the considerably greater expenditure of time. Length and breadth alone could be used for comparisons and are easy to measure and do not involve approximations. Each alone, however, is a poorer indicator of size for the present purpose, although actual test shows that the proportional differences here involved are indeed clearly shown by using length and breadth independently to construct two ratio diagrams.

The nature of the difference between the two forms can be seen in figure 109, which uses *H. sapiens* (American whites [Black 1902]) as a standard of comparison (see also Robinson 1970). It makes no significant difference what form is used as a basis of comparison; with different bases the diagrams look different, but the conclusions are the same. This was tested by constructing a series of such diagrams using different comparison standards. Including substantial samples of the living great apes in the comparisons is particularly illuminating. In order to provide a relatively uncluttered diagram for publication, some of those comparisons included in the original diagram have been left out. For example, the original diagram, using mandibular teeth, included gorilla, orang, and chimpanzee curves based on pooled samples of males and females ($N = 50$, 14, and 50, respectively) as well as curves for males and for females separately ($N = 25$, 7, and 25, respectively), as well as curves for *H. sapiens* samples other than the standard. The diagram presented here, however, is representative: the eliminated *H. sapiens* curves do not deviate markedly from that of American whites, and the orang curves in all cases lay between the curves for the gorilla and the chimpanzee. What the diagram here presented does not show is sexual dimorphism, though this does not affect the purpose for which the diagram is here being used. In all the great apes the female has a proportionately smaller canine than the male; the difference is greatest, and is quite large, in the gorilla and least in the chimpanzee. It is also noteworthy that the gorilla has proportionately the smallest canines of the three great apes. The pongid samples used are from wild-shot specimens only, and these were all dentally mature though not old. All measurements for the pongid and early hominid samples were made by me and are thus closely comparable.

This diagram is very instructive. It shows that with respect to the series P_4–M_3 neither *Paranthropus* nor *H. africanus* differ at all from the living great apes in proportion. Indeed, the chimpanzee is noticeably more like modern man with regard to M_1–M_3 than are either of the early hominids. The proportions for P_4–M_3 are somewhat different in modern *H. sapiens*. The Pekin sample of *H. erectus* does not differ significantly from *H. sapiens* with respect to M_1–M_3—the curve is virtually identical, for instance, to that for the Australian Aborigine (using Campbell's [1925] data). But P_4 is proportionately large compared to that of modern man and resembles that of *H. africanus* more than it does that of *H. sapiens*. With respect to C–M_1, the Pekin sample of *H. erectus* closely resembles *H. africanus* and differs fairly markedly from modern *H. sapiens*.

Although there is a great deal of similarity between the curves for all of the forms from P_4–M_3, this is anything but true for C–P_3. For P_3 there is a group with proportionately large teeth, including the three great apes, a group with teeth of intermediate size that includes *Paranthropus*, *H. africanus*, and *H. erectus*, and a group of small size comprising samples of *H. sapiens*. *Paranthropus*, however, is somewhat anomalous with respect to P_3–P_4. While the curves for *H. erectus* and *H. africanus* are rather close together and parallel, that for *Paranthropus* cuts sharply across the other two so that although it has the largest P_4 of the three, it has the smallest P_3. From P_4 to M_3 the *Paranthropus* curve is almost identical to that of the gorilla, but forward from P_4 it diverges sharply and dramatically from that of the gorilla.

The variation in proportionate size of the canine is very much greater than that of P_3 for the forms being considered. Again, there are three natural groups: one in which the canine is proportionately large, comprising the pongids; one in which the canine is of moderate size, including *H. africanus*, *H. erectus*, and *H. sapiens*; and one in which the canine is proportionately small, including *Paranthropus* alone. The middle size group is compact; *H. africanus* and the Pekin sample of *H. erectus* are almost identical and fall only a short distance from the American white sample of *H. sapiens*. The *Paranthropus* canine is highly anomalous in its size relationship to P_3, and because of the fact that it has a relatively very large P_4 compared to P_3, it is even more anomalous with respect to P_4 size. This sharp reduction in proportionate size of the canine is highlighted by the fact that the *Paranthropus* P_3 is significantly larger proportionately than that of the *H. sapiens* sample, but the canine is very markedly smaller than that of the *H. sapiens* sample. It is further emphasized by the fact that in proportionate size of the canine *H. africanus* lies precisely midway between *Paranthropus* and the mean (male + female) value for a sample of fifty gorilla specimens. Indeed, in proportionate canine size, by the criterion here being used, *H. africanus* and *H. erectus* resemble the female gorilla much more closely than they do *Paranthropus*. The *H. sapiens* sample used lies midway between the *Paranthropus* value and that for the female gorilla sample (of twenty-five) which falls at +2 on the diagram.

Moreover, using pooled samples of all of the specimens known of *Paranthropus*, of *H. africanus*, and of *H. erectus* makes only minor modifications to the diagram: *Paranthropus* remains with a proportionately exceedingly small canine while *H. africanus* and *H. erectus* are closely similar and rather close to *H. sapiens*. As already noted, the criterion of comparison here tends to underestimate differences in canine size. This may not be entirely true of the gorilla, which seems to have a proportionately somewhat shorter canine relative to its girth than is true of the chimpanzee: the cross-sectional area may thus lead one to expect a larger canine than is in fact present.

However that may be, the evidence is clear that *Paranthropus* differs very markedly from *H. africanus* with respect to relative size of the mandibular canine. That this is a very significant difference is demonstrated by the comparisons with pongids. Moreover, the very unusual *Paranthropus* condition endured for a very long time.

The picture presented by the maxillary dentition is in principle the same, but the difference is not so great. The smaller degree of difference between the two early hominids appears to result from the *H. africanus* canine sample being biased to the small side. This is suggested by the fact that the mean size for the maxillary canine sample at Sterkfontein is smaller than that for the mandibular sample. It is very unusual for average values for higher primate—especially hominid—maxillary canines to be less than that for the mandibular canines of the same form. Moreover, in one specimen (Sts 52) both upper and lower canines of the same individual are present and almost unworn, and the maxillary canine is the larger of the two. If the size relationship of the canines in this specimen is used as a basis for estimating a less biased mean size for maxillary canines in relation to the mean for the mandibular sample, a value is obtained that gives a ratio diagram similar to that given by the mandibular teeth.

In the light of all this evidence there can be little doubt that the difference in tooth proportions between *H. africanus* and *Paranthropus* is real and of considerable magnitude. Few would care to doubt that the gorilla differs significantly in canine size from *H. africanus*, and yet, as the evidence shows, *Paranthropus* differs from *H. africanus* in proportionate canine size as much as the latter differs from the gorilla. The occurrence of this anomalous *Paranthropus* condition over at least a million years and over widely different geographical areas leaves no doubt that it was a stable and significant characteristic of this form. In spite of his strong disagreement with this view, Tobias (1967) actually provides good support for it. On p. 226 he compares the size relations of the canine to the two premolars in the two forms and writes, "It may be concluded that there is a real difference between the two taxa in the disparity between the sizes of the canines and the cheek teeth." This statement is fully consistent with the evidence presented above. One is thus at a loss to understand why this conclusion is not referred to again. Instead, on p. 228 he sums up the discussion with the conclusion, ". . . the fundamental morphological basis underlying the . . . dietary hypothesis has largely fallen away—and so, it seems to me, must the hypothesis itself. This conclusion is based solely upon a reassessment of the very facts which the hypothesis was originally erected to explain." And yet the cardinal evidence that led me originally to the hypothesis of dietary, hence also ecological and behavioral, difference between *H. africanus* and *Paranthropus* was the difference in proportionate size of the canines to the cheek teeth—the reality of which he confirms.

But teeth do not exist as independent and isolated entities, and if the setting of the dentition is examined it becomes clear that the differences between the two forms of early hominid are not confined to the teeth. In *Paranthropus* the mandibular ramus is relatively high compared to the corpus length and is set more vertically. Attachment scars indicate very well developed masseter muscles, a conclusion that is confirmed by the massiveness of the jugal arch to which they attach. The scars of origin are well developed, especially anteriorly where the zygomatic bone is actually extended downward toward the mandible in the direction of pull of the anterior muscle fibers. Natural split-lines in the bone follow the direction of pull of the anterior muscle fibers, suggesting a very considerable stressing of the bone in this region. The temporal muscles also were large so that they met to form a bony sagittal crest for some distance on either side of the vertex. Such a crest is present in every fully adult individual so far found in which the relevant region is known. Also, the manner in which the muscle exerted its pull on the mandible appears to have been different from that in *H. africanus* and later hominids. The main insertion appears to have been in much the same position as that in other hominids, and the endocoronoid buttress passes to this region for added strengthening. Whereas in other hominids the end of this buttress coincides with the coronoid itself, in *Paranthropus* the anterior margin of the ramus curves up and over for a considerable distance from the end of the buttress before reaching the coronoid. The coronoid is thus closer to the condyle and the sigmoid notch is narrower than is the case in *H. africanus* and later hominids. As far as I am aware, this situation is unique among primates. The anatomical characteristics of this region suggest that, besides the powerful, mainly vertical component of the muscle, there was also a more than usually important and powerful posterior component that pulled in a more nearly horizontal direction. This suggests that perhaps chewing was not merely powerful but involved also considerable fore and aft, as well as rotatory, movements of the mandible. What evidence there is indicates that the pterygoid muscles were powerfully developed also and that the pterygoid plates were relatively very large. Associated with the proportionate reduction in size of the anterior teeth, the face was relatively vertical with comparatively little maxillary prognathism. The nasal region was completely flat. Because of this flatness and the powerful development of the zygomatic bones already referred to, the face was dished so that the nasal region is completely obscured in true side view. The bony palate also is unusual. It is relatively thick and slopes fairly sharply upward as it passes back from the incisive foramen. Palate thickness can be measured in three specimens from Swartkrans approximately opposite the contact between first and second molars. There is a median sagittal palatal thickening; measured in the midline of this thickening in SK 12, 46, and 52, the thickness is 13.3, 8.0, and 13.2 mm, respectively. To one side of the ridge the thickness is 10.9, 7.5, and 11.3 mm, respectively. By comparison, the *H. africanus* palate, to one side of the midline, is 1.4 and 1.8 mm thick in two specimens. A third cannot be measured, but a hole in the palate reveals thickness comparable to that in the other two. In some cases there is appreciable vaulting from the alveolar margin to the incisive foramen in *Paranthropus*, as in Hominid 5 from Bed I at Olduvai, but usually there is comparatively little or none at all so that the palate midline is quite straight from the anterior alveolar margin to the back end of the palate. The Olduvai Hominid 5 skull is unusual, for *Paranthropus*, in having so much anterior depth. This does not alter the fact that it has the sloping type of palate, typical of *Paranthropus*, which makes a considerable angle with the Frankfurt Plane. The anterior depth is a consequence of an unusual amount of growth in the alveolar part of the maxilla anteriorly. As a consequence the occlusal plane is not approximately parallel to the Frankfurt Plane as it usually is in *Paranthropus* and *H. africanus*, but the anterior end of the tooth row is appreciably farther from the Frankfurt Plane than is the posterior end. This gives the face an unusual jutting forward and downward appearance in side view. Unfortunately, this single specimen cannot indicate whether this was the usual situation in the East African *Paranthropus* of that time period or whether this individual was an unusual variant. That it could be the latter is suggested by a chimpanzee specimen in the Powell-Cotton Museum. This has an exactly similar elongation of the anterior alveolar portion of the maxilla, producing a downward tipped occlusal plane. This is a wild-shot specimen from the Bipindi region, and other specimens from the same region do not exhibit this anomaly. In *H. africanus*, *H. erectus*, and *H. sapiens*, the palatal midline between the incisive foramen and the posterior end is approximately parallel to the Frankfurt Plane in contrast to the condition in *Paranthropus*. A remarkably complete adult *Paranthropus* skull found by Richard Leaky recently in the Lake Baringo region (Leakey 1970), a cast of which he kindly showed me, does not exhibit the downward facial elongation and anterior palatal depth seen in

Olduvai Hominid 5. This tends further to confirm the view that the latter is aberrant rather than typical in these features.

In all of the above features *Paranthropus* differs from *H. africanus*. Although there may be overlap between the two in respect to some of these characters, they are not isolated features but are parts of a single functional complex: as a unit it differs markedly and unmistakably in the two forms.

Since the masticatory apparatus is an integral and fundamental part of a mammal's basic ecological and behavioral adaptation, and following the principle that basic adaptational features will be under very close and precise control of natural selection, the conclusion seems unavoidable that *Paranthropus* and *H. africanus* were not similarly adapted.

Further enquiry provides some insight into the nature of the difference in adaptation with respect to the masticatory apparatus. Clearly the *Paranthropus* apparatus functions with more power: most of the teeth are larger, have thicker enamel, and larger root systems; the bony parts are thicker and more massive and the muscles were larger. This suggests heavy use of the masticatory apparatus. The postcanine teeth were functionally more important, did more work than in *H. africanus*, and the ante-premolar teeth did proportionately less. The postcanine teeth have occlusal surfaces that are large and have very little relief even when unworn. They tend to wear rapidly to a more or less plane surface. These teeth clearly are specialized for crushing and grinding, not slicing and shearing. The piercing, slicing, and gripping function of the anterior teeth must have been relatively unimportant in the function of the masticatory apparatus, otherwise natural selection would not have allowed them to reduce to the extent that they did. The fact that this feature existed unaltered for what seems to have been at least a million years, and in the light of the new evidence from Omo and the Baringo area may well have been very much longer, clearly indicates a stable adaptation closely controlled by selection. That is to say, this unusual proportional arrangement evidently represented an optimal adjustment to the needs of the animal.

I would emphasize again that to assess the significance of the dental differences the dentition should be treated as a unit, not as a series of independent teeth. Changes in adaptation involve shifts in emphasis on the various parts making up the dental unit. The ratio diagram analysis indicated that there is little difference in relative size of the P_4–M_3 portion of the tooth row in *H. erectus*, *H. africanus*, *Paranthropus*, and the great apes. But P_3 varies a good deal more in this group, and with respect to the canine the variation is very great. This may be interpreted as meaning that in the hominoids the ante-premolar teeth, and to some extent P_3, are more sensitive to adaptive change than are the postcanine teeth, and that in the great apes the ante-premolar teeth are more important in the total dental function than they are in *H. africanus* and *H. erectus*. The latter forms, in turn, have ante-premolar teeth that carry a larger share of total dental function than is true of *Paranthropus*. This is not to say that ante-premolar teeth are unimportant in the latter and were in the process of disappearing; rather, they had been adjusted to the needs of the organism by natural selection, but the ecological and behavioral adaptive pattern involved a relatively small amount of work for the ante-premolar teeth.

The combination of features found in *Paranthropus* seems to exclude a diet involving much carnivorousness. Had meat been important in the diet, these teeth would have served too valuable a function, and natural selection would thus have maintained their size rather than favor marked reduction. Moreover, the powerful battery of postcanine teeth with large occlusal surfaces of low relief, which wear flat very rapidly and were used with rotatory chewing movements, are certainly far better suited to crushing and grinding than to the slicing function in specialized carnivores.

In contrast, *H. africanus* had a masticatory apparatus so detailedly resembling that of *H. erectus* and *H. sapiens* with respect to the points discussed above that no grounds appear to exist for supposing it to have had a diet basically different from theirs. That is to say, it evidently was an omnivore.

The argument has been advanced, starting with Darwin, that increased use of culture would cause reduction in anterior tooth size, especially the canines. Most of the function of the canine, it is supposed, would be taken over by tools and weapons. Superficially, this is a seductive hypothesis. It does not, however, survive examination. A general argument against it is that many examples of reduction, or total loss, of canine and other adjacent teeth are known among mammals, none of which involved culture as a factor in the reduction. The phenomenon is widespread and is almost invariably understood in terms of dietary adaptation. But it is, of course, possible that in spite of massive evidence to the contrary, a specific instance of canine reduction could be due to an unusual factor like cultural activity—especially since it is so reasonable to postulate that tools could largely replace canine teeth. But it would follow from this premise that there should be a reasonable degree of negative correlation between canine size and cultural facility. Fig. 109 indicates clearly enough that this is not so; among hominids, culture is most clearly associated with the group with medium-sized canines, not with the small-size group. Conversely, to argue that reduced size reflects cultural activity means that one is committed to the view that *Paranthropus* was far more effectively culture-bearing than is modern man. This, surely, is not likely to have been true.

The evidence thus is that *Paranthropus* differs sharply from *H. africanus* with respect to relative size of the canine to the cheek teeth, though the incisors and third premolar also are relatively small. This condition persisted for a very long period, hence it represents a stable adaptation. Contrary to what some authors have claimed, there is indeed something significant here that needs explanation. Culture appears not to be a satisfactory explanation; dietary differences therefore appear to be the most logical answer. It is worth mentioning that this is the route my thinking followed initially, a decade ago. I was not arguing, as some authors suppose, that examination of a primate's dentition will necessarily allow one to tell exactly what it ate. On the contrary, I believe that higher primate dentitions can be very difficult to interpret dietarily from morphology alone. But since *H. africanus* resembles culture-bearing man so very closely in dentition and the masticatory apparatus as a whole, it is reasonable to attribute to them the rather flexible omnivorousness typical of man in general and especially of those most technologically poorly developed demes of man. Clearly, there are no grounds for asserting that it could *not* have had such an ecology. The problem then is one of attempting to see in what manner the diet of *Paranthropus* may have differed in order to explain why its masticatory apparatus was different. As we have seen, the evidence is much more in favor of more herbivorousness rather than more carnivorousness as compared to the middle-of-the-road omnivorousness of man.

On the other hand, it is not enough merely to hold that *Paranthropus* was a herbivore of the general pongid type, since one would then expect that, like them, it should have had substantial anterior teeth. Presumably its diet will not have included plants with tough outer covering, such as bamboo, which require well-developed anterior teeth to remove the covering. It is not likely to have been a grazer for a number of reasons; from a dental point of view, such specialization usually leads to well-developed cropping teeth, even if the canines disappear entirely. One might suggest, therefore, that the diet probably included fruit, berries, seeds, vegetables, roots and bulbs, tender leaves and shoots, and perhaps nuts that could be cracked open by means of the large cheek teeth.

Dietary differences of the sort here being discussed are not, of course, simply concerned with food; behavior is a very important and integral part of dietary adaptation. The behavior of a primitive hunter-gatherer is vastly different from that of a foraging herbivore such as the chimpanzee or gorilla. *Paranthropus*, according to the conclusions reached here, will have been more like the latter than like a hunter-gatherer behaviorally. Thus the total difference in adaptation between it and *H. africanus* will have been of very considerable magnitude.

Part of the change from the ape grade of organization to the human grade appears to have been a change in diet from the almost completely herbivorous diet of the nonhominid higher primates to the omnivorous diet of the later hominids. It seems therefore that *Paranthropus* had not made this transition while *H. africanus* had, though, as we have seen, *Paranthropus* was evidently not a herbivore of exactly the same type as the pongids. A further part of the change from ape to human grade was a major brain expansion. As we have seen, the fossil evidence suggests that in this respect also *Paranthropus* had not changed whereas *H. africanus* apparently was in the process of changing. The cranial material gives no indication of difference in locomotor habit between the two. We may now turn to a review of the postcranial evidence to see how it amplifies or modifies the above conclusions.

Conclusions from Postcranial Evidence

Postcranial material is rather poorly represented in the known collections of early hominids: in the case of the South African early hominids, less than a tenth of the known specimens are parts of the postcranial skeleton. That this material is so rare is a pity since some very important aspects of the structure and functioning of early hominids can be understood only from noncranial evidence. The relatively small amount of material available, however, is extraordinarily illuminating in diverse ways, as the preceding chapters perhaps suggest. The chief conclusions that have emerged in the present study will be summarized here under two headings: (1) body size and proportions, and (2) locomotor habit. The details are to be found in the preceding chapters.

1. Body Size and Proportions

The basic evidence for general body size comes from a single mature adult female specimen, Sts 14. Much of the spinal column, including all six lumbar vertebrae and the sacrum, a few ribs, almost the complete pelvis, and much of a femur of this individual are known. The small and slender ribs, the extremely small size of the vertebrae—especially their bodies—and the very slender femoral head, neck, and shaft all indicate that the individual was small and very lightly built. The pelvis of this individual gives the impression of having come from a larger individual than actually was the case. This female was evidently small in stature as judged by the length of the spinal column and of the femur. It is not easy to make a reliable estimate of the original height. Estimates were made using as a basis the proportion between different individual lumbar vertebrae body lengths and stature in various living groups of man; the ratio between the sum of the lumbar vertebrae body lengths and stature as well as femur length as a proportion of stature. These gave estimates ranging from just below 4 feet to about 4 feet 6 inches. Finally, using Trotter and Gleser's formulae for estimating stature from the femur alone for Negro and for white females gave a value of 4 feet 3 inches. It therefore seems likely that the height of the individual was between 4 feet and 4 feet 6 inches.

The slenderness of weight-bearing bones such as the femoral shaft and the lumbar vertebrae, the smallness of the sacroiliac articulation, the smallness of the femoral head, and the thinness of acetabular wall all indicate a very lightly built individual. There is no reasonable way of making an accurate estimate of weight; in view of the small stature and lightness of build of this individual, it seems to me likely that its weight would have been approximately 40–60 lbs.

These estimates of size and weight are based on a single female individual. One other adult innominate, the sex of which is uncertain, is of approximately the same size and also has a relatively small auricular area. There is a slight suggestion that this specimen may have belonged to a male individual. If this were so, then it would suggest that males also were small. Whether or not it belonged to a male, it does indicate that the more complete, female specimen was not aberrant in size for the Sterkfontein population.

The single, moderately complete vertebral body (Sts 73) of probably the first or second lumbar vertebra affords evidence of another individual that may have been somewhat heavier than Sts 14 though probably of about the same stature. This is suggested by the fact that the cross-sectional area of the body is somewhat larger but the superior–inferior height is the same as in the Sts 14 equivalent. There is a suggestion of mild sexual dimorphism in the dentition of *H. africanus*. This, coupled with the evidence from Sts 73 of a somewhat more robust individual of roughly the same height as Sts 14, suggests that males and females differed a little in robustness—a not very startling conclusion! This somewhat slender evidence, taken in conjunction with the two incomplete juvenile innominates from Makapansgat, suggests that *H. africanus* was a pygmoid form that was very lightly built and in which males were a little more heavily built than the females.

The Sts 14 female is again the best source of information concerning bodily proportions. A point of considerable interest is whether the lower limb was long, as in modern man, or was relatively short as in the chimpanzee or gorilla. On the basis of a reasonably secure estimate of the femur length in this individual, it appears that the femur was proportionately of about the same length as that of modern man. Using the modified ratio diagram method suggests that the femur in this individual was of the same proportionate length as the average for the Australian Aborigine female. Since the portion of the lower limb from the knee down cannot be too different in length than the femur, the implication is clear that this individual had lower limbs of proportionately about the same length as has modern man.

Unfortunately, this individual provides no evidence about the length of the upper limb. The only relevant evidence is from the single incomplete humerus, Sts 7. This is a robust and relatively long bone. If its size and length are taken as applying to Sts 14, then one would have to conclude that the arm of *H. africanus* was proportionately longer compared to the lower limb than is the case in modern man but not as long as in the pongids. The humerus probably belongs to one of the largest male specimens known, however, since it was directly associated with the most robust *H. africanus* mandible known at present. If this is so, then presumably it is appreciably too large for the small female Sts 14. If the average percentage difference in humerus length between males and females in modern man is used as a basis for estimating humerus length for Sts 14 from Sts 7, a length of 270 mm is obtained. This gives a humero-femoral index of 88, which is intermediate between the average value for man and that of the pongid with the lowest index, the chimpanzee. In modern man the mean index value varies from the upper 60s to about 74; that of the pongids ranges from 101 to 138. Individual index values vary over appreciably greater ranges than these. It is also possible to estimate a femur length for Sts 7 from that of Sts 14. Since the femur in the latter individual indicates a proportionate femur length for *H. africanus* like that of modern man, one may use the average percentage difference between femur lengths of males and females of modern man as a basis for the calculation. This gives an estimated femur length of 337 mm and a humero-femoral index of 89.

These figures suggest that the humerus of *H. africanus* may have been proportionately a little longer than that of modern man on the average, though not outside the range of variation of the latter, but shorter than that of the pongid with the shortest humerus. Obviously, however, this evidence is not good and this conclusion is no more than approximate and tentative.

We may conclude, then, that the bodily proportions in *H. africanus*, with respect to relative limb lengths, were very similar to those of modern man with the possible exception that the arms were relatively a little longer.

It does seem clear, however, that Sts 14 had a rather large pelvis for her size. Furthermore, because of the relatively wide open front of the false pelvis, it is probable that she had a protuberant abdomen. The other ilia available suggest a similar conclusion; hence this was probably a general feature. Since well-developed lumbar curvature was present in this individual, the characteristic set of the trunk on the hips found in man, with a hollowed lower back, was doubtless present, possibly a little exaggerated by the protuberant abdomen. What the buttocks were like is difficult to say because their size and contours depend so heavily on the amount of fat in the superficial fascia.

The lower limbs were probably rather slim; evidence of this can be seen in the smallness of the hamstring attachment area. From the angulation of the femoral shaft it is clear that the knees, and therefore also the feet, were close together during erect standing. If the Olduvai Bed I foot is, as I think likely, representative of *H. africanus*, then it was small compared to that of man, was compact and well arched, and had a fully adducted great toe—a substantial and very human foot.

Although the *H. africanus* skull seems rather small and gracile for a manlike creature, when pictured on the lightly built and small creature just described, it would not be especially small.

H. africanus, then, seems to have had essentially the modern body proportions but had wide hips and a bulging abdomen in a body smaller than that of a pygmy.

The conclusions that can be drawn about *Paranthropus* are not so detailed because less information is available about the postcranial skeleton and much of what is available consists of parts that are not known in *H. africanus*. One thing is clear, however, and that is that *Paranthropus* was a more robust and heavily built creature than *H. africanus*. The ilium, for example, is distinctly larger in the two known specimens than are those of *H. africanus*; in both cases two different sites are involved. Because many of the suitable measuring points in the *Paranthropus* specimens are absent, size comparison is not especially easy. Using what seems to be a good measure of innominate size—width of the iliac shaft at the level of the upper acetabular margin—indicates that the Swartkrans innominate is 22 percent larger than that of Sts 14 from Sterkfontein. The Kromdraai innominate seems to be at least as large. The anterior margin of the ilium is 33 percent longer in the Swartkrans specimen, but this distance seems clearly to be proportionately longer in *Paranthropus* than in *H. africanus*; hence this figure would be exaggerating the difference in general size. The fact that the anterior part of the iliac blade is more protuberant in *Paranthropus* suggests a proportionately thicker torso in this form. Acetabular width differs by about 11 percent in the Swartkrans and Sts 14 innominates. This is a more impressive difference than may appear at first sight because of the remarkably small amount of size difference in this structure in the great apes and man and because of its low coefficient of variation. That is to say, the size of the innominate as a whole varies appreciably more than does the size of the acetabulum—as can readily be appreciated from the fact that the mean absolute width of the acetabulum is almost the same in modern man and the gorilla, but the size of the innominates of the two are very different.

The two *Paranthropus* femora are considerably more robust than is the *H. africanus* (Sts 14 female) equivalent. For example, one has a shaft thickness that is greater than that of the *H. africanus* specimen by 50 percent of the thickness of the latter, and the other is thicker by 75 percent. This is the order of magnitude of the difference found between the smallest and largest femora of modern human racial groups.

Unfortunately, no long bones of *Paranthropus* are available from South Africa that are complete enough to provide any real evidence of their original total length. It might be felt by some that the tibia and fibula from FLK NN I at Olduvai, which are reasonably complete and were associated with the good skull of *Paranthropus* (= "Zinjanthropus"), give evidence of limb length for *Paranthropus*. In my view this would be a mistake because there is no reliable evidence that indicates to which of the two hominids known from the lower levels of Bed I these bones belong. On balance, the length, robustness, and morphology suggest the probability that the two bones belong to *H. africanus*. But in view of the uncertainty about their taxonomic designation, using them to estimate size of either *Paranthropus* or *H. africanus* is an open invitation to error and confusion.

In the circumstances there is no way of assessing whether the two Swartkrans femora had such thick shafts because the bones were long compared to the *H. africanus* equivalent, hence the individuals they belonged to were tall and thus more heavy, or whether the animals were not particularly tall but were heavily built. Putting together the percentage difference in size of the innominates and the femora as compared to those of *H. africanus* suggests to me the probability that *Paranthropus* was a little taller and somewhat more robust than the former. The talus also suggests that *Paranthropus* was at least a moderately heavy animal. If *H. africanus* was approximately 4 to 4½ feet high, then perhaps *Paranthropus* was about 4½ to 5 feet in height. If the estimate of approximately 40–60 lbs is correct for a female *H. africanus*, then perhaps *Paranthropus* weighed about 150–200 lbs. Evidence from vertebrae suggests the possibility that *H. africanus* males and *Paranthropus* females may have been of about the same size.

Also, because of the lack of long bone specimens sufficiently complete to allow estimation of their original total length, it is not possible to judge whether the lower limb was relatively long, as in *H. africanus* and later hominids, or what the proportionate length was of the upper limb. As we saw on p. 155, Schultz's evidence concerning length/robustness relations in primate long bones would indicate a relatively shorter femur, hence lower limb, in *Paranthropus* than in *H. africanus*.

There is some suggestion that *Paranthropus* may have had proportionately large hips, though probably not to the extent found in *H. africanus*. For the same reasons as in the latter, the abdomen was probably protuberant —perhaps even more so than in *H. africanus* because of being more heavily built and with a proportionately more protuberant anterior superior iliac spine region.

2. Locomotor Habit

One of the more significant anatomical changes in the transition from the ape grade of organization to the human one was the complex of changes involved in becoming an erect biped from a quadrupedal condition. This change need not in itself have made a great deal of difference to the population in which it first occurred. Because this change was a prospective adaptation for culture, however, it had very great significance for the emergence of man. For this reason it is of interest to know what was the locomotor habit of *H. africanus* and *Paranthropus*.

Basically there are two major requirements for erectly bipedal posture to be efficient:

(a) The capacity to balance the body effectively. This capacity needs to be effective not only while the individual is standing erect on both feet but also during the phases in walking when one limb only is supporting the trunk.

(b) The capacity to move the legs quickly and through a relatively wide arc so that long steps may be taken. This requires, among other things, ability effectively to extend the lower limb past the vertical.

Both of these capacities involve a number of components and are by no means simple, either anatomically or functionally. It does not seem to me, however, that the change from quadrupedal to erectly bipedal locomotor habit involved basic changes of function on the part of any of the muscles. The change came about through modification of skeletal proportion, orientation, and spatial pattern as well as through changes in muscle size.

(a) Capacity for Balance

One aspect of the capacity for balance is maintaining the trunk erect on the pelvis. Fore and aft balance is maintained by the powerful spinal musculature assisted by the abdominal musculature. Lateral balance is more of a problem; again the powerful spinal musculature is important, and assistance is provided by the lateral trunk musculature that attaches to the brim of the false pelvis. In *H. africanus*, as in man, the widened sacrum increases the separation between the ilia, and the medialward curvature of the iliac blade, along with its relatively great antero-posterior width, results in a wide false pelvis in which the walls extend from the back round to the sides. This is in contrast to the pongid pelvis in which the iliac blades do not extend round to the sides in this manner, except to a small extent in the gorilla. The wide and laterally situated iliac blades in *H. africanus* place the lateral trunk muscles in a more advantageous position to assist the spinal muscles in controlling the trunk attitude. Because the iliac blades lean outward slightly more than is the case in modern man, the *H. africanus* false pelvis may even have had a slight advantage over that of modern man in this respect. With regard to this aspect of balance, *H. africanus* had already achieved the modern human condition.

A more significant aspect of lateral balance control involves control of the attitude of the pelvis, especially during walking. For efficiency in erect-walking, the feet must move in planes that remain close together. The relatively tall body is thus balanced on a small foundation. Moreover, during walking one foot is off the ground much of the time; hence the pelvis and the torso above it are then supported by one limb alone. As is well known, gluteus medius and minimus have acquired the function in man of controlling the pelvis during walking so that it does not flop from one side to the other under the influence of trunk weight as its support alternates from one limb to the other during bipedal movement. On the supporting side these muscles contract, pulling down on the false pelvis on that side and thus preventing it from subsiding to the unsupported side. This cyclic tension and relaxation produces great strain on the iliac blades and has resulted in the development of a supporting buttress, the acetabulo-cristal buttress, in the form of a thickening of the outer table bone of the ilium between the actabulum and the iliac crest. The presence of this buttress is thus evidence of efficient lateral balance control.

A crucial question, therefore, is whether such a buttress existed in *H. africanus*. The general opinion in the literature is that this is not so, that there is no buttress. This study, however, has demonstrated that the acetabulo-cristal buttress is certainly present as a normal part of the *H. africanus* ilium. Indeed, there is a second anteriorly situated buttress, the acetabulo-spinous buttress, which is very well developed. The former serves the same purpose as in modern man; the latter serves, in my opinion, to strengthen the relatively protuberant and laterally flaring anterior superior iliac spine region against strain from tension of the inguinal ligament. The latter is an important structure in erect bipeds that serves in part as the anterior wall of the false pelvis and for attachment of important abdominal muscles. That there is strain in this region caused by the abdominal viscera is demonstrated by the frequency of herniation there. Other structures contribute to place stress on the anterior part of the iliac blade, but the inguinal ligament is probably by far the most important one. During later human evolution, the anterior end of the iliac blade shortened, turned more medialward, and the upper border curved downward more sharply. These changes improved the mechanical efficiency of this region because the line of tension of the inguinal ligament thereby coincided more nearly with the direction of the anterior part of the blade. The need for the acetabulo-spinous buttress thus disappeared, and only the acetabulo-cristal buttress remained, though the other does sometimes still occur in modern man. Because of the greater degree of medialward flexure of the anterior part of the iliac blade, the acetabulo-cristal buttress appears more prominent in modern man. In *H. africanus*, with both present and lying close together, each assisted the other in providing strength; hence neither needed to be as large as it would have been had only one been present.

The evidence thus indicates very clearly that the mechanism for lateral balance control was fully established in *H. africanus*.

There is another aspect of balance control while erect that merits mention. In pongids the center of gravity of the body is well above and forward of the hinge axis of the trunk on the lower limbs. In the erect position the trunk thus tends to fall forward, and this tendency has to be resisted by quite powerful muscular action. In man the center of gravity is very slightly above and behind the hinge axis. This means that the natural tendency of the trunk is to fall backward; that is, the weight of the trunk tends to rotate the pelvis backward or hyperextend it on the femur. Because the center of gravity is so close to the hinge axis, however, less energy is needed to counteract this action, and much of that resistance comes from the ilio-femoral ligament in the specialized form—the Y ligament of Bigelow—in which it occurs in man.

At least three important factors enter into lowering the center of gravity of man and moving it backward. These are:

(1) A shortening of the ilium. Reduction in height of the ilium brings the sacroiliac articulation closer down to the level of the acetabulum. The present study has shown, using the modified ratio diagram method, that by far the most obvious differences in proportion between pongid pelvises and that of modern man are in the relatively great reduction of the ilium height-related dimensions. In this respect *H. africanus* agrees extremely closely with modern man and differs widely from the pongids. Furthermore, a well-developed anterior inferior iliac spine and the suggestion of a femoral tubercle (the area is damaged) indicate that a Y ligament was present.

(2) Lumbar curvature. Since the sacrum cannot rotate in relation to the ilium in order to bring the spinal column into a vertical position without blocking the birth canal, bringing the upper portion of the trunk well back over the pelvis, which also moves the center of gravity back, involves curvature of the lumbar spine. Lumbar curvature was well developed in the only specimen (Sts 14) that is complete enough to allow the point to be checked.

(3) Increased size of the lower limb. The lengthening of the lower limb increases the mass of the lower portion of the body, thus assisting to lower the center of gravity. Reduction of the upper limb contributes to the same end but to a smaller extent. In *H. africanus* the evidence is clear that the lengthening of the lower limb had already occurred and suggests that the upper limb was perhaps only slightly longer proportionately than it is in modern man. The most important reason for lengthening the lower limb, however, concerns quite a different point: the development of a speed-oriented propulsive mechanism, as is discussed elsewhere in this work.

Therefore, it seems clear that the major changes required to lower the center of gravity to the human position had already occurred in *H. africanus* Sts 14. Since these are a complexly interrelated set of changes, it is unlikely that they would all be so well expressed in this one individual if they were not generally present in the population.

This rather detailed evidence from the *H. africanus* material seems to me to demonstrate unequivocally that the whole complexly integrated system of changes in structure, proportion, and spatial arrangement required to produce the uniquely human method of balance control was fully present and operative in *H. africanus*. Indeed, there appear to be no grounds for supposing that it was any less effective in that form than it is in ourselves. The elongated anterior end of the iliac blade, with its own strengthening buttress, would not decrease the efficiency of the balance mechanism—in fact, it may have improved it very slightly.

With respect to *Paranthropus* less evidence in available. For example, nothing is known directly about lumbar curvature or whether the arms were relatively short or whether the lower limbs were relatively long. But it is clear that the reduction of iliac height had proceeded to a point that was at least very similar to that in *H. africanus* and significantly different from the condition in the living pongids. Also, both the acetabulo-cristal and acetabulo-spinous buttresses were well developed on the gluteal face of the ilium. The ilium was expanded backward as well as forward; hence there was a wide base for the lateral trunk muscles. The reduction in height of the ilium presumably means also that the sacrum had already widened. This was part of the compensatory mechanism that maintained birth canal size as the ilium reduced in height. This presumably also means that lumbar curvature had already developed. There is no actual evidence that indicates that the lateral balance mechanism was any differently developed from that in *H. africanus*. If the arms were relatively long and muscular and the lower limbs somewhat short, then the center of gravity would have been less advantageously placed than it is in the latter. Until some evidence of limb proportions is available one cannot be sure whether the two forms were completely alike in respect of lateral balance control.

(b) Capacity for Striding

The capacity to stride, which is essential for efficient bipedal locomotion, is a complex capacity with many contributing components. One of these is effective control of lateral balance; this has already been dealt with and will not be considered further. The aspect that will be considered here is the capacity to extend the femur, hence also the whole limb, past the vertical so that long steps can easily be taken.

An hypothesis proposed by Washburn and accepted by many workers—including myself, for a while—is that the chief difference between the inefficient bipedality of pongids and the efficient bipedality of man concerns the function of gluteus maximus. In pongids this muscle acts over the hip joint, so the hypothesis goes, and is thus an abductor of the thigh at the hip joint. In man its line of action is behind the hip joint, owing to the backward expansion of the iliac blade having carried the line of action of the muscle back of the hip joint; therefore, it functions as a powerful extensor of the thigh at the hip joint. This provides a well-developed source of power to assist the second half of a stride. This extra power in the later phases of the stride, which is not available to pongids, is regarded as the most significant difference between the bipedality of the two forms. On this view, there was a major change in the function of the muscle that occurred after crossing the threshold marked by the point where the line of function passed behind the hip joint when the animal was standing erect.

A major flaw in this hypothesis is the error of its basic assumption that gluteus maximus in pongids is essentially an abductor. Assuming similarity of origin and insertion for pongids and man, and considering the relative positions of these two points in pongids, the assumption appears fully justified. The origin, however, is not the same in pongids and in man; in particular, the chimpanzee and gorilla have a powerful and well-developed origin for the major portion of this muscle from the ischial tuberosity. The Asiatic pongids also have this arrangement, but the ischial origin is less well developed in the orang. The arrangement of this muscle is considered at length by Sigmon (Ph.D. thesis, 1969). The consequence of the powerful attachment to the ischial tuberosity is that the chief action of gluteus maximus in pongids is extension of the thigh at the hip. The major action of this muscle is thus the same in pongids and in man, hence the basis of this hypothesis is destroyed.

A further erosion of the hypothesis comes from the fact that it assumes cardinal importance of gluteus maximus action as a propulsive force during walking in man, and as a postural muscle—the trunk jackknifes if it is paralyzed. That walking was still entirely possible, with only minor compensatory trunk adjustment, with paralyzed gluteus maximus was already known to Duchenne (1867) a century ago. More recent studies have shown that during erect standing at ease this muscle is electrically silent (Joseph 1960) and that during normal walking on a flat surface its only consistent action is at the very beginning of the stance phase, and then primarily for purposes of stabilization (see, for example, Eberhart, Inman, and Bresler 1954). Recent studies of locomotion in man (e.g., Eberhart et al. 1954; Steindler 1955; Kondo 1960; Radcliffe 1962; Battye and Joseph 1966; and Basmajian 1967) confirm that, apart from the small amount of activity early in the stance phase just after heel strike—which is not essential—gluteus maximus serves primarily as a reserve power source to aid in moving the trunk in relation to gravity, as in climbing (e.g., up stairs or an inclined surface), bending, or straightening up from a kneeling or bent-over position, or in rapid locomotion where it is concerned with moving the limb.

Where then does the difference lie between pongid and hominid bipedal walking? At least three major factors appear to be involved.

(1) *Orientation of the pelvis.* When in the quadrupedal position, the pelvis of a pongid is so orientated that the ischial shaft is directed backward and downward. The hamstring muscles and the hamstringlike ones, such as gluteus maximus and part of adductor magnus, can then provide power to extend the femur at the hip joint until it is well past the vertical before near-parallelism of the ischial and femoral shafts robs them of capacity for further extension. When the same animal rears the trunk up and assumes erect posture, the pelvis is so orientated that the ischial shaft is directed straight down toward the ground. The extensors of the thigh, functioning exactly as they previously had, run out of power before the femur is fully vertical, again because of near-parallelism of the ischial and femoral shafts. Hence, only short steps can be taken; striding is not possible. In modern man, however, the pelvis is orientated in roughly the same manner when the individual is fully erect as it is in a pongid when it is in the quadrupedal position. This is part of the reason for lumbar curvature in man: to maintain the desirable pelvis orientation while having the spinal column erect. Hence, when the body is erect, the ischial shaft is directed backward and downward; therefore, the extensors of the thigh at the hip can extend the thigh past the vertical, just as a pongid can when in the quadrupedal position. The resulting capacity, in man, to move the thigh through a large arc is a very important component of the ability to stride.

In *H. africanus* well-developed lumbar curvature was present, and the relationship of the sacrum to the innominates was exactly as it is in man. We may conclude, therefore, that the pelvis was orientated as it is in modern man. This aspect of the capacity to stride was thus fully developed.

(2) *Relationship between hamstrings, moment arm, and lever length.* In the great apes the femur is relatively short; hence the lever length in this case, the distance from the acetabulum to the ground, is relatively short. The maximum moment arm of the muscles that extend the lower limb on the trunk is relatively long. That is, the distance from the center of the acetabulum to the end of the ischium is relatively long. The extensor muscles are relatively short but are comparatively powerfully developed. The consequence of this arrangement is that the extensors are capable of exerting a great deal of power; the specialization of this mechanism is more for power than for speed. Moreover, the greatest length of the moment arm, including that of gluteus maximus, occurs at the beginning of the stance phase. For all of these muscles the mechanical advantage reduces as the thigh moves back.

In modern man the lever length is proportionately longer than it is in pongids and the moment arm length is proportionately shorter. The effect of this is to approximately halve the ratio between the maximum moment arm length and the lever length in man as compared to the pongids. This arrangement results in less power but greater speed of action at the end of the lever. The longer lever length also means that steps are longer for the same angular displacement than would be true for a shorter lever. This mechanism is clearly less advantageous than the pongid one for a climber but is more advantageous for bipedal walking and running. The moment arm length of the hamstrings is greatest early in the stance phase, hence their mechanical advantage reduces as the thigh is extended. On the other hand, the moment arm length of gluteus maximus is greatest near the end of the stance phase, hence its mechanical advantage increases as the thigh moves back. Thus, as the mechanical advantage of the hamstrings reduces, that of gluteus maximus increases. The application of extensor power can thus be more even over the whole of the stance phase in man than in the pongids.

As we have seen, *H. africanus* already had the relatively long lower limb of modern man. According to the views expressed in the literature, however, it had a long apelike ischium. Usually such statements give no indication of what is meant by ischium length. But Washburn did define his use of the term as the distance between the acetabular margin and the nearer edge of the ischial tuberosity; that is, only a part of the ischium, the shank. I cannot see what functional meaning this definition of ischium length has.

The full length of the ischium has very real functional meaning, however, since it is also the maximum moment arm for the hamstrings. Comparison of the whole ischium with those of pongids and of modern man, as this study demonstrates, shows unequivocally that *H. africanus* has an ischium that is not only relatively distinctly shorter than that of the pongids but also is clearly shorter than the average relative length in modern man. This has been shown to be true by using the logarithmic ratio diagram method, by comparing it to femur length, and by comparing it to the greatest diameter of the body surface of the sacrum that articulates with the last lumbar vertebra. In all cases the *H. africanus* ischium is shown to be shorter even than is true of modern man. Because of the relatively small size of the ischial tuberosity, it seems likely that the hamstrings were not large and powerful muscles, as they are in the pongids, but were slender as in man.

We must therefore conclude that the hamstring, moment arm, and lever length complex in *H. africanus* was specialized for speed rather than power and for long steps to an extent at least equal to, and perhaps even greater than, the average condition in modern man.

(3) *The foot.* An important feature of effective use of long steps in walking or running is the push-off given by the foot at the end of one step and the start of the next. A very flexible, rather handlike foot is not very effective for this purpose. The more compact, less flexible, and well-arched foot of man, with its fully adducted and large hallux, is very effective in providing a springy push-off, assisted by the powerful calf muscles.

No *H. africanus* foot is available from South Africa; hence, there is no actual direct evidence of what the foot was like that was associated with the sophisticated locomotor apparatus of this form as known from Sterkfontein. If the foot from Olduvai Bed I does belong to *H.* "*habilis*," as is generally supposed, and if this form is an *H. africanus*, which I believe is so for reasons not concerned with the foot, then the *H. africanus* foot is of the sort that would be well suited to provide the sort of push-off found in man. Unfortunately, however, the true owner of this foot is not known for certain; hence, the above is no more than an attractive possibility. On the other hand, since the foot is so extremely important a part of the locomotor apparatus, and in view of the advanced and sophisticated nature of the rest of the *H. africanus* locomotor apparatus, including many details of femur and knee joint dealt with in the body of this work but not repeated here, it seems highly improbable that the *H. africanus* foot could have been any less manlike than that from Olduvai.

Hence, although secure evidence of the foot is missing, the available evidence clearly indicates that the pelvic and lower limb anatomy was such that locomotor specialization was primarily for speed rather than power of limb movement and allowed long steps to be used.

In the case of *Paranthropus* the evidence again is less complete. There is no direct evidence of whether lumbar curvature was present, nor do we have a sacrum. Some evidence is available in the form of the auricular area, however, to suggest that the sacrum was related to the innominate in the same way as it was in *H. africanus*. One auricular area is very poorly preserved, however, and the other is incomplete; hence this evidence is suggestive but not good. But the fact that the ilium was shortened and expanded backward implies that the sacrum had already widened and was orientated in such a way that lumbar curvature probably was already in existence.

There was a well-developed anterior inferior iliac spine, and both the femora, which are of two different individuals, have well-developed femoral tubercles. This evidence indicates that the ilio-femoral ligament was in the form of a Y ligament.

The ischium length can be determined accurately for the Swartkrans innominate, and it is long. By the modified logarithmic ratio diagram method it can be shown to be appreciably longer proportionately than that of *H. africanus* or the average for modern man, falling very close to the pongids. Since we do not have a femur length estimate for this form, it is not possible to determine whether this relatively long and apelike ischium was associated with a long lower limb, comparable to that of man or *H. africanus*, or a short one such as is found in the great apes. Only in the event that the lower limb was actually considerably longer proportionately than that of modern man or of *H. africanus* could the conclusion be avoided that the *Paranthropus* mechanism was specialized for greater power and less speed than was true of *H. africanus*. Since so elongate a limb is not likely to have been present, we must conclude that, at best, *Paranthropus* was intermediate between *H. africanus* and modern man, on the one hand, and the living pongids on the other with respect to this character. If the *Paranthropus* femur was short, as possibly it was because of Schultz's finding that primate femora with thick shafts are also relatively short, then it would have to be ranged with the great apes regarding propulsive mechanism specialization. Because the ilium was more or less of the *H. africanus* type, with evidence of a well-developed capacity for lateral balance control, it seems improbable that the limb extensor mechanism would have been purely of the pongid type. I suspect, therefore, that the lower limb was at least a little elongated compared to the great ape condition but not as much as in *H. africanus* or man.

The *Paranthropus* condition was thus probably a compromise between the essentially quadrupedal climber pattern of the great apes and the efficiently bipedal pattern of *H. africanus* and modern man. This implies that it combined bipedality of a kind that was somewhat less efficient than that of *H. africanus* with some locomotor activity that required power use of the limb extensor apparatus. The most likely activity in the circumstances would seem to have been climbing trees.

There is a small amount of information available about the foot. This slim evidence, derived from a talus, suggests a relatively mobile, flexible foot that perhaps had an incompletely adducted hallux. As in the case of the lower limb extensor apparatus, there is here a suggestion of a compromise, in this case between the fully human foot type and that of a great ape such as a chimpanzee. This is consistent with the conclusion reached above.

Synthesis: *Homo africanus* and *Paranthropus*

The foregoing brief survey of some of the broader conclusions that have emerged from the study of the early hominid postcranial material supports the conclusion drawn from the cranial material that *H. africanus* and *Paranthropus* were not similarly adapted. There is nothing in the postcranial material available, or in its interpretation as set out in this work, that contradicts or renders improbable the conclusions based on the cranial material. Instead, the latter conclusions are amplified and supported.

The picture of *H. africanus* that emerges is fairly clear. It had essentially the same body proportions as has modern man and evidently was, for all practical purposes, basically as well adapted to erectly bipedal posture and locomotion as we are, although some slight improvements in efficiency still had to be made here and there. Adaptation to this locomotor habit was so advanced that it seems improbable that *H. africanus* could have been noticeably more effective at using quadrupedal posture than we are. It seems to me that even occasional knuckle-walking is improbable in the extreme, not only because of the generally high level of adaptation to erect bipedalism but also because of the lengthened lower limb.

The animal was small and gracile and was probably capable of running and walking fast, as the anatomy of its propulsive mechanism clearly suggests. Those anatomical features that allowed it to have so well developed a mechanism for static and dynamic balance control, as well as its posture and limb proportions, would also have given it considerable capacity to use the body for such actions as wielding tools or weapons, throwing, and so on.

The evidence from the skull indicates that it was an omnivore of much the same sort as earlier and later forms of true man. The type of brain expansion associated with the emergence of man seems already to have been under way. The geological evidence indicates that in the late Lower

Pleistocene it was essentially a dweller in rather dry plains or upland conditions. Putting together the locomotor, dietary, and climatic evidence suggests that *H. africanus* was a hunter-gatherer living in fairly dry regions.

Almost certainly it was at least a fairly skilled tool-user that gradually became, at least in some demes such as in East Africa, a maker of tools who used stone, among other materials. Communication was probably more efficient than is true of any living nonhominid animals, but we have no means of knowing whether there was anything that could be regarded as a true language.

The one suggestion of a not very human structure is in the scapula and perhaps in a slightly long arm. Unfortunately, the one available specimen is so damaged and incomplete that the picture is anything but clear, and it is not known what, if any, functional differences were involved.

The adaptive pattern seems to have been in principle the same as that of man. Indeed, it seems to me that this *is* man, anatomically, ecologically, and behaviorally; for which reason I originally proposed that the generic name "*Australopithecus*" (*sensu stricto*) be sunk into *Homo* and have now adopted that proposal in practice. This is a logical consequence of the belief that a genus represents a clearly defined adaptive zone or way of life, within which various species can occur that represent no more than variations in detail on the basic adaptive theme. The basic adaptation of *H. africanus* seems to me to be the same as that of man, hence the two should be in the same genus.

It is worth digressing at this point from the main theme to take note of views in the literature that are relevant to the above conclusion. The first is the firm objection of Simons (1967) to the point of view expressed in the preceding paragraph:

> However, if Robinson means to suggest . . . that time-successive genera . . . include only succeeding species in the same "adaptive zone," we are then making adaptive zone a new measure for defining genera. Other taxonomists have seldom considered this a factor in generic definition, and, inasmuch as a full knowledge of the way in which extinct species occupied their adaptive zones can seldom be obtained, the criterion is vague and is simply not applicable to most fossils.

That no objective criteria exist for recognizing the genus category is well recognized; the species is the only rank in the hierarchy that comes close to objectivity based on the presence or absence of genetic isolation. Once clear genetic isolation is present, it becomes a matter of personal interpretation what level of the hierarchy is being dealt with. But the modern taxonomist works in an evolutionary context—he is classifying the products of evolutionary activity, regardless of his views on the relative merits of traditional versus numerical taxonomy—and it is difficult to avoid the conclusion that the one obligatory rank above the species, the genus, rests on an adaptive basis. Insofar as a classification is reflecting any biological reality, it must surely also be reflecting adaptation. It is commonly held that a genus should, if possible, be monophyletic. If so, it contains a number of species that speciated from a common stock. Had any of the species diverged adaptively from the others in a very obvious manner, it would no longer be placed in the same genus because it would be too different. Even if a genus is polyphyletic, the species assembled in it were put there because they resemble each other more than any resembles species in other genera. Again, this implies similarity of adaptation.

That this is, indeed, how some able taxonomists view the genus is clear from the literature. Mayr, Linsley, and Usinger (1953), in a widely used textbook on taxonomy, write (pp. 49–50):

> The genus, as seen by the evolutionist, is a group of species that has descended from a common ancestor. It is a phylogenetic unit. . . . The genus, however, has a deeper significance. Upon closer examination, it is usually found that all the species of a genus occupy a more or less well-defined ecological niche. The genus is thus a group of species adapted for a particular way of life. . . . On this theoretical basis, it is probable that all generic characters are either adaptive or correlated with adaptive characters. Lack (1947) has made a particularly convincing analysis of the adaptive significance of the genera of Galapagos finches.

This book has been revised (Mayr 1969), and the quoted passage has been rewritten. But it still expresses exactly the same point of view and is even a bit more strongly worded, e.g., "it is usually found" is now "Almost invariably . . .".

Another classic text of the rather small number of general texts on the principles of classification—that of Simpson (1961)—does not deal as specifically with the genus as such, but the discussion of higher categories expresses the same basic idea:

> Breadth and retention of adaptation form a continuous scale which can be related to categories and put in those terms. Thus there are species-specific, genus-specific, family-specific, and so on even to

> kingdom-specific adaptations and corresponding somatic characteristics, although for various reasons these become somewhat more liable to exceptions as the taxonomic scale is ascended.

All of this is not to say that Simons is wrong in holding a non-adaptive view of the genus, since every taxonomist is entirely at liberty to formulate his own taxonomic framework. The International Code of Zoological Nomenclature governs nomenclature alone and specifically does not concern itself with the reasons for a taxonomic decision, as is clearly and specifically stated in the Preamble to the 1961 edition of the Code. I wish only to point out that I believe Simons to be mistaken in asserting that "other taxonomists have seldom considered this a factor in generic definition" when referring to adaptation. The view of the genus to which I subscribe is in fact a very commonly used one.

Simons is also mistaken in writing that I sidestepped the serious problem of genetic continuity by placing two forms in one genus because I believed the one to be ancestral to the other ("*Australopithecus*" and *Homo* in Robinson 1967). In that paper I considered the problem of within-lineage and between-lineage classification at some length. Moreover, it is clear in that paper, as it is in the present work, that my reason for advocating the sinking of "*Australopithecus*" (*sensu stricto*) into *Homo* is that the former consists of a species that has the same adaptive characters basically as do species in the genus *Homo* and should therefore be placed in that genus. On the other hand, I believe *Paranthropus* to be the parent lineage of *H. africanus* —that is, that the first population of *H. africanus* was a genetically remodeled local population of *Paranthropus*. But since I also believe that the basic adaptive pattern of *Paranthropus* was very different from that of *Homo*, I do not advocate that *Paranthropus* be sunk into *Homo* in spite of believing that a population of *Paranthropus* was directly ancestral to *H. africanus*. The logic underlying this view of the genus is fully consistent with, and a consequence of, the beliefs that (a) evolution is a fact; species are products of evolutionary activity; and (b) every natural population possesses some degree of adaptedness to its natural habitat.

The views of Campbell (1968) concerning "*Australopithecus*" and *Homo* are relevant here. He believes that adaptive differences between these two groups are sufficiently clear to warrant retention of two genera. He believes the adaptive zone of "*Australopithecus*" to be characterized by the following:

> *Australopithecus* is a bipedal hominid living in open savannah plains of East and South Africa, on vegetables and small animals, and

> other scavenged food. Olduvai has revealed remains of birds, lizards and rodents, pebble tools and a windbreak, alongside the Bed I remains of the *Homo habilis* and *Zinjanthropus boisei* fossils, which we consider should be classified in the genus *Australopithecus*.

Before making this statement, Campbell expresses the opinion that it is not clear to him whether the gracile and robust early hominids represent two or a single species. He believes the two populations "... were more or less contemporary and more or less sympatric for about 1 million years ..." If they were sympatric and recognizably distinct for so long, there are three possibilities: (1) they represent dimorphic forms of the same species, i.e., males and females; (2) they were subspecies or racial variants of the same species; or (3) they were genetically isolated lineages, i.e., two good species at least. The first alternative can be eliminated on the basis of the South African evidence. It is extremely unlikely that at Sterkfontein over an accumulation period of probably hundreds of thousands of years only females existed or only female remains were preserved while at nearby Swartkrans males alone existed or were preserved for thousands of years. The dry period deposits—Taung, Sterkfontein, and Makapansgat—on this interpretation, would have had, or preserved, only females over a very long period of time; then followed the two wetter period deposits—Swartkrans and Kromdraai—with males only. The second alternative, that the two forms were subspecies of the same species, is also highly improbable. By definition, subspecies are not sympatric except in very unusual circumstances such as in the overlapping ends of a ring distribution or in the case of modern man where advanced means of transport can bring different racial variants together. The former could not be involved since one is unlikely to have the overlapping ends of a ring distribution occupying the major portion of the African continent. The other possibility is also unlikely; even had transport means existed to bring different races together, since there would have been genetic continuity between them the distinctions between the two could hardly have been maintained undiminished for a million years or more if they were sympatric. Of the three alternatives only the third is reasonable—that two genetically isolated lineages existed. The distribution in the South African sites is reasonably interpreted as a reflection of ecological difference between the two since both were present elsewhere in Africa during that period.

Turning now to Campbell's reasons for retaining both *Homo* and "*Australopithecus*": if one starts with the assumption that all the early hominid material must be placed in a single genus, then it follows logically that that genus is distinguishable from *Homo*. The features that make it distinct, however, are due to *Paranthropus*. If the analysis of the anatomical and functional aspects of the cranial and postcranial skeleton presented in this book has any validity, then it is clear that if the robust and gracile forms are compared with each other and with *Homo*, the robust form can easily be differentiated from both the gracile form and *Homo*. The distinction between the latter two, however, is of a wholly different order; indeed, the former grades fairly smoothly into the latter, but there is no intergrading between *Paranthropus* and either "*Australopithecus*" or *Homo*. So I agree with Campbell that the gracile and robust early hominids and *Homo* all cannot reasonably be placed in the same genus; the odd man out in this trio is *Paranthropus*.

Of some relevance here is Tobias's suggestion (e.g., 1968) that a species may reasonably be placed in two different genera at once. He proposes this as a possible solution for coping with the fact that *Homo* "*habilis*" closely resembles *Australopithecus africanus*; thus, *Australopithecus/Homo habilis*. It would seem clear enough that a species cannot logically be placed in two validly different genera simultaneously any more than one child can simultaneously have two sets of biological parents. The above situation does not suggest to me the need for nomenclatural reform or innovation. Instead, it suggests that the two genera providing the horns of the dilemma are not validly distinct; that the very need for a hybrid nomenclature is evidence that the two genera are not soundly based. The problem completely disappears with respect to the known specimens when "*Australopithecus*" (*sensu stricto*) is sunk into *Homo* and *Paranthropus* retained as a valid genus. The different but related difficulties of both Campbell and Tobias disappear once it is recognized that "*Australopithecus*" and *Homo* belong to the same adaptive zone—efficiently bipedal, omnivorous hunter-gatherers who have cultural facility as a basic part of their adaptive pattern—while *Paranthropus* occupied a quite different adaptive zone. Of course, it is obvious that where there is genetic continuity through time, a very full fossil record will still provide cases that are awkward nomenclaturally simply because the bases of nomenclature are discrete, discontinuous, nonoverlapping taxonomic units while such discontinuity does not exist within lineages but only between different lineages. If careful attention is paid to differentiating between sequential taxa in the same lineage and taxa that belong to different lineages (see Robinson 1967), and if genera are based as firmly as possible in terms of well-defined adaptive zones, most of the difficulties of the type here discussed disappear.

Returning now to the main theme after this digression, *Paranthropus* seems clearly not to fit the adaptive pattern of the *Homo* lineage. In every respect it appears to represent a creature that was less advanced in the direction of man than was *H. africanus*. It apparently was an herbivore that still had this specialization at a time when *H. erectus* already was present, with a large brain, in the same region. There is no evidence in *Paranthropus* of the human type of brain expansion; hence the absence of a forehead and the presence of a brain evidently not significantly larger than that found in the smaller *H. africanus*. Although it clearly must have had appreciable capability as an erect biped, it was not as effectively specialized in this direction as was *H. africanus* since it still had a long ischium and therefore still used its propulsive mechanism at least partly in a power-specialized manner. This suggests a compromise adaptation—perhaps not completely efficient bipedality on the ground, coupled with spending some time in the trees. The small amount of evidence of the hand suggests that this was powerful but elongate with a short thumb. The evidence of the foot suggests mobility and flexibility and perhaps an incompletely adducted great toe—also a compromise between ape and human characteristics. Such hands and feet would be consistent with some climbing activity. The dietary habit is also consistent with such an interpretation. Climatic evidence, where it is clear, indicates that *Paranthropus* lived in reasonably wet conditions. It may be argued that this is not consistent with the Olduvai evidence that indicates that Bed I times were rather dry. But Hay (1965) has concluded that in spite of being rather dry there nevertheless was a fair amount of flowing water coming from higher ground. It seems likely that then, as today, high mountains had considerably different climates on their slopes than obtained on the plains nearby. It may well be, therefore, that the general Olduvai region contained areas of wetter climate interspersed in the generally drier plains.

Paranthropus was evidently a rather apelike form in its way of life and was stably adapted to that way of life. It probably lived in woodland conditions; that is, a combination of wooded and grassy conditions. Separate generic status from *H. africanus* seems suitable in order to reflect the quite different adaptive patterns of the two, which involve differences of anatomy, ecology, and behavior.

The rather vague suggestion from the one very incomplete *H. africanus* scapula known, that the shoulder of this form may have had some quadrupedal climber characteristics, is of interest. It certainly has not been previously supposed that *H. africanus* may so recently have had an ancestor that regularly still climbed trees. If the conclusion that *Paranthropus* did so is correct, and if this form was indeed ancestral to *H. africanus*, then it would not be unreasonable to find a partially apelike shoulder and upper limb in a form otherwise transformed into an efficient erect biped. This state of affairs would soon change, however, once tree-climbing was abandoned and cultural activity actively engaged in during the change to a plains-dwelling, hunter-gatherer way of life, as is postulated here for *H. africanus*.

Paranthropus seems to be a suitable ancestral type for *H. africanus;* elsewhere (Robinson 1962 and 1963, for example) I have presented an interpretation of how desiccation in Africa could have brought about a change to omnivorous diet along with the emergence of culture under the control of natural selection, to give rise to the culture-bearing *Homo* line in drier regions, while *Paranthropus* persisted in wetter islands in East and Central Africa. Subsequently, with the Pleistocene climatic amelioration, *Paranthropus* spread again, and the two lineages could be found occupying the same regions. During this phase *Paranthropus* is found at Omo, Olduvai, Swartkrans, and at Sangiran, Java, living in the same region as representatives of the *Homo* lineage. The newer evidence from Omo and the Baringo region has extended the hominid record much further back in time to probably about four million years (Howell 1969). There is good evidence of both *Paranthropus* and the gracile form through almost all of this period. The information is better for *Paranthropus*, of which some mandibles with teeth are known. The teeth of the gracile form do not appear to be distinguishable from those of *H. africanus*. The position concerning cultural activity in relation to the Omo material is not yet clear; no stone tools have yet been reported from there in direct association with early hominid material. The investigations in this area are still in an early stage, however, and further work is needed before the situation is clarified. By early Bed I time at Olduvai the *Homo* line appears to have been well launched into active cultural evolution.

Manifestly, it is not a clear-cut case which hominid made the tools since representatives of both forms are known from before the time of the first tools. It seems to me that on balance the evidence favors the view that *Paranthropus* was not actively culture-bearing and that all of the known cultural material was the work of the *Homo* lineage. The reasons for this conclusion are these:

1. Analysis of the dentition and masticatory apparatus indicates that association of small canines with cultural activity in *Paranthropus* is very improbable since relating the two in an inverse causal relationship leads to the improbable conclusion that *Paranthropus* was more actively culture-bearing than any phase of the *Homo* lineage, including the present.
2. Anatomically, *Paranthropus* is demonstrably less like either *H. erectus* or *H. sapiens* than is *H. africanus*. The total evidence suggests a creature that was distinctly more apelike in anatomy, ecology, and behavior than any known phase of the *Homo* lineage and included herbivorous diet, incomplete development of erect posture, and incomplete emancipation from the trees.
3. There is now evidence of *Paranthropus* over some millions of years and no indication during that time of significant change in its

characteristics. During this time there was active modification going on in the *Homo* lineage and also in the stone tool tradition.

4. Cultural activity represents a powerful means of adaptation. It seems improbable that two forms, one of which was distinctly more manlike than the other, could both have been developing it and yet be sympatric over a long period of time. The more advanced one would presumably have displaced or killed off the less advanced one.
5. We know that the *Homo* line did in fact become actively culture-bearing.

In other words, that the *Homo* line had culture is clear; the evidence in favor of *Paranthropus* having had it, or been likely to have had it, is at best dubious.

Subsequently, in what appears to have been the early Middle Pleistocene, *Paranthropus* disappears from the record, evidently by extinction. *H. africanus* disappears from the record a little earlier, in the late Lower Pleistocene. On the present record there appears to be no reasonably clear-cut case of sympatry of *H. africanus* and *H. erectus*, though there is of *Paranthropus* and *H. erectus*. The evidence suggests that *H. africanus* and *H. erectus* grade into each other; there appear to be no grounds for making an absolute distinction between them, as can be made between *Paranthropus* and either of them. Cultural activity appears in the record in association with the *Homo* line before there is evidence of brain capacity in excess of that found among modern great apes. *H. africanus* disappears from the record at about the time that *H. erectus* appears in it. These lines of reasoning suggest that the *Homo* line was a genuine ancestral-descendant lineage. Since *H. africanus* was quite widely spread in Africa—there is no clear evidence of it outside of Africa, as there is for *Paranthropus*—it is unreasonable to expect that every deme at the same time transformed into *H. erectus*. It is more likely that one deme, probably in East Africa, made the transition first in conditions most suitable for evolutionary change. As the new *H. erectus* spread it would have displaced or killed off less progressive demes by being competitively superior. Very short-term sympatry of *H. erectus* and *H. africanus* is entirely probable, it seems to me, but not long-term sympatry. Indeed, if we had a full fossil record from Sterkfontein Lower Breccia time through to early Swartkrans time, I believe that we would have seen *H. erectus* move in and destroy *H. africanus*, after which *Paranthropus* arrived and lived sympatrically with *H. erectus*.

Deriving *H. africanus* from the *Paranthropus* lineage seems to me reasonable. In every definable respect *Paranthropus* is less manlike than *H. africanus*, except for the canine teeth—but even in that respect *Paranthropus* is not intermediate between *H. africanus* and either *H. erectus* or *H. sapiens*. Moreover, *Paranthropus* already had a compact hominid dentition and a hominid occiput and had completed most of the transition to erect posture. By economy of hypothesis, the principle of William of Ockham's razor, it is preferable to assume that these features, and in combination, arose once rather than two or more times, unless compelling evidence to the contrary is uncovered. The sequence of the changes involved will be discussed later in this chapter.

The view that *Paranthropus* could be the ancestor of *H. africanus* has not been favorably received. This seems to be due primarily to the belief that *Paranthropus* is very specialized. Certainly it is the most aberrant hominid known at present. This does not exclude it, however, from ancestry to later hominids; presumably the ancestor to the hominids was not a typical hominid. Tobias (1967) has argued that *Paranthropus* is specialized in having massive cheek teeth and supporting structures. The ratio diagram analysis discussed earlier shows, however, that in proportions along the tooth row *Paranthropus* does not differ significantly from *H. africanus* or the pongids with respect to almost the whole cheek tooth row, P_4–M_3. Given that it is a large animal—hence the absolute size of the cheek teeth is somewhat greater than in *H. africanus*—its cheek teeth are not out of character with those of the great apes or other early hominids. The specialization does not lie in this distinction but in the small anterior teeth. According to Tobias, stripping *Paranthropus* of the dental specialization and the skull features associated with it would leave a creature like *H. africanus*. This point of view does not seem to me to be valid. Increasing the canine size would increase the similarity to *H. africanus* in that respect, but there would still be almost all of the more apelike skull features of *Paranthropus*, including a low, forehead-less braincase, along with other less manlike features such as incomplete adaptation to erect posture, herbivorous diet, and so on.

From the evidence now available from East Africa, it seems probable that *Paranthropus* was becoming smaller from the late Pliocene to the Middle Pleistocene. Projecting this trend backward in time, one may thus postulate that the forerunner of *Paranthropus* was a larger animal that had an anterior dentition somewhat between the proportionate size that we know and that typical of the great apes. The latter point is based on the fact that canines that are proportionately larger than those of hominids are usual in non-hominid primates. Interestingly enough, a fossil form with just those characteristics is in fact known—*Gigantopithecus*. A modified ratio diagram based on the teeth in the three mandibles of this form from China (Woo's data, 1962) indicates the close similarity of the three (Fig. 110). All have M_1–M_3 curves closely fitting those of the great apes, *Paranthropus*, *H. africanus*, and

H. erectus. The canines fit in proportionate size with those of *H. africanus* and *H. erectus* and are roughly midway between those of the pongids and of *Paranthropus*. As interesting is the fact that P_3 and P_4 are proportionately very large and of roughly equal size: in mandibles II and III, P_3 is slightly smaller than the corresponding P_4; in the other it is slightly larger. This ratio diagram profile fits extraordinarily well an intermediate stage in a sequence of change from an animal with a pongid ratio diagram profile to one with a *Paranthropus* profile. This is particularly striking with respect to the large premolars, with P_3 tending to become smaller than P_4. Members of the *Homo* lineage do not form a stage of such a sequence of change. With the recent discovery of a Middle Pliocene mandible of this form from India, *Giganto-pithecus bilaspurensis*, there is now a representative of this lineage that is evidently earlier than the earliest known early hominid.

Simons (1970), following suggestions of Jolly (1970), concludes that *Gigantopithecus* was "granivorous"—ate grass seeds, stems, rhizomes, etc.—and lived in an open savannah-woodland habitat. This corresponds closely with what I have long believed of *Paranthropus*, except that I have not stressed grass seeds in the diet and believe that berries and leaves and perhaps nuts probably were also part of the diet. On balance, it seems to me that *Giganto-pithecus* is a most attractive potential ancestor for *Paranthropus* and, through the latter, of the other hominids also. It is the fashion at present to believe *Ramapithecus* to be ancestral to the hominids, and this is advanced as grounds for the belief that *H. africanus* is nearer to the ancestral hominid since *Rama-pithecus* is small and is more readily seen as an ancestor of *H. africanus* than of *Paranthropus*. Nevertheless, examination of the *Ramapithecus* material, including *Kenyapithecus wickeri*, has convinced me that there is not one specifically hominid feature known in this form and that in all respects it is more pongid than hominid. I do not believe that it is a hominid or that it was ancestral to hominids. The reasons for this view will be dealt with more fully elsewhere; let it suffice here to say that the canine, upper incisor, P_3, the upper precanine diastema, and the nature of the wear on the upper and lower teeth all indicate closer resemblance to pongids than to hominids.

More specimens are needed of these various forms before certainty can be reached about their affinities. To me, by far the most reasonable hypothesis at present is that *Gigantopithecus* represents a stream of evolution in which a shift was made from the forests to savannah-woodland habitat that also involved a shift in diet away from the pongid type of herbivorousness. These were large animals, hence they were probably largely immune from predators; for this reason, and because the diet did not require heavy use of them, the anterior teeth reduced in size. One line of this stream gave rise to the Pleistocene *Gigantopithecus* of China; another, perhaps in Africa but not necessarily, gave rise to *Paranthropus*. Probably before the canines had reduced completely, speciation in *Paranthropus* gave rise to *H. africanus*.

The Origin and Evolution of Erect Posture

The evidence concerning the anatomy of the postcranial skeleton of *H. africanus* and of *Paranthropus* that is presented in this book throws some light on the evolution of erect posture from the background of a broadly apelike grade of organization. The known fossil material does not include the earliest stages of the change. By the time *H. africanus* as now known was in existence the process was complete except for minor refinements. *Paranthropus* apparently represents a stage that was not as advanced as that of *H. africanus*, but even there some of the major steps to erect bipedality had already been taken.

According to a widely held current opinion, the process of remodeling the locomotor apparatus was set into motion by the backward expansion of the iliac blade, which caused the line of action of gluteus maximus to come to lie behind the hip joint when the individual was standing erect. The crossing of this threshold brought a different selection pattern into existence that rapidly reorganized the locomotor apparatus into bipedal form.

As we have seen, this explanation is certainly incorrect. There appears to have been no such threshold or sharp changes in gluteus maximus function. Instead, a series of gradual and smooth modifications of proportion and orientation seems to have brought about the transformation. Gluteus medius was apparently the only muscle that changed in function: in pongids it is primarily an extensor of the thigh while in hominids it is primarily an abductor of the thigh.

Analysis of the postcranial material of the early hominids leads me to suppose that the change to the erectly bipedal type of skeleton was initiated by a tendency to shift downward the center of gravity of the body. A number of reasons lead me to this conclusion. As already noted, the high and forward position of the center of gravity of a pongid when standing erect is very disadvantageous because it makes the animal use a great deal of muscular energy to maintain the erect position. It follows that if erect posture came into increasingly frequent use in such a creature, natural selection would increasingly favor reduction of the heavy energy cost involved. An important effect of such selection is likely to have been reduction of iliac height. This seems a likely first point of attack, as it were, because (1) a long ilium contributes to the raised position of the center of gravity, (2) a long ilium produces an unstable situation in the erect position because the points of weight support of the vertebral column by the ilia are relatively high above the hinge axis of the trunk on the limbs, (3) the most conspicuous difference between pongid and hominid pelvises is the much reduced iliac height in hominids, and (4) the height reduction of the ilium occurred in the earlier stages of the evolution of erect posture because it was already complete in *Paranthropus* even though other important phases (reduction of ischium length, elongation of the lower limb) had either not yet started or were still incomplete. Since the ilium is proportionately shorter in hominids than in any other higher primates, it is likely that the ancestors of the hominids also had relatively long ilia, though not necessarily as elongate as in the modern pongids.

The essential point in the hypothesis of how erect posture evolved, which is being advanced here, is that erect posture came to be used with increasing frequency and as a consequence brought into action selection favoring the lowering of the center of gravity. One may speculate that, in a savannah-woodland setting with tall grass, standing erect would have very real advantage for observational purposes. This sort of behavior is common among many different animals, and anyone with experience of these conditions in Africa will have observed such behavior many times. Keeping constantly aware of what is going on about them is the price of staying alive for most animals. Such behavior would be especially advantageous to a higher primate with its relatively poorly developed sense of smell and heavy reliance on its visual sense. Using the hands for getting food from shrubs and the lower branches of trees, or carrying food, could have contributed to the advantage of erect posture. Any one of these, and other, activities is not likely to have been sufficient alone to have caused frequent enough use of erect posture for selection to have improved its efficiency. That is to say, it is probably an error to look for *the* reason why erect posture came to be used. There are many reasons why erect posture would be used, and contribute to survival ability, in woodland-savannah conditions. The fact that it was frequently used is more important than why it was used.

Another important aspect is the absence of strong selection for some other locomotor habit. For example, chimpanzees use erect posture but the demands for quadrupedal-climber activity are more important, therefore selection maintains the mechanisms suitable for that purpose. Another relevant point may have been competition. A number of different nonmonkey higher primates were in existence in Africa during the Miocene and Pliocene. Perhaps competition with forms better adapted to forest life pushed the hominid ancestors into the grassy woodland habitats that were emerging due to the slow drying of much of the African continent during the later Miocene and Pliocene. The shrinking forests and expanding savannah-woodland habitats, on the one hand, would have increased competition for space and food in the forests and, on the other hand, provided new opportunities in the savannah-woodland regions.

The quadrupedal-climber locomotor habit, evidently widely used among nonmonkey higher primates, includes moderate capacity to use erect posture. Presumably the hominid ancestors will also have had that capacity. All that would be needed, then, is increased stimulus to use that capacity; no matter what reasons led to increased use of erect posture, the mere fact that it was being used frequently would bring into action selection favoring lowering of the center of gravity. As suggested, probably erect posture was used for a variety of reasons, no one of which is the key to understanding why erectly bipedal posture came into regular use. The shift to grassy woodland conditions was probably the most significant factor in causing erect posture to be used more frequently. As Kortlandt (1962) and Van Lawick-Goodall (1968) have observed, chimpanzees that live in more open conditions use erect posture more frequently than those in more dense forest.

Once erect posture came to be used more frequently, then, a pattern of selection favoring the lowering of the center of gravity would come into play, one consequence of which would be reduction of iliac height. Reduction of the height of the ilium would have far-reaching consequences. Very soon such a trend would have been met by counter-selection to maintain birth canal size in females. About the only way for both requirements to be met would be for the ilium to expand backward as it reduced in height and for the sacrum to expand in breadth. Both of these changes are readily apparent in *H. africanus*. Indeed, the expansion of the sacrum is especially impressive since the sacrum is proportionately as wide as is that of modern man; but, because the vertebral bodies are relatively small in this form, the relatively great width was achieved by the development of proportionately very large lateral masses. The backward expansion of the iliac blade assisted in moving the center of gravity backward, which was also desirable for efficient use of the trunk in the erect position. The latter process would have been assisted greatly and easily by the rotation of the sacrum in relation to the ilium so that the spinal column became vertical. This, however, would have caused obstruction of the birth canal by the coccyx. On the other hand, the pelvis could not rotate as a whole on the hinge axis to get the spinal column erect since, as we have seen, this would result in inability to extend the femur past the vertical because the ischium would then point straight downward. The latter is a prime contributory factor to inefficient bipedal walking in pongids. The compromise achieved was that the pelvis remained orientated almost as it had been, the sacrum actually rotated a little in the opposite direction to that required to make the spinal column erect and straight, and strong lumbar curvature developed in getting the spinal column erect.

These seem to have been the essential skeletal changes involved in the initial readjustment from quadrupedality to erect bipedality. It seems probable that, once started, any change in the direction of erect bipedality would necessarily have gone this far. One can assume that a necessary part of this change was the establishment of a proper lateral balance control mechanism. The expanded iliac blades that were more or less laterally situated, at least in part because of sacrum expansion, would have placed the lateral trunk muscles in a position to function as they do in man. The two gluteal muscles on the hip similarly would have been placed in a manner that allowed the human type of pelvic balance control to operate. Hence, it is to be expected that the acetabulo-cristal buttress would have come into existence then also. It is to be expected that the medialward flexure of the anterior part of the iliac blade would not yet have occurred, as is indicated by the pelvises of both *H. africanus* and *Paranthropus*. This change represents a later refinement. The evidence from the latter specimens indicates that this region was protuberant; hence, we can conclude that it is probable that the acetabulo-spinous buttress came into existence at the same time as the acetabulo-cristal buttress.

This entire complex of changes can thus be regarded as the necessary basic remodeling involved in the establishment of erect bipedality. At this stage, as is suggested by *Paranthropus*, the ischium was still of ape length and the foot was only partly modified for bipedality and was still partly adapted for climbing. It is probable, in my opinion, that the lower limb was not as elongated as it is in *H. africanus* and in modern man, but more so than it is in the great apes. The ability to extend the femur past the vertical will have been present because of the orientation of the pelvis, but the long ischium, moderately short lower limb, and compromise foot would have meant that the efficient striding of *H. africanus* and modern man was not yet present. Bipedal locomotion, however, will have been distinctly more efficient than that of pongids. *Paranthropus* evidently was in this stage and, apparently, became extinct as a lineage in the Middle Pleistocene without significant change. That selection still maintained a power-oriented propulsive mechanism probably means that, like modern pongids, *Paranthropus* climbed trees to sleep at night and perhaps also at other times for protection or feeding.

The next major stage, which involved much less change than the previous one, was the modification of the proportions of the hamstring moment arm and lever lengths to bring the propulsive mechanism into the state of specialization for speed, rather than power, found in modern man. This evidently involved also some reduction of the robustness of the hamstrings. Longer steps would then have been possible with the longer lower limb, and balance would have been improved by the elongation of the lower limb, thus lowering still further the center of gravity. It is difficult to avoid the conclusion that the essentially human type of foot, such as that from Olduvai Bed I, was achieved as part of the process of developing effective striding in this stage.

This is the stage of *H. africanus*, a very effective biped. These changes presumably developed as part of the change to a plains-dwelling, hunter-gatherer way of life.

The further evolution of the locomotor apparatus since the *H. africanus* stage has involved no major changes but only relatively minor refinements. The medialward bending of the anterior part of the iliac blade improved the mechanical efficiency of the anterior part of the false pelvis; the acetabulo-spinous buttress disappeared and the acetabulo-cristal buttress became more conspicuous as a result. The outward lean of the ilia was reduced slightly. The femoral head and acetabulum sizes increased, which reduced the weight stress per unit area of bearing surface, and the acetabular wall increased in thickness. The mechanical efficiency of the sacroiliac joint was similarly improved by an expansion in its surface area. These changes were probably associated with, but were not entirely due to, increasing size and weight of the body. Part of this change was an increase in robustness of the bodies of the vertebrae and probably a reduction of the lumbar vertebrae to a modal number of five. The attachment area of the hamstrings on the ischium spread, thus improving versatility in adjusting power needs to the task in hand.

Thus it seems to me that the main reorganization of the locomotor apparatus to the bipedal type can be explained as a direct consequence of a tendency to lower the center of gravity of the body. The latter tendency was almost certainly a result of increased use of erect bipedality in grassy woodland conditions by a quadrupedal climber. If the interpretation given in this work is correct, then it is unlikely that cultural activity was involved as a contributory factor. Culture apparently emerged later as a part of a dietary change that occurred after *Paranthropus* had developed its partly bipedal, partly quadrupedal-climber adaptation, presumably well back in the Pliocene. There is no obvious reason why this early phase of adaptation to erect posture should have been rapid; it probably occurred quite slowly.

The next step, to the fully human type of locomotor adaptation as seen in *H. africanus*, which probably occurred late in the Pliocene, was part of the process of adjusting to a hunter-gatherer way of life in a rather dry territory. The emergence of culture was an integral part of this change, which produced the human adaptive type.

The early hominid material thus suggests, at least to me, that man did not come into existence from the ape grade of organization in a single major change, nor did erect bipedality of the fully developed sort do so. In both cases the first phase was necessary in order for the second to occur, but the second phase was not a necessary and inevitable consequence of the first. In both cases the changes were initiated by behavioral changes stemming primarily from changed environmental conditions, thus emphasizing again the importance of behavioral flexibility and change as a key factor in initiating major adaptational change. The early hominid material has been extraordinarily fruitful in illuminating the history of the emergence of man and the factors that operated to bring about this momentous transition. In a very real sense the transition from ape to the fullness of the stature of man is not yet complete and we are no more than refined ape-men. This is chiefly because we have not recognized clearly that with the change to the human adaptive type a new mode of evolution of vast potential came into existence that gives man powers incomparably greater than those of any other organism. Thus far we have not learned the rules of the new mode very well and have tended to misuse more often than use well and wisely the powers that we possess. It seems to me that we should think of man not so much as an advanced mammal but as a primitive representative of a wholly new type of organism with a new type of evolutionary process and vast evolutionary potential that has only just begun to be explored.

Appendix

Metrical Data

Table 21
Innominate Dimensions of Hylobates

	N	$\bar{X}$	Observed Range	Standard Pop. Range	s	$s_{\bar{X}}$	$s_{\bar{X}}$ as % of $\bar{X}$	V
a	26	36.5	31–44	24.8–48.2	3.89	0.76	2.1	10.7
b	26	44.0	34–53	27.4–60.6	5.53	1.08	2.5	12.6
c	26	57.3	41–73	36.4–82.2	8.30	1.63	2.8	14.5
d	26	46.5	39–52	34.7–58.3	3.94	0.77	1.7	8.5
e	26	49.6	42–58	35.7–63.5	4.62	0.91	1.8	9.3
f	26	23.4	17–29	15.6–31.2	2.60	0.51	2.2	11.1
g	26	47.3	40–55	35.2–59.4	4.03	0.79	1.7	8.5
h	26	20.5	19–22	17.7–23.3	0.93	0.18	0.9	4.5
j	26	33.4	25–40	23.5–43.3	3.30	0.65	1.9	9.9
k	26	17.5	14–20	12.9–22.1	1.52	0.30	1.7	8.7
l	26	33.3	29–38	26.4–40.2	2.30	0.45	1.7	8.8
*m**	22	91.00	77–120	61.3–120.7	9.91	2.10	2.3	10.9
*o**	22	39.09	35–48	29.9–48.3	3.07	0.65	1.7	7.9

* After Schultz (1930). See figure 111 for key to dimensions used in this and the following tables.

Table 22
Innominate Dimensions of Pan

	N	$\bar{X}$	Observed Range	Standard Pop. Range	s	$s_{\bar{X}}$	$s_{\bar{X}}$ as % of $\bar{X}$	V
a	20	85.8	66–104	52.6–129.0	10.06	2.25	2.6	11.7
b	20	116.6	102–130	89.2–144.0	9.13	2.04	1.7	7.8
c	20	117.6	102–138	91.4–143.8	8.73	1.95	1.7	7.4
d	20	114.8	94–132	87.3–142.3	9.15	2.05	1.8	8.0
e	20	87.7	76–102	68.7–106.7	6.33	1.42	1.6	7.2
f	20	54.9	46–67	37.9–66.6	5.91	1.32	2.4	10.8
g	20	87.5	73–98	71.1–103.9	5.46	1.22	1.4	6.2
h	20	40.7	34–46	30.8–50.6	3.31	0.74	1.8	8.1
j	20	48.9	39–66	26.5–71.3	7.48	1.67	3.4	15.3
k	20	29.1	22–36	17.9–40.3	3.73	0.83	2.9	12.8
l	20	66.4	59–76	52.2–80.6	4.74	1.06	1.6	7.2
m	10	178.3	145–187	141.8–214.8	12.16	3.84	2.2	6.8
o	10	79.9	64–94	55.7–104.1	8.07	2.55	3.2	10.1

Table 23
Innominate Dimensions of Gorilla

	N	$\bar{X}$	Observed Range	Standard Pop. Range	s	$s_{\bar{X}}$	$s_{\bar{X}}$ as % of $\bar{X}$	V
a	20	119.1	87–146	74.6–163.6	14.84	3.32	2.8	12.5
b	20	198.5	162–243	129.3–267.7	23.05	5.16	2.6	23.4
c	20	125.3	93–146	82.9–167.7	14.12	3.16	2.5	11.3
d	20	167.4	134–200	110.7–224.1	18.89	4.23	2.5	11.3
e	20	91.6	74–109	63.5–119.7	9.37	2.09	2.3	10.2
f	20	75.8	52–100	38.3–113.3	12.51	2.80	3.7	16.5
g	20	103.0	77–120	69.6–136.4	11.13	2.49	2.4	10.8
h	20	55.5	42–69	34.6–76.4	6.95	1.56	2.8	12.5
j	20	58.8	30–79	11.1–106.5	15.91	3.56	6.1	27.1
k	20	41.0	32–52	20.3–61.7	6.91	1.54	3.8	16.9
l	20	96.1	80–112	69.2–123.0	8.95	2.00	2.1	9.3
*m**	13	229.38	191–268	156.0–302.8	24.47	6.8	3.0	10.7
*o**	10	119.3	95–145	67.1–171.5	17.40	5.5	4.6	14.6

* After Schultz (1930).

Table 24
Innominate Dimensions of Pongo

	N	$\bar{X}$	Observed Range	Standard Pop. Range	s	$s_{\bar{X}}$	$s_{\bar{X}}$ as % of $\bar{X}$	V
a	20	66.40	55–78	39.8–93.0	8.86	1.98	3.0	13.3
b	20	120.15	96–148	73.2–167.2	15.67	3.50	2.9	13.0
c	19	99.6	81–115	75.7–123.5	7.98	1.83	1.8	8.0
d	20	109.05	85–129	66.4–151.8	14.22	3.18	2.9	13.0
e	20	76.20	67–93	55.4–97.0	6.93	1.55	2.0	9.1
f	20	45.65	30–62	19.1–72.3	8.86	1.98	4.3	19.4
g	20	81.85	69–95	60.7–103.1	7.06	1.58	1.9	8.6
h	20	41.75	35–50	26.3–57.3	5.18	1.16	2.8	12.4
j	20	44.60	27–66	9.0–80.2	11.85	2.65	5.9	26.6
k	20	24.85	18–34	10.2–39.6	4.89	1.09	4.4	19.6
l	20	73.05	61–83	55.6–90.6	5.83	1.30	1.8	8.0
m	10	152.9	132–77	112.4–193.4	13.50	4.27	2.8	8.8
o	10	74.3	62–84	48.4–100.2	8.64	2.73	3.7	11.6

Table 25
Innominate Dimensions of Homo sapiens

	N	$\bar{X}$	Observed Range	Standard Pop. Range	s	$s_{\bar{X}}$	$s_{\bar{X}}$ as % of $\bar{X}$	V
a	50	34.7	16–52	9.5–59.9	8.35	1.18	3.4	24.1
b	50	150.2	119–84	105.2–195.2	14.98	2.12	1.4	10.0
c	50	57.1	42–71	37.9–76.3	6.35	0.90	1.6	11.1
d	50	131.4	107–62	92.7–170.1	12.94	1.83	1.4	9.8
e	50	54.6	35–66	34.8–74.4	6.57	0.93	1.7	12.0
f	50	31.3	16–45	16.6–46.0	4.91	0.69	2.2	15.7
g	30	100.3	79–120	69.4–131.2	10.34	1.87	1.9	10.3
h	50	53.9	45–60	41.9–65.9	3.95	0.56	1.0	7.3
j	50	39.3	30–63	20.1–58.5	6.36	0.90	2.3	16.2
k	50	27.8	20–34	18.2–37.4	3.20	0.45	1.6	11.5
l	50	65.0	49–82	46.4–83.6	6.16	0.87	1.3	9.5
m	40	123.9	103–44	97.2–150.6	8.92	1.41	1.1	7.2
o	40	80.6	66–93	58.4–102.8	7.36	1.18	1.5	9.1

Table 26
Innominate Dimensions of Homo africanus—Sts 14

	a	*b*	*c*	*d*	*e*	*f*	*g*	*h*	*j*	*k*	*l*	*m*	*o*
Right	29.0	109.0	45.0	94.0	44.0	23.0	65.0	39.2	—	15.0	54?	102.0	45.0
Left									28.0		61.0		

Table 27
Innominate Ratios of Hylobates

	N	$\bar{X}$	Observed Range	Standard Pop. Range	s	$s_{\bar{X}}$	$s_{\bar{X}}$ as % of $\bar{X}$	V
$\frac{b \times 100}{c}$	26	79.1	49.3–117.1	23.6–134.6	18.5	3.6	4.6	23.4
$\frac{a \times 100}{c}$	26	64.6	46.5–87.8	36.1–93.1	9.5	1.9	2.9	14.7
$\frac{c \times 100}{d}$	26	123.0	97.6–144.9	83.4–162.6	13.2	2.6	2.1	10.7
$\frac{a \times 100}{e}$	26	74.5	54.4–102.4	34.6–114.4	13.3	2.6	3.5	17.9
$\frac{f \times 100}{b}$	26	53.7	40.4–66.7	31.8–75.6	7.3	1.4	2.6	13.6
$\frac{h \times 100}{c}$	26	36.4	29.6–46.3	22.6–50.2	4.6	0.9	2.5	12.6
$\frac{g \times 100}{c}$	26	84.2	65.6–114.6	43.1–125.3	13.7	2.7	3.2	16.3
$\frac{a \times 100}{d}$	26	78.5	65.3–97.7	57.2–99.8	7.1	1.4	1.8	9.0
$\frac{e \times 100}{g}$	26	104.8	97.7–114.3	91.3–118.3	4.5	0.9	0.9	4.3
$\frac{h \times 100}{b}$	26	47.3	37.7–60.0	27.5–67.1	6.6	1.3	2.7	14.0
$\frac{e \times 100}{c}$	26	88.2	66.7–118.8	44.4–132.0	14.6	2.9	3.3	16.6
$\frac{e \times 100}{b}$	26	114.4	89.6–155.6	59.5–169.3	18.3	3.6	3.1	16.0
$\frac{h \times 100}{l}$	26	61.7	55.3–69.0	49.1–74.3	4.2	0.8	1.3	6.8
$\frac{g \times 100}{d}$	26	102.6	80.0–125.0	61.1–144.1	13.84	2.7	2.6	13.5

Table 28
Innominate Ratios of Pan

	N	$\bar{X}$	Observed Range	Standard Pop. Range	s	$s_{\bar{X}}$	$s_{\bar{X}}$ as % of $\bar{X}$	V
$\frac{b \times 100}{c}$	20	134.7	113.3–154.5	93.9–175.5	13.6	3.0	2.2	10.1
$\frac{a \times 100}{c}$	20	73.1	64.5–95.4	49.1–97.1	8.0	1.8	2.5	10.9
$\frac{c \times 100}{d}$	20	102.8	90.2–125.5	74.9–130.7	9.3	2.1	2.0	9.0
$\frac{a \times 100}{e}$	20	98.6	80.0–131.2	52.7–144.5	15.3	3.4	3.4	15.5
$\frac{f \times 100}{b}$	20	47.4	39.3–58.2	27.9–66.9	6.5	1.4	3.0	13.7
$\frac{h \times 100}{c}$	20	34.8	28.3–39.3	23.7–45.9	3.7	0.8	2.3	10.6
$\frac{g \times 100}{c}$	20	74.7	63.8–88.6	55.5–93.9	6.4	1.4	1.9	8.6
$\frac{a \times 100}{d}$	20	74.8	63.6–88.1	53.5–96.1	7.1	1.6	2.1	9.5
$\frac{o \times 100}{m}$	10	44.8	40.9–52.1	34.9–54.7	3.3	1.0	2.2	7.4
$\frac{e \times 100}{g}$	20	100.3	86.4–114.6	81.4–119.2	7.0	1.6	1.5	6.9
$\frac{h \times 100}{b}$	20	35.1	28.6–42.4	24.0–46.2	3.7	0.8	2.3	10.5
$\frac{e \times 100}{c}$	20	74.9	60.1–90.5	53.0–96.8	7.3	1.6	2.1	9.7
$\frac{e \times 100}{b}$	20	75.6	61.1–90.3	51.6–99.6	8.0	1.8	2.4	10.6
$\frac{h \times 100}{l}$	20	61.5	53.1–67.8	46.2–76.8	5.1	1.1	1.8	8.3
$\frac{g \times 100}{d}$	20	76.7	61.9–88.3	54.5–98.9	7.4	1.7	2.2	9.7
$\frac{h \times 100}{o}$	10	49.3	45.6–55.3	40.4–58.2	3.0	0.9	1.9	6.0
$\frac{h \times 100}{m}$	10	22.1	19.5–24.6	16.6–27.6	1.8	0.6	2.6	8.3

Table 29
Innominate Ratios of Gorilla

	N	$\bar{X}$	Observed Range	Standard Pop. Range	s	$s_{\bar{X}}$	$s_{\bar{X}}$ as % of $\bar{X}$	V
$\frac{b \times 100}{c}$	20	159.3	131.0–206.5	106.8–211.8	17.5	3.9	2.4	11.0
$\frac{a \times 100}{c}$	20	95.6	67.4–120.4	62.9–128.3	10.9	2.4	2.5	11.4
$\frac{c \times 100}{d}$	20	75.0	65.0–84.3	59.4–90.6	5.2	1.2	1.6	6.9
$\frac{a \times 100}{e}$	20	130.6	96.7–151.4	85.3–175.9	15.1	3.4	2.6	11.6
$\frac{f \times 100}{b}$	20	38.1	31.3–43.9	26.4–49.8	3.9	0.9	2.3	10.2
$\frac{h \times 100}{c}$	20	44.5	36.4–54.8	30.1–58.9	4.8	1.1	2.5	10.8
$\frac{g \times 100}{c}$	20	82.5	70.6–91.4	64.2–100.8	6.1	1.4	1.7	7.4
$\frac{a \times 100}{d}$	20	71.3	56.9–81.4	51.5–91.1	6.6	1.5	2.1	9.3
$\frac{e \times 100}{g}$	20	89.2	77.6–101.3	69.7–108.7	6.5	1.5	1.6	7.3
$\frac{h \times 100}{b}$	20	28.0	25.3–30.4	23.5–32.5	1.5	0.3	1.1	5.4
$\frac{e \times 100}{c}$	20	73.3	63.7–82.6	58.9–87.7	4.8	1.1	1.5	6.5
$\frac{e \times 100}{b}$	20	46.4	38.5–55.8	33.2–59.6	4.4	1.0	2.2	9.5
$\frac{h \times 100}{l}$	20	57.8	48.0–65.7	43.7–71.9	4.7	1.1	1.8	8.1
$\frac{g \times 100}{d}$	20	61.6	56.3–73.2	49.1–74.1	4.2	0.9	1.5	6.8

Table 30
Innominate Ratios of Pongo

	N	$\bar{X}$	Observed Range	Standard Pop. Range	s	$s_{\bar{X}}$	$s_{\bar{X}}$ as % of $\bar{X}$	V
$\frac{b \times 100}{c}$	20	118.2	98.1–146.5	79.5–156.9	12.9	2.9	2.5	10.9
$\frac{a \times 100}{c}$	20	65.3	48.7–76.5	42.5–88.1	7.6	1.7	2.6	11.6
$\frac{c \times 100}{d}$	20	94.3	80.2–119.4	63.7–124.9	10.2	2.3	2.4	10.8
$\frac{a \times 100}{e}$	20	87.9	60.6–108.3	43.8–132.0	14.7	3.3	3.8	16.7
$\frac{f \times 100}{b}$	20	38.2	26.8–49.6	17.2–59.2	7.0	1.6	4.2	18.3
$\frac{h \times 100}{c}$	20	41.1	31.8–48.5	27.6–54.6	4.5	1.0	2.4	10.9
$\frac{g \times 100}{c}$	20	80.9	53.9–92.7	54.2–107.6	8.9	2.0	2.5	11.0
$\frac{a \times 100}{d}$	20	61.2	47.3–77.6	41.4–81.0	6.6	1.5	2.5	10.8
$\frac{o \times 100}{m}$	10	48.6	42.2–55.2	36.3–60.9	4.1	1.3	2.7	8.4
$\frac{e \times 100}{g}$	19	94.7	81.7–112.3	64.1–125.3	10.2	2.3	2.4	10.8
$\frac{h \times 100}{b}$	20	34.8	31.0–40.0	27.3–42.3	2.5	0.6	1.7	7.2
$\frac{e \times 100}{c}$	19	77.4	66.4–97.9	47.4–107.4	10.0	2.3	3.0	12.9
$\frac{e \times 100}{b}$	20	64.6	46.4–85.4	30.1–99.1	11.5	2.6	4.0	17.8
$\frac{h \times 100}{l}$	20	57.1	48.7–64.5	43.0–71.2	4.7	1.1	1.8	8.2
$\frac{g \times 100}{d}$	20	75.8	60.8–93.7	51.4–100.2	8.1	1.8	2.4	10.7
$\frac{h \times 100}{o}$	10	55.3	51.0–57.9	46.5–64.2	2.95	0.9	1.7	5.3
$\frac{h \times 100}{m}$	10	26.8	24.1–28.4	22.1–31.5	1.57	0.5	1.8	5.9

Table 31
Innominate Ratios of Homo sapiens

	N	$\bar{X}$	Observed Range	Standard Pop. Range	s	$s_{\bar{X}}$	$s_{\bar{X}}$ as % of $\bar{X}$	V
$\frac{b \times 100}{c}$	50	264.7	205.8–323.8	183.7–345.7	27.0	3.8	1.4	10.2
$\frac{a \times 100}{c}$	50	60.8	30.2–96.3	20.0–101.6	13.6	1.9	3.1	22.4
$\frac{c \times 100}{d}$	50	43.6	34.0–53.9	31.0–56.2	4.2	0.6	1.4	9.6
$\frac{a \times 100}{e}$	50	63.8	33.3–88.6	20.9–106.7	14.3	2.0	3.1	22.4
$\frac{f \times 100}{b}$	50	20.9	15.9–28.5	12.5–29.3	2.8	0.4	1.9	13.4
$\frac{h \times 100}{c}$	50	95.2	79.4–119.0	67.3–123.1	9.3	1.3	1.4	9.8
$\frac{g \times 100}{c}$	30	178.4	139.1–221.4	119.0–237.8	19.8	3.6	2.0	11.1
$\frac{a \times 100}{d}$	50	26.4	13.7–39.1	9.0–43.8	5.8	0.8	3.0	22.0
$\frac{o \times 100}{m}$	40	65.1	52.1–73.3	49.9–80.2	5.1	0.8	1.2	7.8
$\frac{e \times 100}{g}$	30	53.2	44.3–65.2	38.2–68.2	5.0	0.9	1.7	9.4
$\frac{h \times 100}{b}$	50	36.1	31.2–43.8	27.7–44.5	2.8	0.4	1.1	7.8
$\frac{e \times 100}{c}$	50	96.4	64.8–126.2	56.2–136.6	13.4	1.9	2.0	13.9
$\frac{e \times 100}{b}$	50	36.5	28.5–45.0	24.2–48.8	4.1	0.6	1.6	11.2
$\frac{h \times 100}{l}$	50	83.7	61.0–100.0	57.9–109.5	8.6	1.2	1.4	10.3
$\frac{g \times 100}{d}$	30	77.5	66.9–91.5	60.4–94.6	5.7	1.0	1.3	7.4
$\frac{h \times 100}{o}$	40	68.4	61.0–78.5	55.5–81.3	4.3	0.7	1.0	6.3
$\frac{h \times 100}{m}$	40	44.4	39.6–49.6	35.8–53.0	2.9	0.5	1.0	6.4

Table 32
Innominate Ratios of Homo africanus—Sts 5 (right innominate)

$\frac{b \times 100}{c}$	$\frac{a \times 100}{c}$	$\frac{c \times 100}{d}$	$\frac{a \times 100}{e}$	$\frac{f \times 100}{b}$	$\frac{h \times 100}{c}$
242.3	64.5	47.9	47.6	21.1	97.8
$\frac{g \times 100}{c}$	$\frac{a \times 100}{d}$	$\frac{h \times 100}{o}$	$\frac{h \times 100}{m}$	$\frac{o \times 100}{m}$	$\frac{e \times 100}{g}$
144.5	30.9	87.1	38.4	44.1	67.7
$\frac{h \times 100}{b}$	$\frac{e \times 100}{c}$	$\frac{e \times 100}{b}$	$\frac{h \times 100}{l}$	$\frac{g \times 100}{d}$	
36.0	97.8	40.4	64.3	69.2	

Table 33
Sacrum Dimensions

	N	$\bar{X}$	Observed Range	Standard Pop. Range	s	$s_{\bar{X}}$	$s_{\bar{X}}$ as % of $\bar{X}$	V
Pan								
a	8	37.1	32–42	27.0–47.2	3.36	1.19	3.2	9.0
b	8	23.4	21–25	19.5–27.3	1.31	0.46	2.0	5.6
e	8	70.3	67–77	57.7–82.8	4.17	1.47	2.1	5.9
f	8	47.3	43–51	38.2–56.3	3.01	1.06	2.2	6.4
Pongo								
a	14	35.4	27–45	20.5–50.2	4.95	1.32	3.7	14.0
b	14	22.4	18–27	13.7–31.1	2.90	0.78	3.5	12.9
e	14	75.9	61–91	53.3–98.5	7.53	2.01	2.6	9.9
f	14	46.5	37–54	31.1–61.9	5.14	1.37	2.9	11.1
Gorilla								
a	14	45.1	35–55	25.1–65.2	6.68	1.79	4.0	14.8
b	14	27.2	21–33	14.2–40.3	4.35	1.16	4.3	16.0
e	14	90.6	77–102	67.5–113.7	7.70	2.06	2.3	8.5
f	14	62.5	49–76	37.8–87.3	8.25	2.21	3.5	13.2
Homo sapiens								
a	40	53.5	31–70	31.3–75.7	7.40	1.17	2.2	13.8
b	40	32.6	24–38	22.3–42.8	3.41	0.54	1.7	10.5
e	40	115.4	93–133	85.0–145.9	10.16	1.61	1.4	8.8
f	40	55.2	43–65	41.7–68.8	4.51	0.71	1.3	8.1
Homo africanus—Sts 14								
a	1	27.0						
b	1	17.1						
e	1	76.0						
f	1	35.6						

Key to sacral dimensions used in this and the following tables:

a. Maximum transverse diameter of the superior surface of the body of the first sacral element.

b. Mid-sagittal diameter of the superior face of the body of the first sacral element.

e. Maximum transverse width of the first sacral element including the lateral masses.

f. Mid-sagittal length of the first two sacral elements measured on the anterior or pelvic surface.

Table 34
Sacrum Ratios

	N	$\bar{X}$	Observed Range	Standard Pop. Range	s	$s_{\bar{X}}$	$s_{\bar{X}}$ as % of $\bar{X}$	V
Pan								
$\frac{f \times 100}{e}$	8	67.5	59.7–72.9	50.8–84.1	5.55	1.96	2.9	8.2
Pongo								
$\frac{f \times 100}{e}$	14	61.4	54.8–70.4	47.5–75.3	4.63	1.24	2.0	7.5
Gorilla								
$\frac{f \times 100}{e}$	14	69.0	59.0–77.3	48.8–89.2	6.72	1.80	2.6	9.7
Homo sapiens								
$\frac{f \times 100}{e}$	40	48.0	41.9–56.6	37.1–59.0	3.65	0.58	1.2	7.6
Homo africanus—Sts 14								
$\frac{f \times 100}{e}$	1	46.8						

Table 35
The Angle of Pelvic Torsion

	N	$\bar{X}$	Standard Pop. Range	s	$s_{\bar{X}}$	$s_{\bar{X}}$ as % of $\bar{X}$
Homo sapiens	47	65.6	48.2–83.0	5.8	0.8	1.3
Gorilla	33	108.1	95.8–120.4	4.1	0.7	0.7
Pan	51	118.7	100.1–137.3	6.2	0.9	0.7
Pongo	11	111.1	94.3–127.9	5.6	1.7	1.5
Hylobates	2	98.5	92–105*	—	—	—
Papio	6	90.8	81.8–99.8	3.0	1.2	1.4
Cercopithecus	13	72.2	50.0–94.4	7.4	2.0	2.8
Macaca	18	76.5	51.3–101.7	8.4	2.0	2.6
Homo africanus	1	86.5	—	—	—	—

* Observed range. Data adapted from Chopra (1962).

Literature Cited

Alexander, R. M. 1968. *Animal Mechanics*. London: Sidgwick and Jackson.

Ashton, E. H., M. J. R. Healy, and S. Lipton. 1957. The descriptive use of discriminant functions in physical anthropology. *Proc. Roy. Soc. Lond.* (B) 146(925):552.

Ashton, E. H., and C. E. Oxnard. 1964. Functional adaptations in the primate shoulder girdle. *Proc. Zool. Soc. Lond.* 142:49.

Ashton, E. H., C. E. Oxnard, and T. F. Spence. 1965. Scapular shape and primate classification. *Proc. Zool. Soc. Lond.* 145:125.

Ashton, E. H., and S. Zuckerman. 1951. Some cranial indices of *Plesianthropus* and other primates. *Amer. J. Phys. Anthrop.* 9:283.

Basmajian, J. V. 1962. *Muscles Alive: Their functions revealed by electromyography*. Baltimore: Williams and Wilkins Co.

———. 1967. *Muscles Alive: Their functions revealed by electromyography*. 2d ed. Baltimore: Williams and Wilkins Co.

Battye, C. K., and J. Joseph. 1966. An investigation by telemetering of the activity of some muscles in walking. *Med. and Biol. Engng.* 4:125.

Black, G. V. 1902. *Descriptive Anatomy of Human Teeth*. 4th ed. Philadelphia: S. S. White Mfg. Co.

Blumenbach, J. F. 1776. *De Generis Humani Varietate Nativa Liber Cum Figuris Aeri Incisis*. Goettingae, V.A. Vandenhoeck.

———. 1865. *The anthropological treatises of. . . .* Thomas Bandyshe (ed.). London: Longmans, Green.

Boné, Edouard. 1955. Une clavicule et un nouveau fragment mandibulaire D'*Australopithecus prometheus*. *Palaeontologica Africana*, III:87.

Boné, Edouard L., and R. A. Dart. 1955. A catalog of the australopithecine fossils found at the Limeworks, Makapansgat. *Amer. J. Phys. Anthrop.* 13(4):621.

Brain, C. K. 1958. The Transvaal Ape-Man-Bearing Cave Deposits. *Transvaal Museum Memoir* no. 11.

———. 1967. Transvaal Museum fossil project at Swartkrans. *S. Afr. J. Sci.* 63:378.

———. 1970. New finds at the Swartkrans Australopithecine site. *Nature* 225:1112.

Broom, R. 1918. The evidence afforded by the Boskop skull of a new species of primitive man (*Homo capensis*). Amer. Mus. of Nat. Hist., New York, *Anthropological Papers*, vol. 23, pt. 2:63.

———. 1936. New anthropoid skull from South Africa. *Nature* 138:486.

Broom, R., and J. T. Robinson. 1949. The lower end of the femur of *Plesianthropus*. *Ann. Transv. Mus.* vol. 21, pt. 2.

———. 1950. Notes on pelves of fossil ape-man. *Amer. J. Phys. Anthrop.* 8(4):489.

———. 1952. Swartkrans Ape-Man. *Transvaal Museum Memoir* no. 6.

Broom, R., J. T. Robinson, and G. W. H. Schepers. 1950. *Sterkfontein Ape-Man. Transvaal Museum Memoir* no. 4.

Broom, R., and G. W. H. Schepers. 1946. The South African Fossil Ape-Men —the Australopithecinae. *Transvaal Museum Memoir* no. 2.

Buettner-Janusch, J. 1966. *Origins of Man.* New York: John Wiley & Son.

Campbell, B. G. 1966. *Human Evolution.* Chicago: Aldine.

———. 1968. The evolution of the human hand. In Yehudi A. Cohen (ed.), *Man in Adaptation: The Biosocial Background.* Chicago: Aldine. Pp. 128–130.

Campbell, T. D. 1925. Dentition and palate of the Australian aboriginal. Adelaide, S. Australia. *Publication under the Keith Sheridan Foundation for Medical Research* no. 1.

Chopra, S. R. K. 1958. A "pelvimeter" for orientation and measurements of the innominate bone. *Man* no. 171.

———. 1962. The innominate bone of the Australopithecinae and the problem of erect posture. *Bibl. Primat.* 1:93.

Dart, R. A. 1949a. Innominate fragments of *A. prometheus. Amer. J. Phys. Anthrop.* 7(3):301.

———. 1949b. First pelvic bones of *A. prometheus. Amer. J. Phys. Anthrop.* 7:255.

———. 1956. Cultural status of the South African man-apes. *Smithsonian Report for 1955* 317.

———. 1957. The second adolescent (female) ilium of *Australopithecus prometheus. J. of Pal. Soc. of India* (Lucknow) 2:73.

———. 1958. A further adolescent Australopithecine ilium from Makapansgat. *Amer. J. Phys. Anthrop.* 16:473.

Davivongs, V. 1963a. The pelvic girdle of the Australian aborigine: sex differences and sex determination. *Amer. J. Phys. Anthrop.* 21(4):443.

———. 1963b. The femur of the Australian aborigine. *Amer. J. Phys. Anthrop.* 21(4):457.

Day, M. H. 1967. Olduvai Hominid 10: A multivariate analysis. *Nature, London* 215:323.

———. 1969. Femoral fragment of a robust Australopithecine from Olduvai Gorge, Tanzania. *Nature* 221:230.

Day, M. H., and J. R. Napier. 1964. Fossil foot bones. *Nature* 201:969.

Day, M. H., and B. A. Wood. 1968. Functional affinities of OH8 talus. *Man* 3(3):429.

Duchenne, G. B. A. 1867. *Physiologie des mouvements.* Transl. by E. B. Kaplan (1949) London: W. B. Saunders.

Dwight, T. 1904–5. The size of the articular surfaces of the long bones as characteristic of sex; an anthropological study. *M. J. Anat.* 4:19.

Eberhart, H. D., V. T. Inman, and B. Bresler. 1954. The principal elements in human locomotion. In Klopsteg, P. E., and P. D. Wilson (eds.). *Human Limbs and Their Substitutes.* New York: McGraw-Hill.

Elftman, H., and J. Manter. 1935. Chimpanzee and human feet in bipedal walking. *Amer. J. Phys. Anthrop.* 20:69.

Erickson, G. E. 1963. Brachiation in New World monkeys and in anthropoid apes. *Symp. Zool. Soc. Lond.* no. 10:135.

Fick, R. 1926. Massverhältnisse an den oberen Gliedmassen des Menschen und den Gliedmassen der Menschenaffen. *Sitzungsber d. Preuss. Akad. d. Wissensch., Phys.-mathem.* Kl. 30:417.

Frazer, J. E. 1965. *The anatomy of the human skeleton*, 6th ed. Ed. by S. A. Breathnach. London: Churchill.

Frey, H. 1923. Untersuchungen über die Scapula, speciell über ihre äussere Form und deren Abhängigkeit von der Funktion. *Zeitschr. f. Anat. u. Entwicklungsgesch.* 68:277.

———. 1924. Weitere Untersuchungen über die Scapula, speziell über die Scapula scaphiodes. *Zeitschr. f. Anat. u. Entwicklungsgesch.* 74:240.

Gregory, W. K. 1912. Notes on the principles of quadrupedal locomotion and on the mechanism of the limbs in hoofed animals. *Annals N. Y. Acad. Sci.* 22:267.

Gutgesell, V. J. 1970. "Telanthropus" and the single species hypothesis: A reexamination. *Amer. Anthropologist* 72(3):565.

Hay, R. L. 1965. Preliminary notes on the stratigraphy of Beds I–IV, Olduvai Gorge, Tanganyika. *Olduvai Gorge 1951–61* 1:94.

Holloway, R. 1970. New endocranial values for the Australopithecines. *Nature* 227:199.

Howell, F. C. 1965. *Early Man.* New York: Time-Life.

———. 1969. Remains of hominidae from Pliocene/Pleistocene formations in the Lower Omo Basin, Ethiopia. *Nature* 223:1234.

Huxley, J. S. 1947. *Touchstone for Ethics.* New York: Harper & Bros.

———. 1955. Evolution, cultural and biological. In W. Thomas (ed.) *Yearbook of Anthropology.* New York: Wenner-Gren Foundation for Anthropological Research.

Jolly, C. J. 1970. The seed-eaters: A new model of hominid differentiation based on a baboon analogy. *Man* 5:5.

Joseph, J. 1960. *Man's Posture Electromyographic Studies.* Springfield, Ill.: Charles C. Thomas.

Kajava, Y. 1924. Veber dem Schultergürtel der Finnen. *Annales Acad. Scientiarum fennicae* (Helsinki) serie A, t. 21:1–69.

Kern, H. M., and W. L. Straus, Jr. 1949. The femur of *Plesianthropus transvaalensis. Amer. J. Phys. Anthrop.* 7 (1):53.

Kondo, S. 1960. Anthropological study on human posture and locomotion. *J. Faculty of Sci. Univ. Tokyo.* Sec. 5, vol. 2, pt 2:189.

Kortland, A. 1962. Chimpanzees in the wild. *Sci. Amer.* 205(5):128.

———. 1963. Protohominid behavior in primates. *Symp. Zool. Soc. Lond.* 10:61.

Lack, D. 1947. *Darwin's finches.* Cambridge: Cambridge Univ. Press.

Leakey, L. S. B. 1960. The affinities of the new Olduvai Australopithecine. *Nature* 186:458.

Leakey, L. S. B., P. V. Tobias, and J. R. Napier. 1964. A new species of the genus *Homo* from Olduvai Gorge. *Nature* 202:7.

Leakey, R. 1970. Fauna and artefacts from the Koobi Fora Area. *Nature* 226:228.

Le Gros Clark, W. E. 1947a. Observations on the anatomy of the fossil Australopithecinae. *J. Anat., London* 81:300.

———. 1947b. The importance of the fossil Australopithecinae in the study of human evolution. *Sci. Prog.* 35:377.

———. 1947c. Anatomical studies of fossil hominoidea from Africa. *Proc. of Pan-African Congress on Prehistory.*

———. 1950. New Palaeontological evidence bearing on the evolution of the Hominoidea. *Quart. J. Geol. Soc. Lond.* 105:225.

———. 1955. The os innominatum of the recent ponginae with special reference to that of the Australopithecinae. *Amer. J. Phys. Anthrop.* 13 (1):19.

———. 1955. *The Fossil Evidence for Human Evolution: Introduction to the Study of Paleoanthropology.* Chicago: Univ. of Chicago Press.

———. 1957. *History of the Primates: An Introduction to the Study of Fossil Man.* Chicago: Univ. of Chicago Press.

———. 1964. *Fossil Evidence of Human Evolution: Introduction to the Study of Paleoanthropology.* 2d ed. Chicago: Univ. of Chicago Press.

———. 1967. *Ape-Man or Man-Ape.* New York: Holt-Rinehart.

Le Gros Clark, W. E., and D. P. Thomas. 1951. Associated jaws and limb bones of *Limnopithecus macinnesi. Fossil Mammals of Africa,* no. 3. London: Brit. Mus. (Nat. Hist.).

Lisowski, F. P. 1966. Growth changes and comparison of the primate talus. *Ethiopian Medical Journal* 4:173–79.

Lovejoy, C. O., and K. G. Heiple. 1970. A reconstruction of the femur of *A. africanus, Amer. J. Phys. Anthrop.* 32(1):33.

Manter, J. T. 1946. Distribution of compression forces in the joints of the human foot. *Anat. Rec.* 96(3):313.

Martin, R., and K. Saller, 1957. *Lehrbuch der Anthropologie,* 4 bds. Stuttgart: G. Fischer. Verlag.

Martin, R., and K. Saller. 1959. *Lehrbuch der Anthropologie*. Stuttgart: G. Fischer.

Mayr, E. 1950. Taxonomic categories in fossil hominids. *Cold Spring Harbor Symp. Quant. Biol.* 15:109.

———. 1969. *Principles of Systematic Zoology*. New York: McGraw-Hill.

Mayr, E., E. G. Linsley, and R. L. Usinger. 1953. *Methods and Principles of Systematic Zoology*. New York: McGraw-Hill.

Mednick, L. W. 1955. The evolution of the human ilium. *Amer. J. Phys. Anthrop.* 13(2):203.

Morton, D. J. 1926. Evolution of man's erect posture. *J. of Morphology* 43:147.

———. 1952. *Human Locomotion and Body Form*. Baltimore: Williams and Wilkins.

Napier, J. R. 1959. Fossil metacarpals from Swartkrans. *Fossil Mammals of Afr.* no. 17. London: Brit. Mus. (Nat. Hist.).

———. 1962a. Fossil hand bones from Olduvai. *Nature* 196:409.

———. 1962b. Olduvai Gorge and human antiquity. *Antiquity* 36:41.

———. 1964. The evolution of bipedal walking in the hominids. *Arch. de Biol.* (Liège) 75: suppl. 673–708.

———. 1967. The antiquity of human walking. *Sci. Amer.* 216:56.

Napier, J. R., and P. R. Davis. 1959. The fore-limb skeleton and associated remains of *Proconsul africanus*. In *Fossil Mammals of Africa*, No. 16. London: Brit. Mus. (Nat. Hist.).

Oxnard, C. E. 1963. Locomotor adaptations in the primate forelimb. *Symp. Zool. Soc. Lond.*, no. 10:165.

———. 1967. Aspects of the mechanical efficiency of the scapula in some primates. *Anat. Rec.* 157:296.

———. 1968. A note on the Olduvai clavicular fragment. *Amer. J. Phys. Anthrop.* 29:429.

Parsons, F. G. 1914. The characters of the English thigh-bone. *J. Anat. and Physiol. Lond.* 48:238.

Pauwels, F. 1948. Die Bedeutung der Bauprinzipien für die Beanspruchung der Röhrenknochen. *Z. Anat. Entw. Ges.* 114:129.

Pearson, K., and J. Bell. 1919. A study of the long bones of the English skeleton. Part 1. The femur. *Draper's Co. Research Mem.*, Biometric series 10. Cambridge: Cambridge Univ. Press.

Peters, H. B. 1936. Ergänzendes zur Primateneinteilung. *Z. Rassenk.* 3 and 4:205.

Pilbeam, D. 1970. *The evolution of man*. New York, Funk & Wagnalls.

Radcliffe, C. W. 1962. The biomechanics of below knee prostheses in normal level bipedal walking. *Artif-limbs* 6:16.

Raven, C. H. 1950. Regional anatomy of the Gorilla. In Gregory, W. K. (ed.). *The Anatomy of The Gorilla, Raven Memorial Volume*. New York: Columbia Univ. Press.

Reynolds, E. 1931. The evolution of the human pelvis in relation to the mechanics of erect posture. *Papers of the Peabody Museum of American Archaeology and Ethnology*, Harvard Univ. vol. 11, no. 5:255.

Ripley, S. 1967. The leaping of langurs. *Amer. J. Phys. Anthrop.* 26:149.

Robinson, J. T. 1954. The genera and species of the Australopithecinae. *Amer. J. Phys. Anthrop.* 12:181.

———. 1956. The dentition of the Australopithecinae. *Transvaal Museum Memoir* no. 9.

———. 1961. The Australopithecines and their bearing on the origin of man and of stone tool making. *S. Afr. J. Sci.* 57:3.

———. 1962. The origin and adaptive radiation of the Australopithecines. In G. Kurth (ed.). *Evolution und Hominisation*. Stuttgart: G. Fischer Verlag.

———. 1963. Adaptive radiation in the Australopithecines and the origin of man. In F. C. Howell & F. Bourliere (eds.). *African Ecology and Human Evolution*. Chicago: Aldine.

———. 1964. Some critical phases in the evolution of man. *S. Afr. Archaeol. Bull.* 19 (73):3.

———. 1965a. Comment on "New Discoveries in Tanganyika" by Tobias. *Current Anthropology* 6:403.

———. 1965b. *Homo habilis* and the Australopithecines. *Nature* (*Lond.*) 205:121.

———. 1966. The distinctiveness of *Homo habilis*. *Nature* (*Lond.*) 209:953.

———. 1967. Variation and the taxonomy of the early hominids. In *Evolutionary Biology*. Th. Dobzhansky, M. K. Hecht, and Wm. Steere (eds.). New York: Appleton-Century-Crofts. Pp. 69–99.

———. 1969. Dentition and adaptation in early hominids. *Proceedings VIIIth International Congress of Anthropological and Ethnological Sciences* 1: *Anthropology*.

———. 1970. Two new early hominid vertebrae from Swartkrans. *Nature* 225:1217.

Schaller, G. B. 1961. The Orangutan in Sarawak. *Zoologica* 46(2):73.

———. 1963. *The Mountain Gorilla*. Chicago: Univ. of Chicago Press.

Schofield, G. 1959. Metric and morphological features of the femur of the New Zealand Maori. *J. Roy. Anthrop. Inst.* 89:89.

Schultz, A. H. 1930. The skeleton of the trunk and limbs of higher primates. *Human Biol.* 3:303.

———. 1937. Proportions, variability, and asymmetries of the long bones of the limbs and the clavicles in man and apes. *Human Biol.* 9:281.

———. 1949. Ontogenetic specializations of man. *Arch. Jul. Klaus-Stift.* 24:197.

———. 1953. The relative thickness of the long bones and the vertebrae in primates. *Amer. J. Phys. Anthrop.* 11(3):277.

———. 1960. Age changes in primates and their modification in man. In J. M. Tanner (ed.). *Human Growth.* London: Pergamon Press.

Schultz, A. H., and W. L. Straus, Jr. 1945. The numbers of vertebrae in primates. *Proc. Amer. Philos Soc.* 89:601.

Sergi, S. 1904. Le variazioni dei solchi centrale e la loro origine segmentale nell'Hylobates. *Ricerche Lab. Anat. Norm. R. Universita.* Rome.

———. 1908. Di una classificatione razionale dei gruppi umani. *Atti Soc. Ital. Progr. Sci.* Roma.

Sigmon, B. A. 1969. Anatomical structure and locomotion habit in anthropoidea with special reference to the evolution of erect bipedality in man. Ph.D. thesis, Univ. of Wisconsin.

Simons, E. L. 1967. The significance of primate paleontology for anthropological studies. *Amer. J. Phys. Anthrop.* 27(3):307.

Simons, E. L., and S. R. K. Chopra. 1969. A preliminary announcement of a new *Gigantopithecus* species from India. In H. O. Hofer (ed.). *Recent Advances in Primatology, Proceedings of the Second International Congress of Primatology, vol. 2.*

Simons, E., and P. Ettel. 1970. *Gigantopithecus. Scientific American* 222:76.

Simpson, G. G. 1941. Large Pleistocene felines of North America. *Amer. Mus. Novitates.* 1136:1.

———. 1949. *The Meaning of Evolution.* New Haven: Yale Univ. Press.

———. 1961. *Principles of Animal Taxonomy.* New York: Columbia Univ. Press.

———. 1963. The meaning of taxonomic statements. In S. L. Washburn (ed.). *Classification and Human Evolution.* Viking Fund Publications in Anthropology, no. 37.

Simpson, G. G., A Roe, and R. Lewontin. 1960. *Quantitative Zoology.* New York: Harcourt, Brace.

Smith, J. M., and R. J. G. Savage. 1956. Some locomotory adaptations in mammals. *J. Linn. Soc.* (*Zool.*) 42:603.

Steindler, A. 1955. *Kinesiology of the Human Body under Normal and Pathological Conditions.* Springfield, Ill.: Charles C. Thomas.

Straus, W. E., Jr. 1929. Studies on primate ilia. *Amer. J. Anat.* 43:403.

———. 1948. The humerus of *Paranthropus robustus. Amer. J. Phys. Anthrop.* 6:285.

Teilhard de Chardin, P. 1959. *The Phenomenon of Man.* New York: Harper-Row.

Tobias, P. V. 1965. New discoveries in Tanganyika: Their bearing on hominid evolution. With comments by M. H. Day, F. C. Howell, C. H. R. von Koenigswald, J. R. Napier, and J. T. Robinson. *Curr. Anthrop.* 6:391.

———. 1967. Pleistocene deposits and new fossil localities in Kenya. *Nature.* 215:479.

———. 1968. Mid and upper Pleistocene members of the genus *Homo* in Africa. In G. Kurth (ed.). *Evolution und Hominisation.* Stuttgart: Fischer.

Tuttle, R. H. 1967. Knuckle-walking and the evolution of hominid hands. *Amer. J. Phys. Anthrop.* 26:171.

Vallois, H. 1928–1946. L'Omoplate humaine: Étude anatomique et anthropologique. *Bulletins et Mémoires de la Société D'Anthropologie.*

Van. Lawick-Goodall, J. 1968. Behaviour of free-living chimpanzees of the Gombe Stream area. *Anim. Behav. Monogr.* 3.

Von Eikstedt, E. 1937. Geschichte der anthropologischen Namengebung und classifikation. *Z. Rassenk.* 5 and 6:209.

Waddington, C. H. 1947. *Science and Ethics.* London: Allen & Unwin.

———. 1960. *The Ethical Animal.* Chicago: Univ. of Chicago Press.

Washburn, S. L. 1950. The analysis of primate evolution with particular reference to the origin of man. *Cold Spring Harbor Symposium on Quant. Biol.* 16:67.

———. 1963. Behavior and human evolution. In S. L. Washburn (ed.). *Classification and Human Evolution.* Viking Fund Publications in Anthropology, no. 37.

Waterman, H. C. 1929. Studies on the evolution of the pelvis of man and other primates. *Bull. Amer. Mus. Nat. Hist.* 55:585.

Weidenreich, F. 1913. Über das Hüftbein und das Becken der Primaten und ihre Umformung durch den aufrechten Gang. *Anatom. Anzeiger.*

———. 1941. The extremity bones of *Sinanthropus pekinensis. Palaeont. Sinica,* whole no. 116, n.s. D no. 5.

Wheatley, M. D., and W. D. Jahnke. 1951. Electromyographic study of the superficial thigh and hip muscles in normal individuals. *Arch. Phys. Med.* 32:508.

Wolpoff, M. H. 1968. "Telanthropus" and the single species hypothesis. *Amer. Anthropologist* 70:477.

Woo, J. K. 1962. Mandibles and Dentition of *Gigantopithecus. Palaeont. Sinica,* whole no. 146, n.s. D no. 11.

Wood Jones, F. 1942. *Anatomy of the Hand.* 2d ed. Baltimore: Williams and Wilkins.

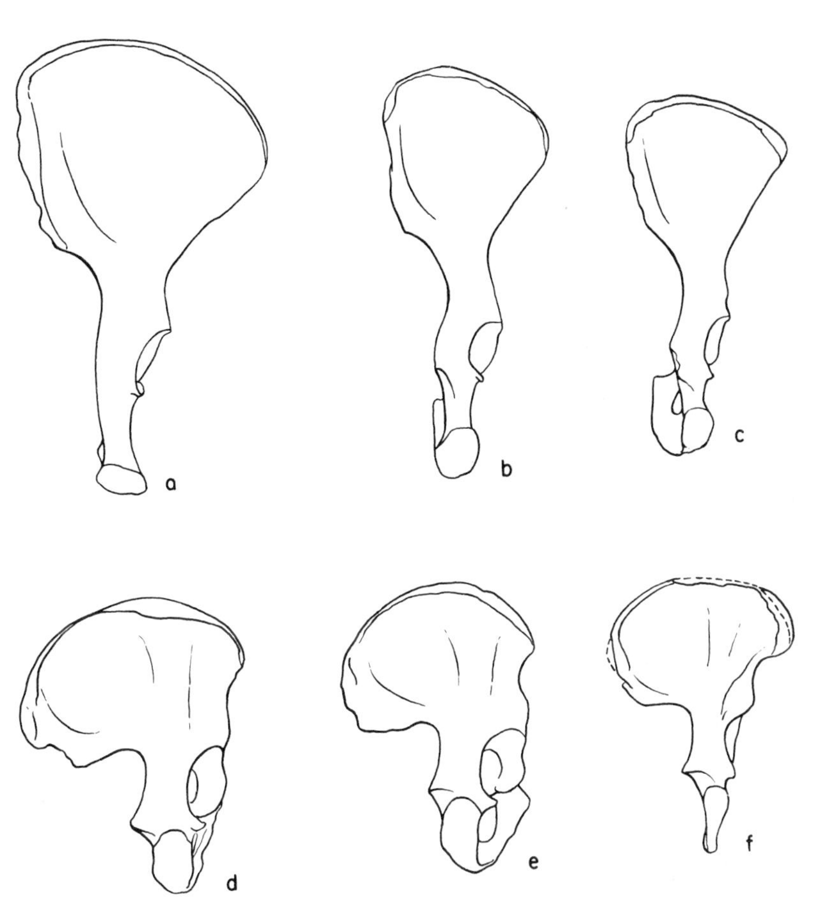

Fig. 1. Comparison of some pongid and hominid innominate bones: a, gorilla; b, orang; c, chimpanzee; d, *Homo sapiens* (American white); e, *H. sapiens* (Bush); f, *H. africanus* (Sts 14). All to same scale. The iliac blades are all similarly orientated.

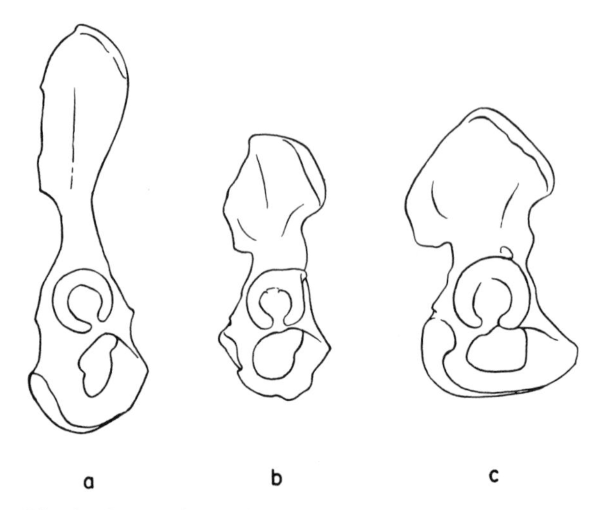

Fig. 2. Comparison of some hominoid innominate bones: a, orang; b, *Homo africanus* (Sts 14) without any reconstruction; c, *H. sapiens* (Bush). All to same scale. The acetabula are all similarly orientated.

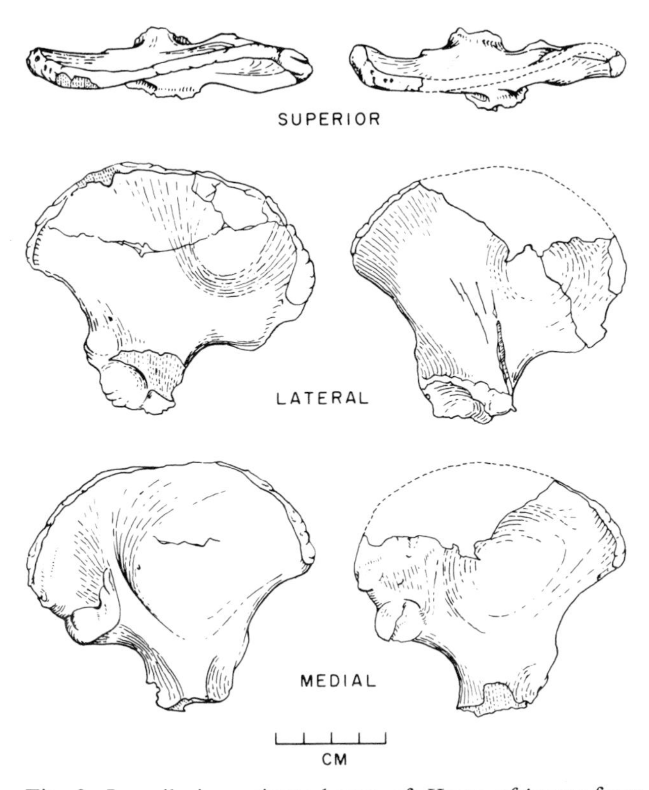

Fig. 3. Juvenile innominate bones of *Homo africanus* from Makapansgat. Redrawn after Dart.

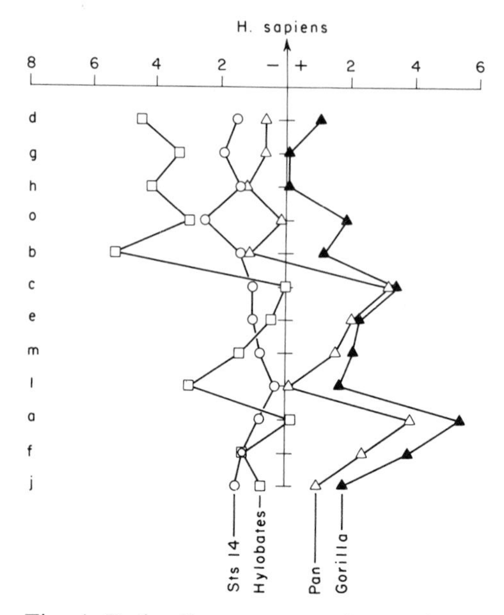

Fig. 4. Ratio diagram comparing twelve innominate bone dimensions of some hominoids, using *Homo sapiens* as standard. The more nearly lines representing two forms parallel each other, the more alike the two are in proportion in the dimensions being compared. The dimensions here used are defined in Fig. 111.

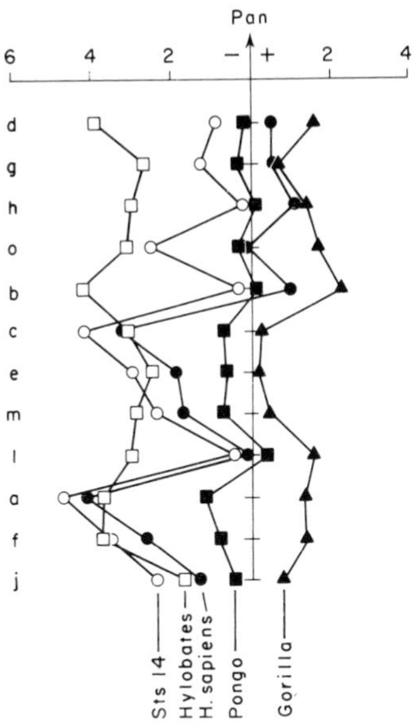

Fig. 5. Ratio diagram comparing twelve hominoid innominate dimensions. *Pan* is here used as the standard of comparison. See caption to Fig. 4.

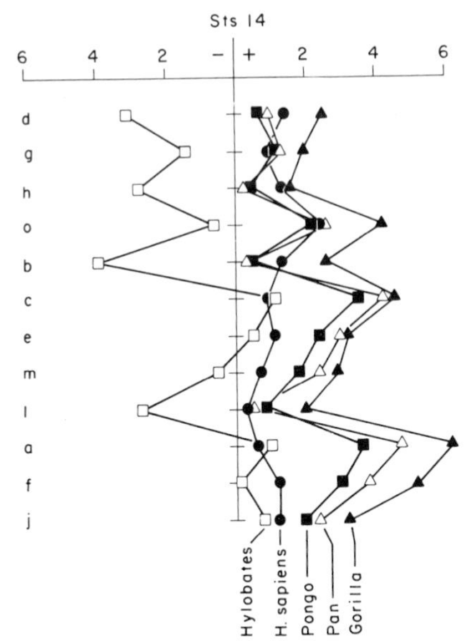

Fig. 6. Ratio diagram comparing twelve hominoid innominate dimensions. *Homo africanus* (Sts 14) is used as the standard of comparison. See caption to Fig. 4.

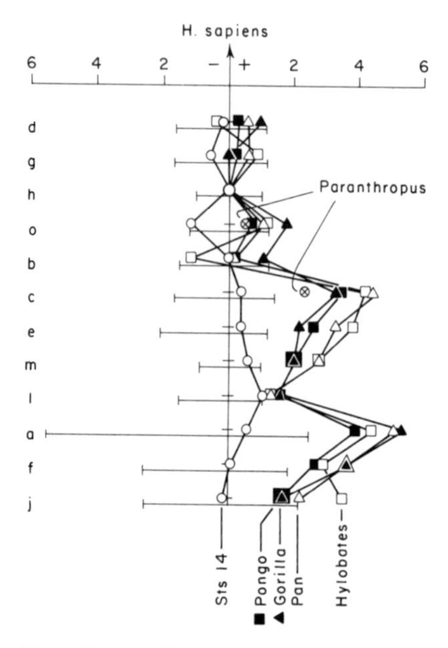

Fig. 7. Modified ratio diagram comparing the same twelve innominate dimensions used in Figs. 4–6. Acetabular width is used as a standard for coincidence; that is, in each form all dimensions are related to a standard acetabular size equal to that of the standard of comparison—*Homo sapiens* in this case. The horizontal bar for each dimension was obtained by taking the *H. sapiens* mean ± 3 standard deviations and then plotting these two values in the same way as the other values were plotted. Key to dimensions used is given in Fig. 111.

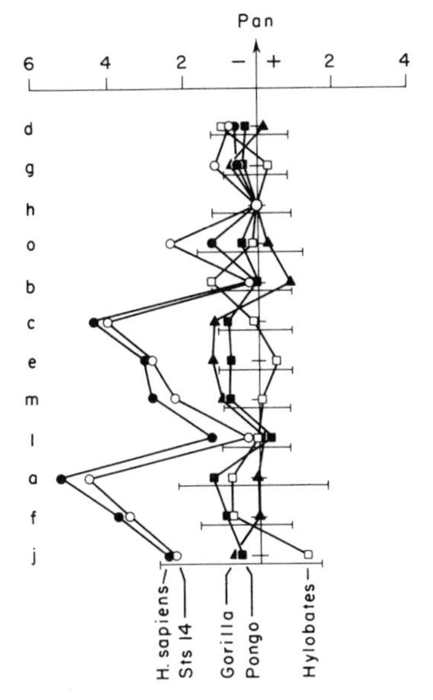

Fig. 8. Modified ratio diagram comparable to that in Fig. 7 using *Pan* as the standard of comparison.

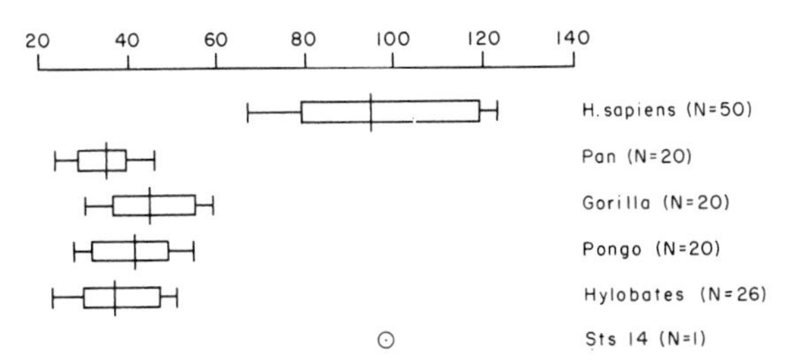

Fig. 9. Comparison of the ratio $100h/c$ in a number of hominoids. In each case the center crossbar represents the mean, the inclosed area represents the observed range, and the maximum length represents the standard population range (mean ± 3 s.d.). Sample size is given in brackets in each case.

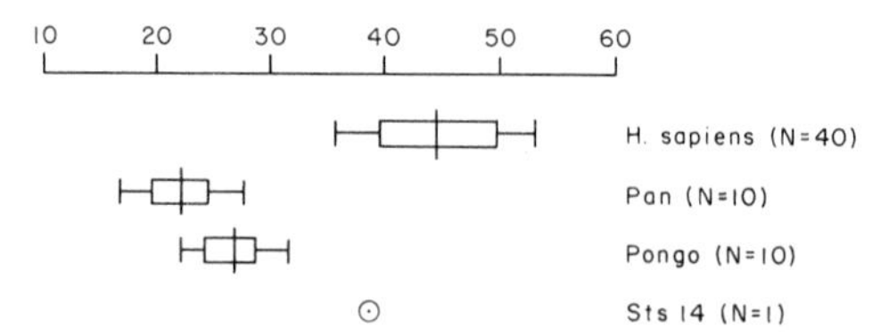

Fig. 10. Comparison of the ratio $100h/m$ in a number of hominoids. In each figure the center bar represents the mean, the box represents the observed range, and the total length the standard population range (mean ± 3 s.d.). Sample size is given in brackets.

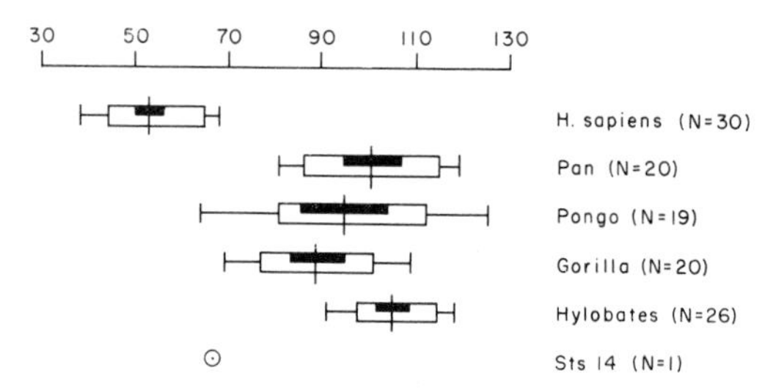

Fig. 11. Comparison of the ratio $100e/g$ in some hominoids. The center bar represents the mean, the boxed area the observed range, the total length the standard population range (mean ± 3 s.d.), and the black bar the confidence interval for the mean where $p = .001$. The ratio diagrams (Figs. 7–8) show that g is proportionately very similar in all of the forms compared; the separation between *H. sapiens* and *H. africanus* (Sts 14) on the one hand and the pongids on the other is thus due to differences in e in the two groups.

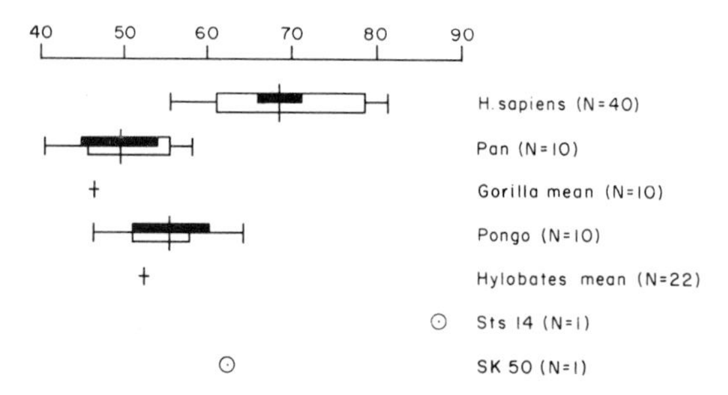

Fig. 12. Comparison of the ratio $100h/o$ in some hominoids. The conventions used are as in Fig. 11. The values of o for *Gorilla* and *Hylobates* were taken from Schultz 1930, where insufficient data was provided to allow full diagrams to be constructed, hence in those cases mean index values only are given. Sts 14 represents *H. africanus* and SK 50 *Paranthropus*.

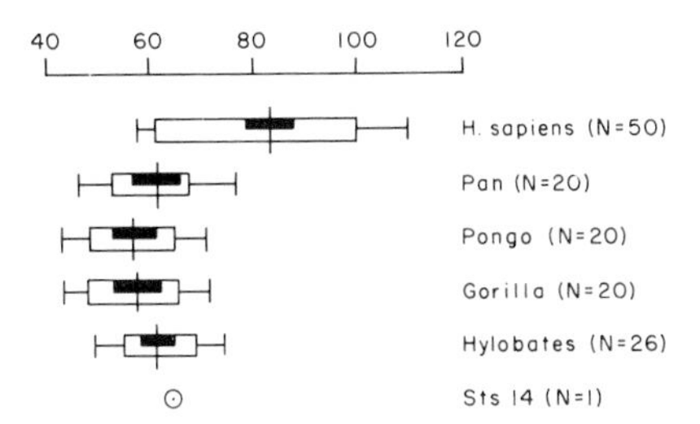

Fig. 13. Comparison of the ratio $100h/l$ in some hominoids. The conventions used are as in Fig. 11.

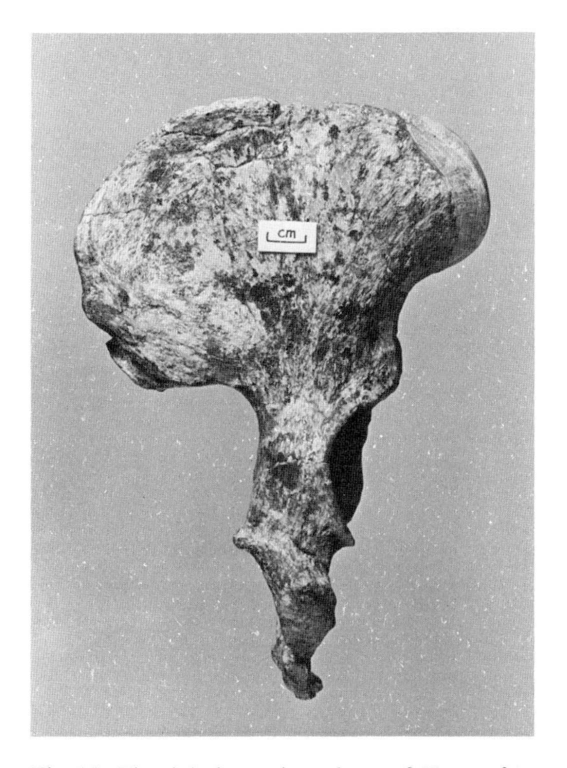

Fig. 14. The right innominate bone of *Homo africanus* (Sts 14) in lateral view. Portion of the anterior superior iliac spine region was missing and has been restored from the left innominate of the same pelvis. There has been minor damage and warping of portion of the iliac crest. See also Fig. 15.

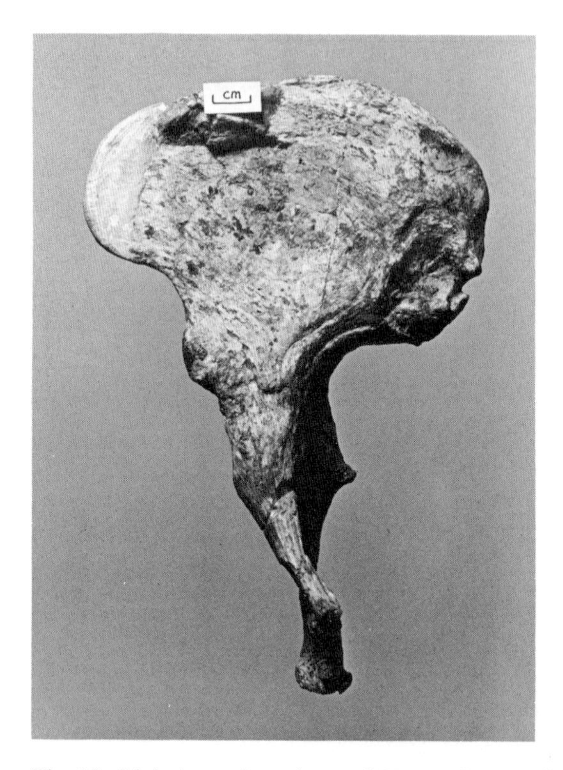

Fig. 15. Right innominate bone of *Homo africanus* (Sts 14) in medial view. The distal end of the pubis is missing.

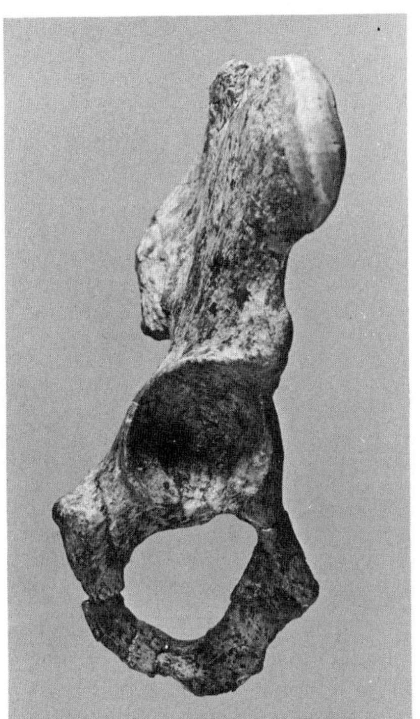

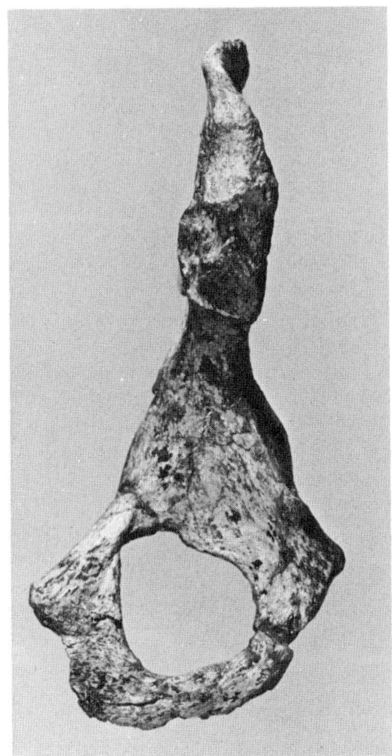

Fig. 16. Right innominate bone of *Homo africanus* (Sts 14) orientated so that the observer is looking vertically into the acetabulum. Part of the ischial tuberosity surface is missing; the small bulge on the end of the ischial shaft consists of original tuberosity surface. The ischial and pubic rami are not complete and have been moderately warped. A restored view of the bone is given in Fig. 19.

Fig. 17. Right innominate bone of *Homo africanus* (Sts 14) orientated so that the observer is looking vertically onto the back of the acetabulum. All of the pubic symphysis is missing. Cracks in the pubic ramus testify to some distortion.

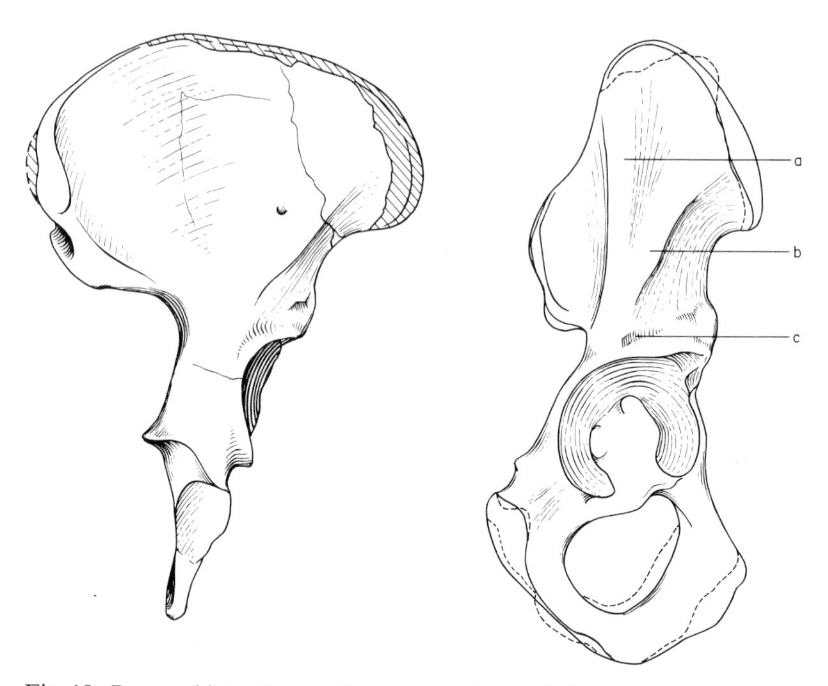

Fig. 18. Restored lateral view of innominate bone of *Homo africanus* (Sts 14). Restored areas shown hatched; they are present on the left innominate of the same pelvis. The well-defined, crestlike anterior margin of the acetabulo-spinous buttress is clearly shown trending upward toward the anterior superior iliac spine region.

Fig. 19. Restored right innominate of *Homo africanus* (Sts 14). The margins of the specimen as it now is are indicated by broken lines where damage or distortion is present: a, acetabulo-cristal buttress; b, acetabulo-spinous buttress; c, scar of attachment of the reflected head of the rectus femoris muscle.

Fig. 20. Right innominate bone of *Homo africanus* (Sts 65) lighted to demonstrate the fan-shaped area of thickened outer table bone comprising the acetabulo-cristal and acetabulo-spinous buttresses and the shallow hollow that partly separates the two. Their directions are also indicated by the natural cracks due to pre-fossilization weathering. Note that some of the terminal cracks of the acetabulo-cristal buttress curve sharply over to the right.

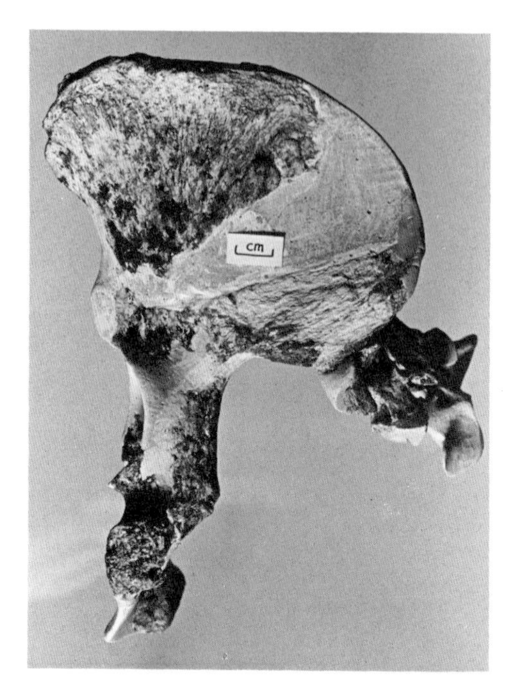

Fig. 21. Left innominate bone of *Homo africanus* (Sts 14). The missing portions have been restored from the right innominate of the same pelvis. The raised iliac tubercle is visible in the middle of the preserved portion of the iliac crest. It is better seen in Fig. 22.

Fig. 22. Detail of portion of the left ilium of *Homo africanus* (Sts 14) shown in Fig. 21. The thickened iliac tubercle is clearly visible.

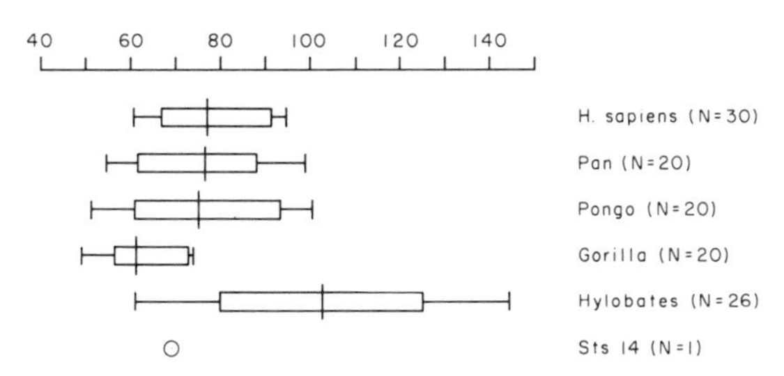

Fig. 23. Comparison of the index $100g/d$ in some hominoids. See text for interpretation of this diagram. In each case the center bar represents the mean, the box the observed range, and the total length the standard population range (mean ± 3 s.d.).

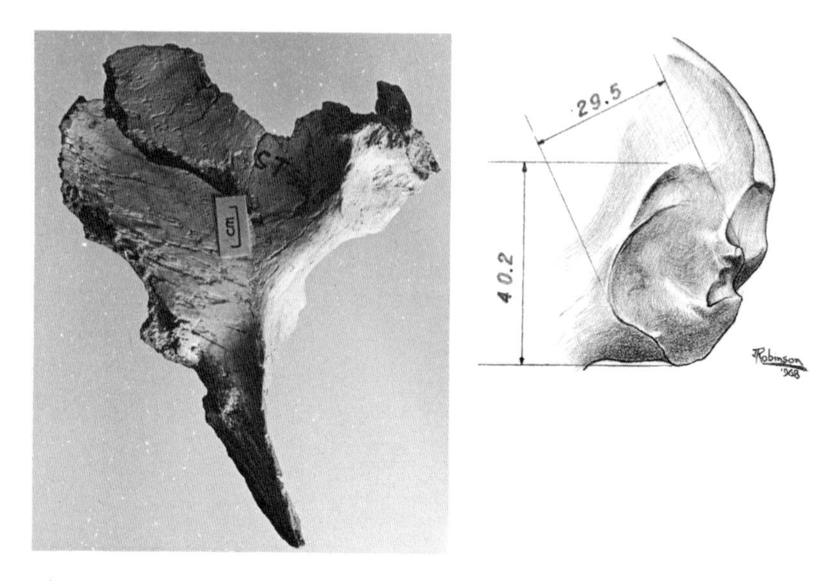

Fig. 24. Pelvic surface of incomplete right innominate bone of *Homo africanus* (Sts 65).

Fig. 25. Auricular surface of right innominate bone of *Homo africanus* (Sts 14).

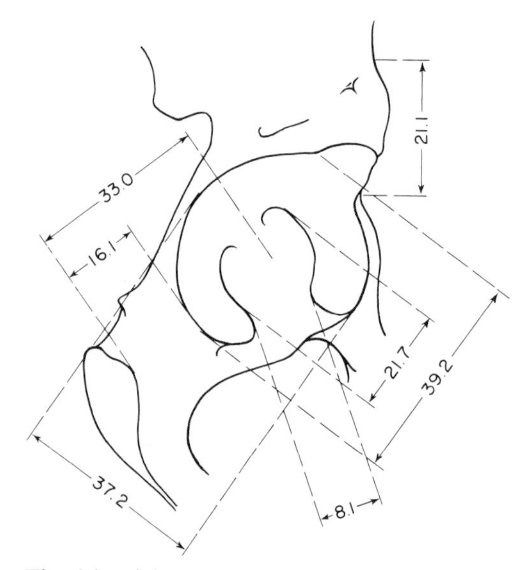

Fig. 26. Right acetabulum of *Homo africanus* (Sts 14). Dimensions are in millimeters.

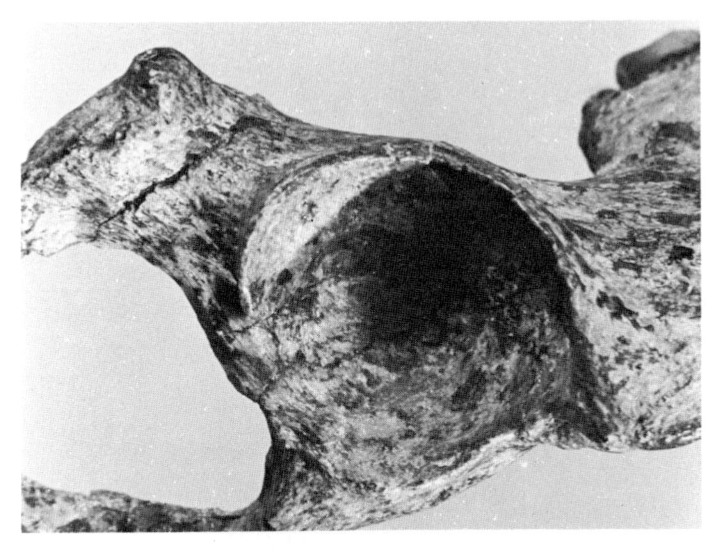

Fig. 27. Acetabulum and ischial shaft of *Homo africanus* (Sts 14, right). The tuberclelike structure at top left is all that remains of the surface of the ischial tuberosity; the area to the left and below the "tubercle" is composed of cancellous bone filled with matrix where the original tuberosity surface was removed before fossilization.

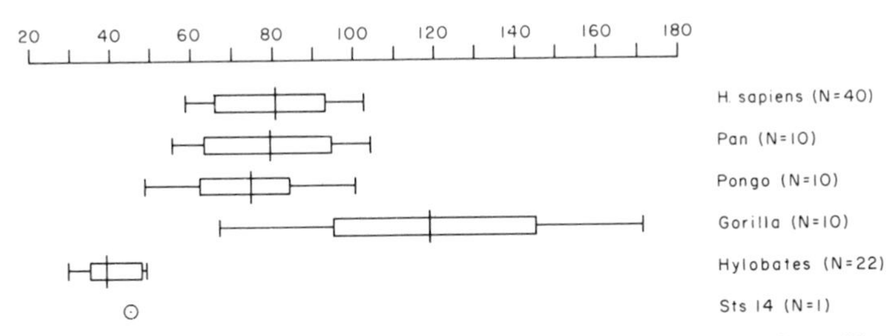

Fig. 28. Comparison of absolute ischium length (*o*) in a number of hominoids. The center bar represents the mean, the box represents the observed range, and total length the standard population range (mean ± 3 s.d.). Data for *Gorilla* and *Hylobates* taken from Schultz 1930.

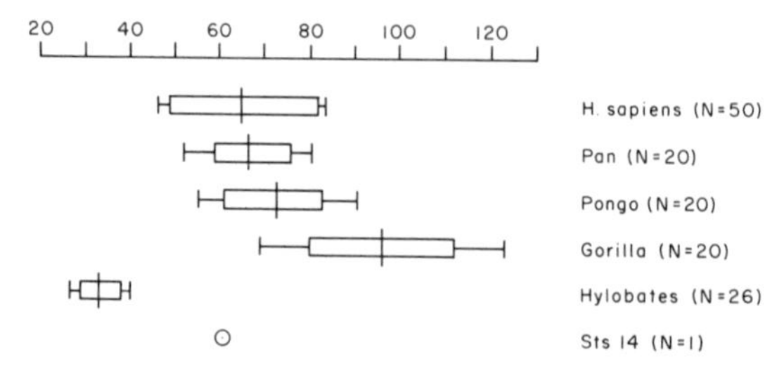

Fig. 29. Absolute length of the pubis (l) from the acetabular margin to the distal extremity compared in a number of hominoids. The center bar represents the mean, the box the observed range, and total length the standard population range (mean ± 3 s.d.).

Fig. 30. Portion of left pubis of *Homo africanus* (Sts 14). A small patch of bone is missing in the region where the pubic tubercle would have been had one been present. The symphysis is not damaged; see Fig. 31.

Fig. 31. Pubis symphyseal surface of left pubis of *Homo africanus* (Sts 14).

Fig. 32. Lateral view of right innominate bone of *Homo africanus* (Sts 65) to show the natural split-line pattern produced by weathering prior to fossilization. One set of cracks trends up in a narrow fan along the acetabulo-cristal and acetabulo-spinous buttresses, while another curves away sharply to the right and follows the acetabulo-sacral buttress. There are a number of cracks due to pressure that should not be confused with split-lines. The split-lines near the anterior top portion of the acetabulo-cristal buttress curve sharply over to the right—these are better seen in Fig. 20.

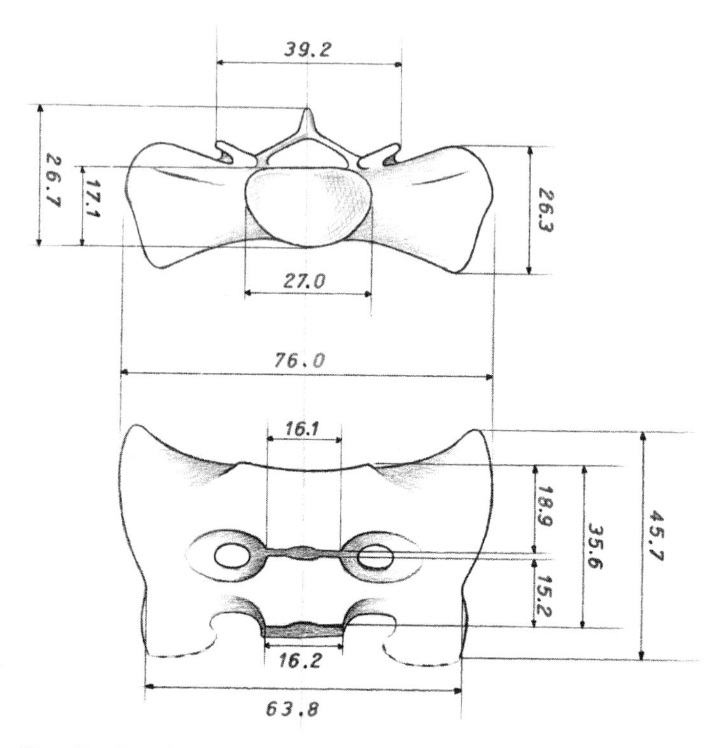

Fig. 33. Superior and anterior views of the incomplete sacrum of *Homo africanus* (Sts 14).

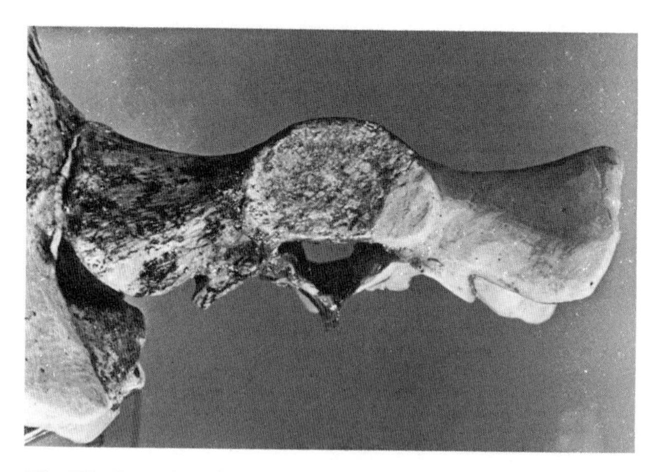

Fig. 34. Superior view of the *Homo africanus* sacrum. Approximately natural size.

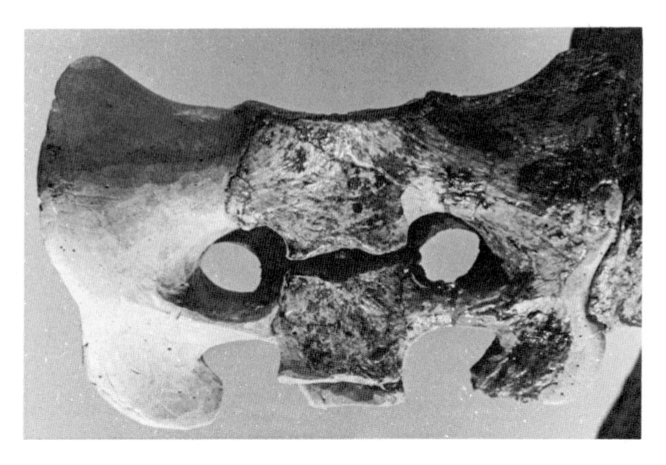

Fig. 35. Anterior view of the *Homo africanus* sacrum. Approximately natural size.

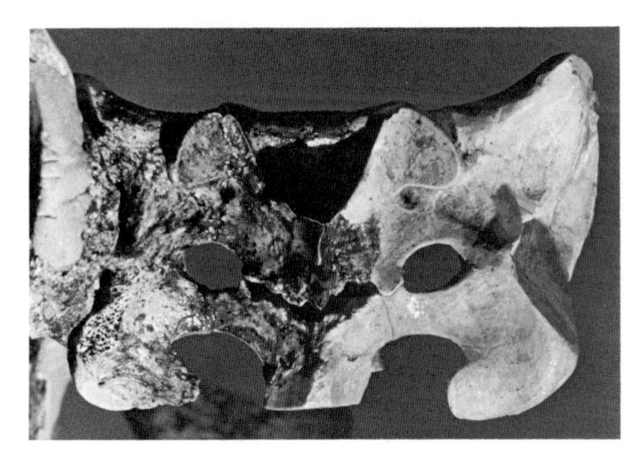

Fig. 36. Posterior view of *Homo africanus* sacrum. Approximately natural size.

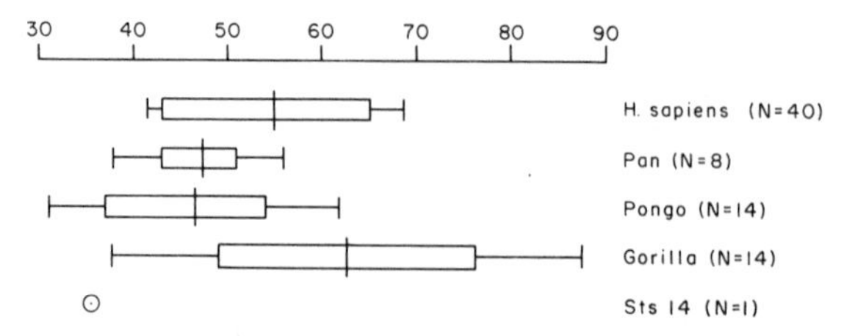

Fig. 37. Comparison of the absolute length (f) of the first two sacral elements, measured in the midline on the anterior surface, of some hominoids. The center bar represents the mean, the box represents the observed range, and total length the standard population range (mean ± 3 s.d.).

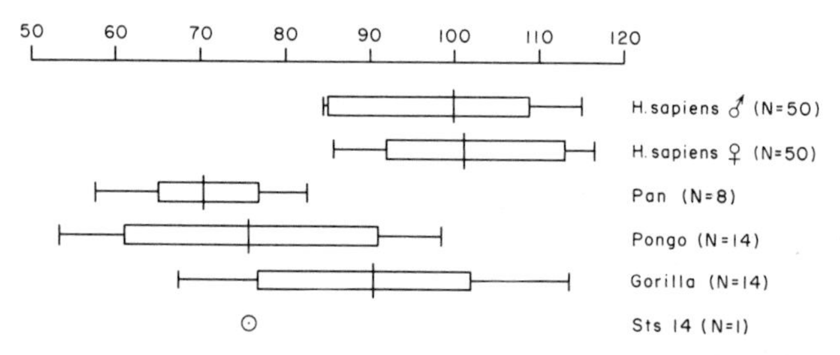

Fig. 38. Comparison of the absolute maximum breadth (e) of the sacrum in some hominoids. Conventions as in Fig. 37.

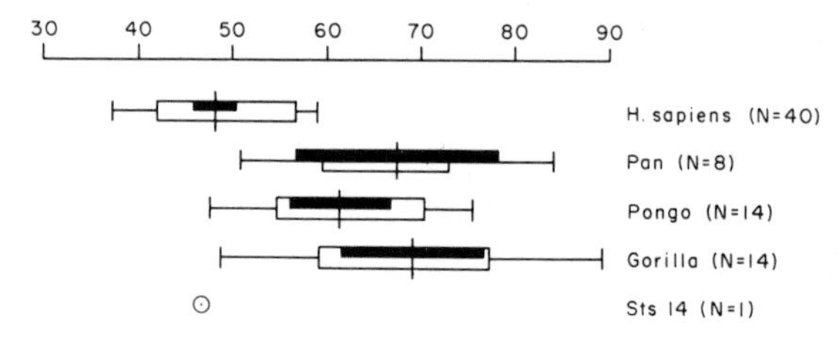

Fig. 39. Comparison of the ratio $100f/e$ for sacra of some hominoids. Conventions as in Fig. 37. The black bar represents the confidence interval for the mean where $p = .001$.

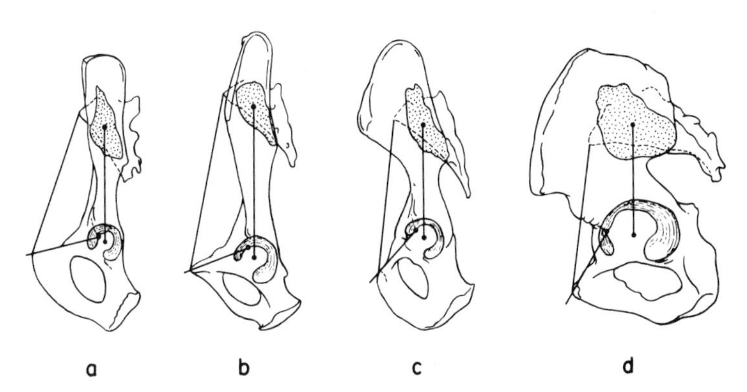

Fig. 40. Dioptrographic tracings of the left side views of the pelvises of some adult primates, reduced to approximately the same height. Redrawn from Schultz 1930: a, rhesus monkey female; b, gibbon male; c, gorilla male; d, human male.

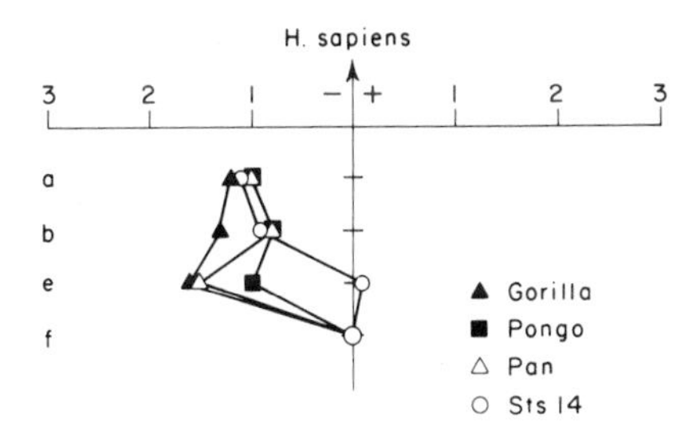

Fig. 41. Modified ratio diagram comparing four sacral dimensions in some hominids. *Homo sapiens* is the standard of comparison, and *f* (length of first two sacral vertebrae) is used as the basis for coincidence.

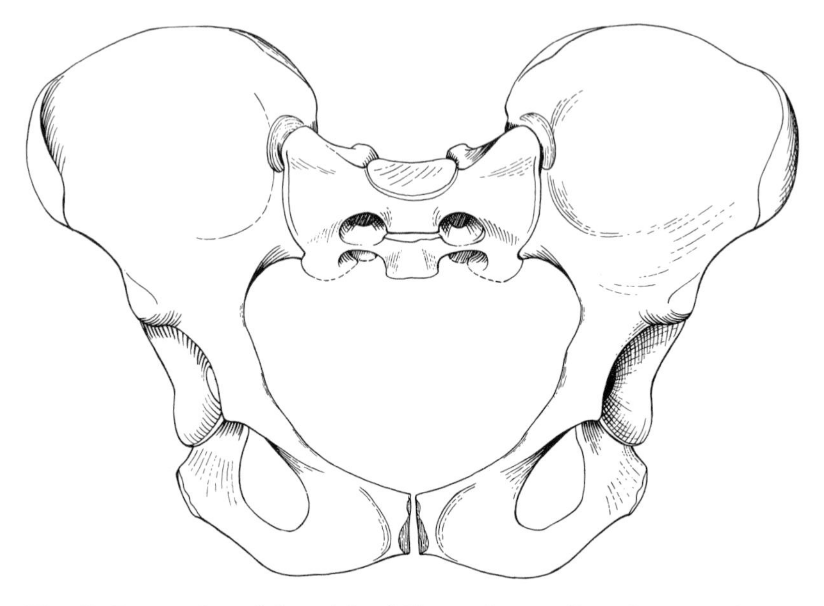

Fig. 42. Restoration of the pelvis of *Homo africanus* (Sts 14). Except for the distal portion of the sacrum, this pelvis is essentially complete and restoration involved primarily the correction of some relatively minor distortions (see also Fig. 43). Approximately half natural size.

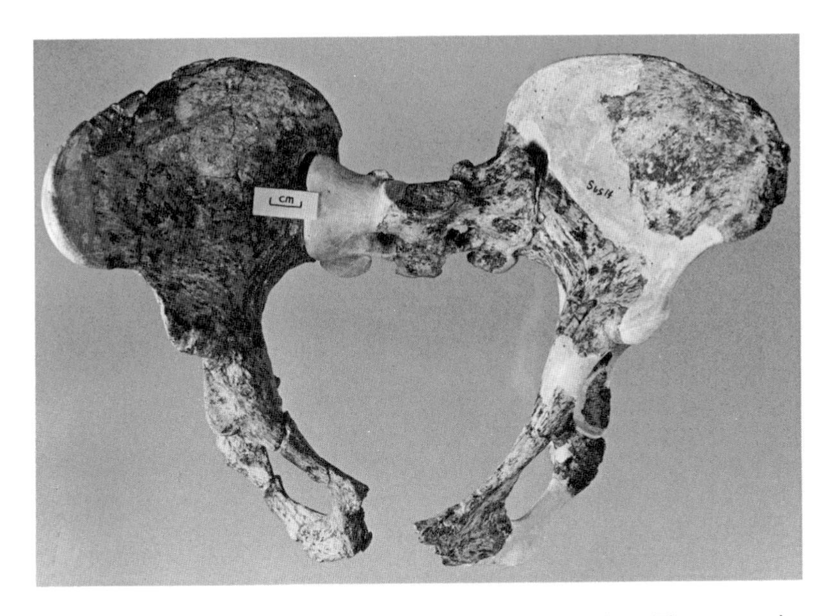

Fig. 43. Pelvis of *Homo africanus* (Sts 14) in anterior view. The restoration in Fig. 42 is based on this specimen, but the specimen is not identically orientated in the two figures. Approximately half natural size.

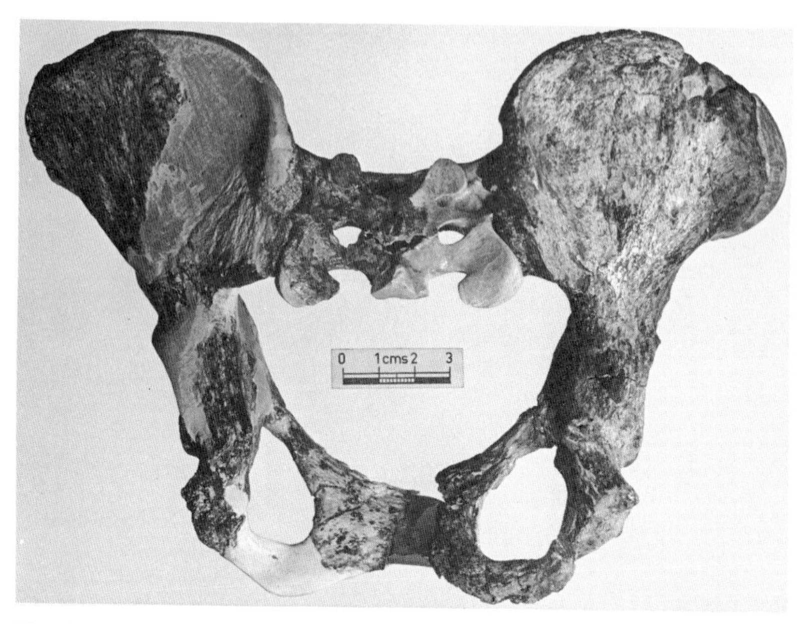

Fig. 44. Pelvis of *Homo africanus* (Sts 14) in rear view. Approximately half natural size.

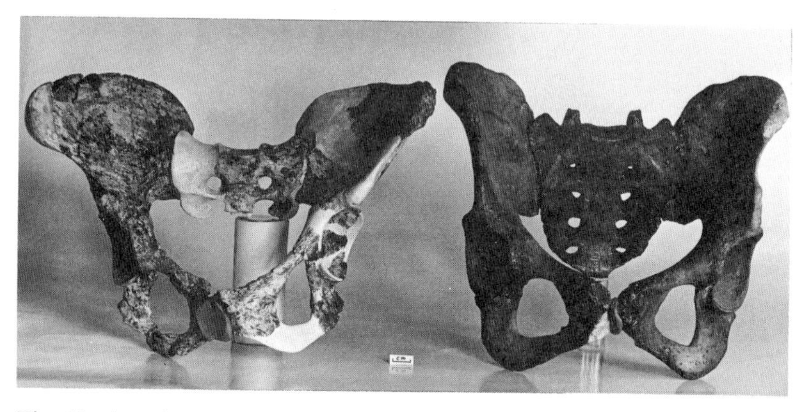

Fig. 45. Anterior view of pelvis of *Homo africanus* (Sts 14), left, and of *H. sapiens* (Bush male), right.

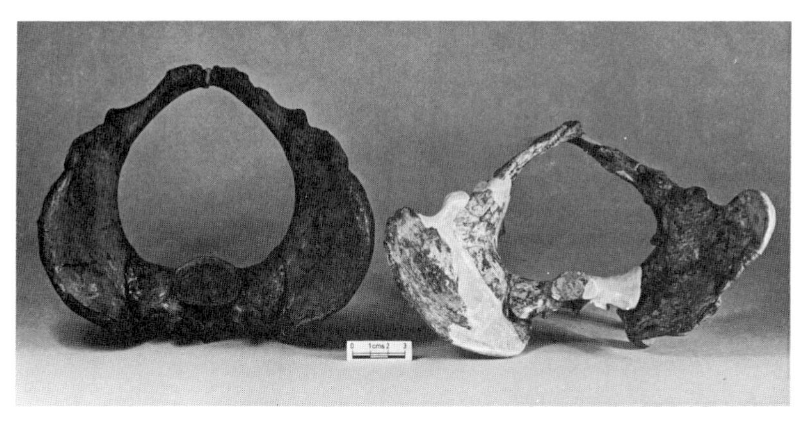

Fig. 46. Superior views of the pelvis of *Homo sapiens* (Bush male), left, and of *H. africanus* (Sts 14), right.

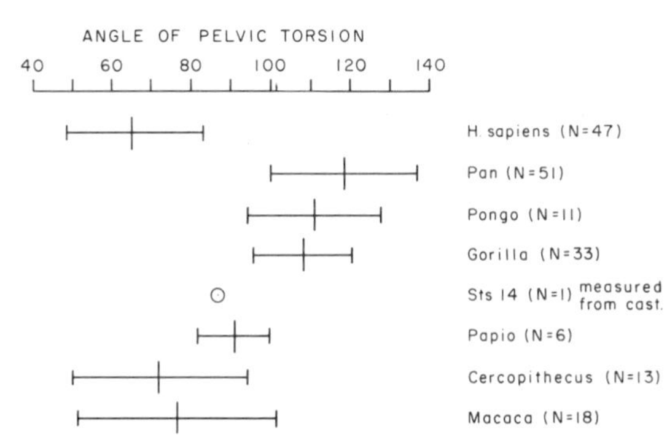

Fig. 47. Comparison of the angle of pelvic torsion in a number of primates. The center bar represents the mean and the total length the standard population range (mean ± 3 s.d.). Data adapted from Chopra (1962).

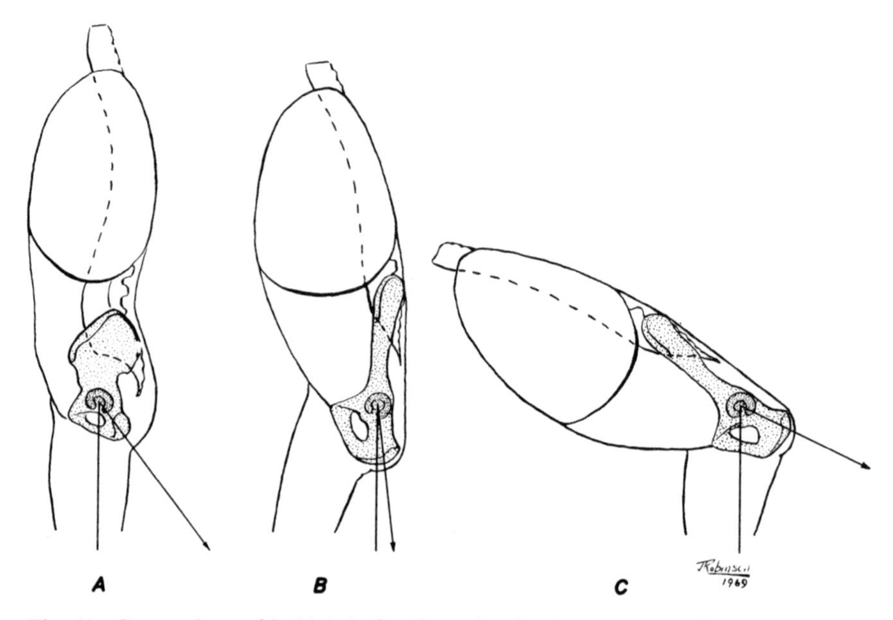

Fig. 48. Comparison of ischial shaft orientation in: A, a man standing erect; B, a chimpanzee standing erect; and C, a chimpanzee in normal quadrupedal position. The arrow indicates ischial shaft orientation. In B the ischial shaft is almost vertical and precludes extension of the thigh past the vertical. In part modified after Schultz (1930).

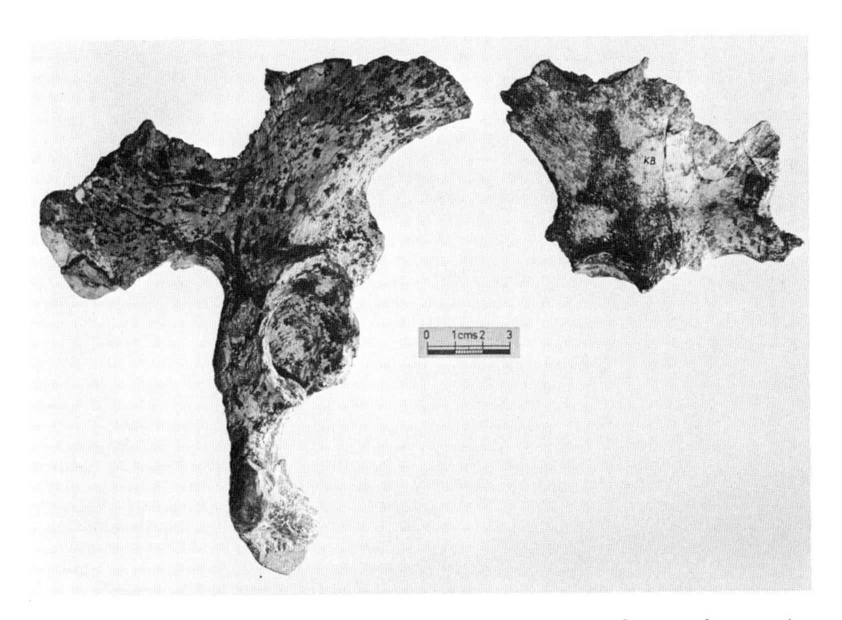

Fig. 49. Lateral views of incomplete innominate bones of *Paranthropus*. At left is SK 50 from Swartkrans; on the right is TM 1605 from Kromdraai.

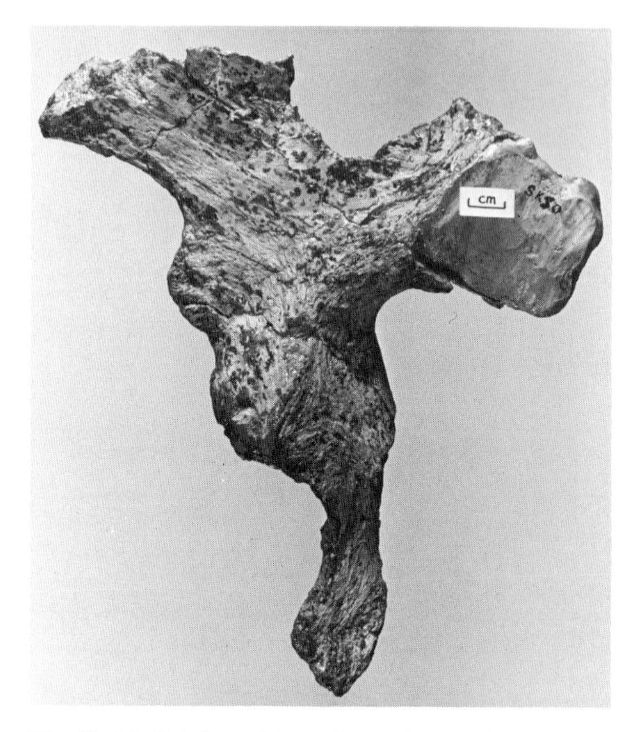

Fig. 50. Medial view of *Paranthropus* (SK 50) innominate bone. The auricular surface is damaged and is here covered with plaster to strengthen this much cracked region.

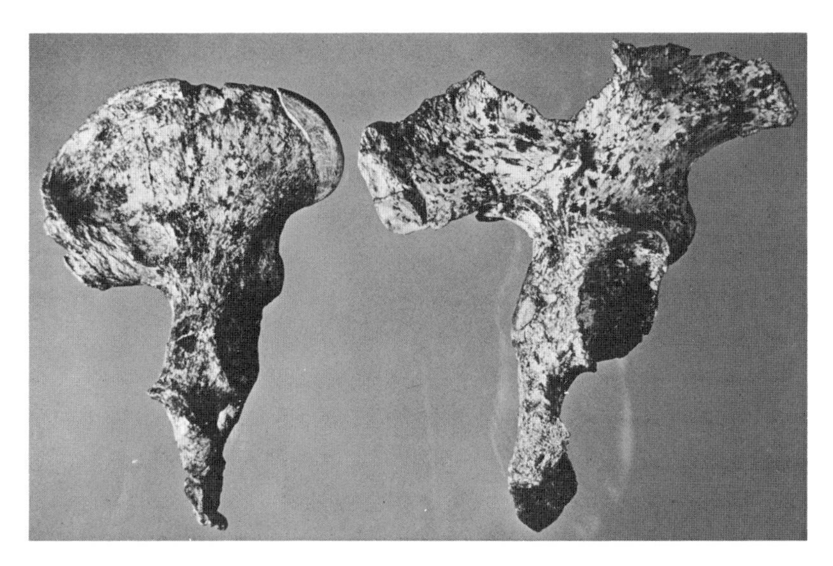

Fig. 51. Lateral views of innominate bones of *Homo africanus* (Sts 14), left, and *Paranthropus* (SK 50), right.

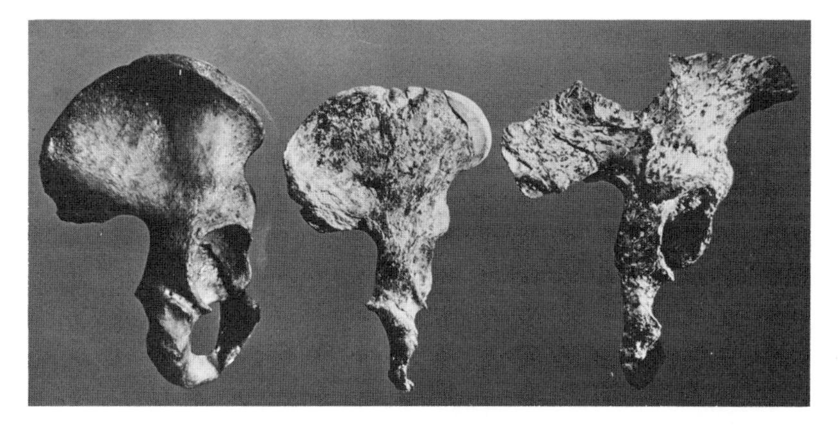

Fig. 52. Lateral views of innominate bones of *Homo sapiens*, left; *H. africanus*, center; and *Paranthropus*, right.

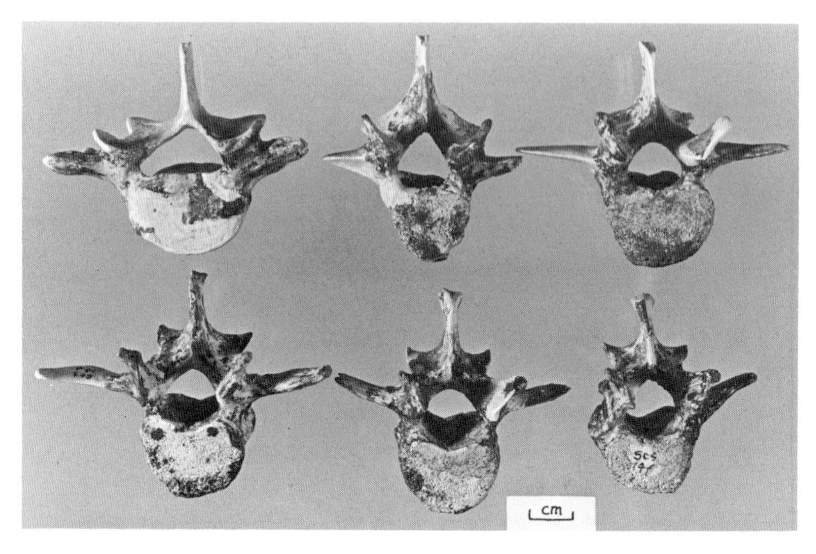

Fig. 53. Superior views of the six lumbar vertebrae of *Homo africanus* (Sts 14). The specimens are arranged in order from the last lumbar at top left to the first lumbar at bottom left, going from left to right along the top and then bottom rows.

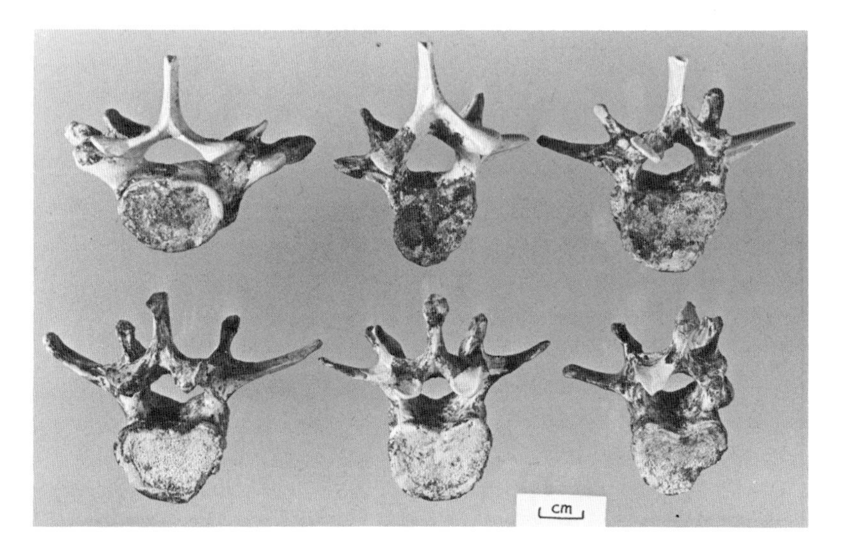

Fig. 54. Inferior views of the six lumbar vertebrae of *Homo africanus* (Sts 14). The specimens are arranged as in Fig. 53.

Fig. 55. Posterior view of the articulated lumbar spine of *Homo africanus* (Sts 14).

Fig. 56. View from the right side of the articulated lumbar spine of *Homo africanus* (Sts 14) attached to the sacrum and left innominate bone.

Fig. 57. Posterior view of articulated lumbar spine of *Homo africanus* (Sts 14) attached to sacrum and left innominate bone.

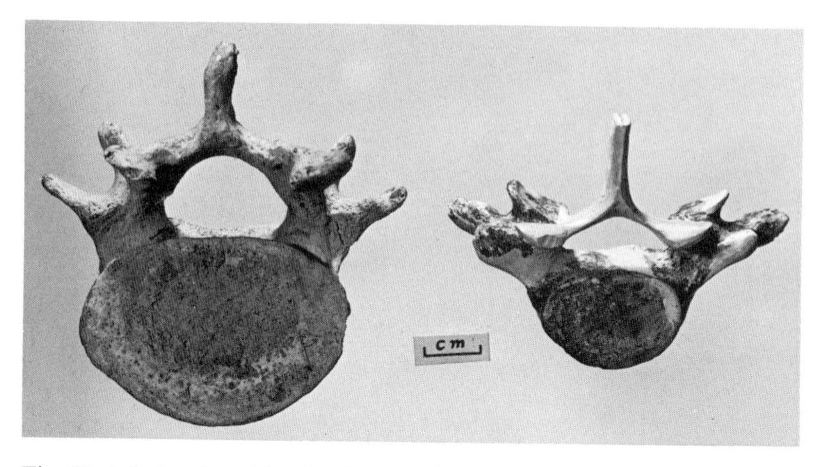

Fig. 58. Inferior view of last lumbar vertebrae of *Homo sapiens* (Bush female), left, and *H. africanus* (Sts 14 female). The disparity in size of the inferior surface of the body in the two specimens is obtrusive. Approximately natural size.

Fig. 59. Lateral view of the second lumbar vertebrae of *Homo africanus* (Sts 14), left, and *H. sapiens* (Bush female).

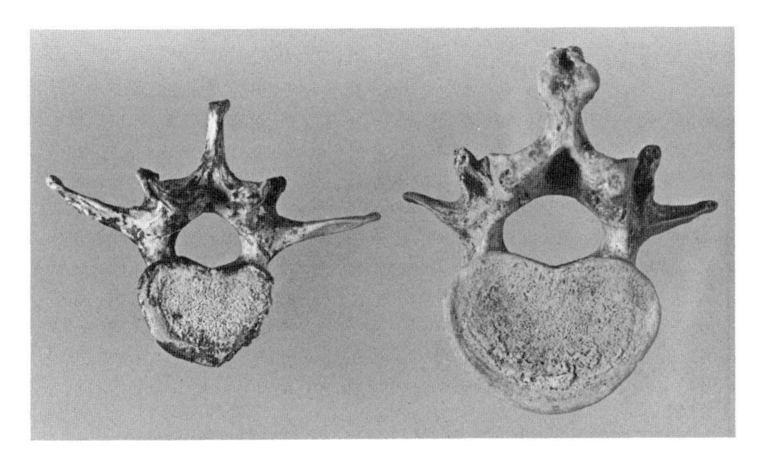

Fig. 60. Inferior view of second lumbar vertebrae of *Homo africanus* (Sts 14), left, and *H. sapiens* (Bush female).

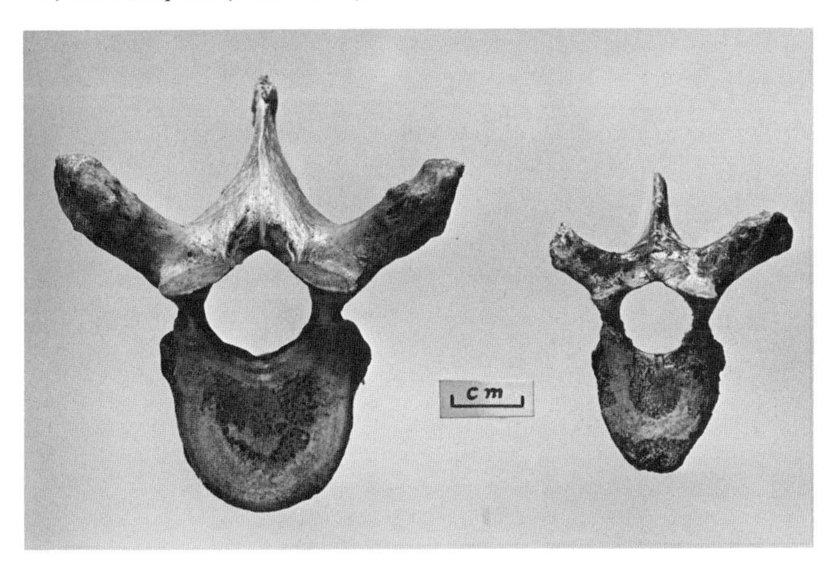

Fig. 61. Superior view of sixth last thoracic vertebrae of *Homo sapiens* (Bush female), left, and *H. africanus* (Sts 14). Approximately natural size.

Fig. 62. Lateral view of sixth last thoracic vertebrae of *Homo africanus* (Sts 14), left, and *H. sapiens* (Bush female). Approximately natural size.

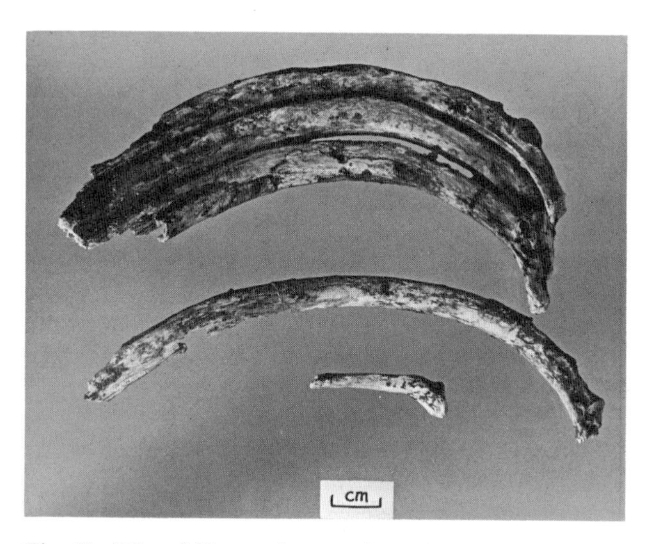

Fig. 63. Ribs of *Homo africanus* (Sts 14).

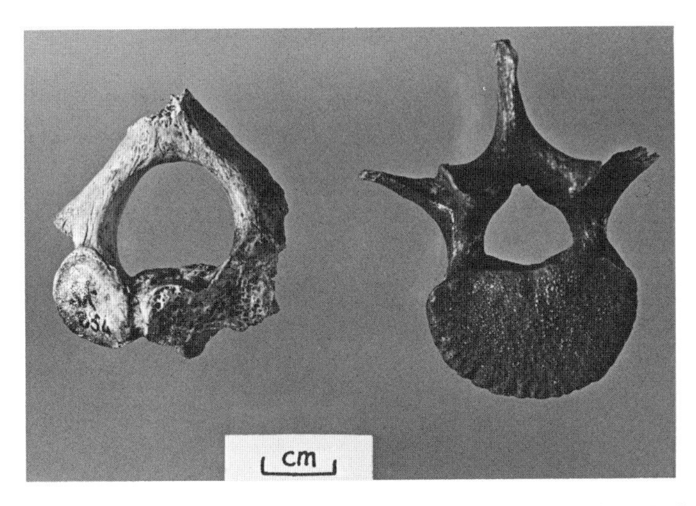

Fig. 64. Two vertebrae from Swartkrans in superior view. Left is an axis vertebra (SK 854), which probably belongs to *Paranthropus*. The other (SK 853) is a not fully mature lumbar vertebra, which probably belonged to *H. erectus* (= "Telanthropus"). Approximately natural size.

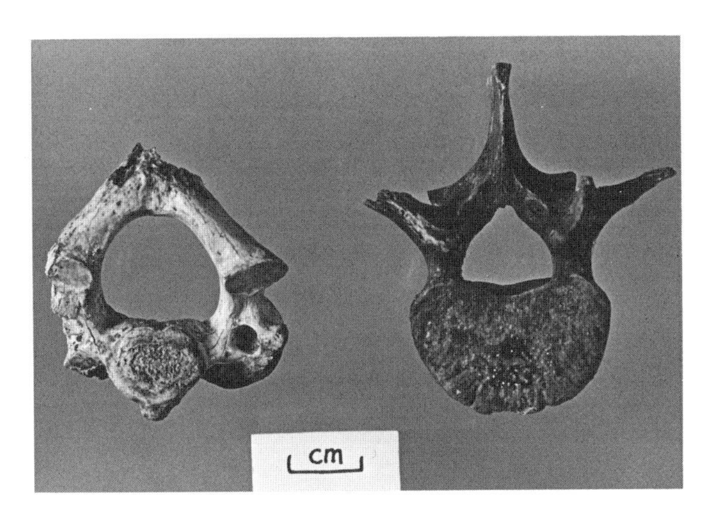

Fig. 65. Two vertebrae from Swartkrans in inferior view. Legend as for Fig. 64.

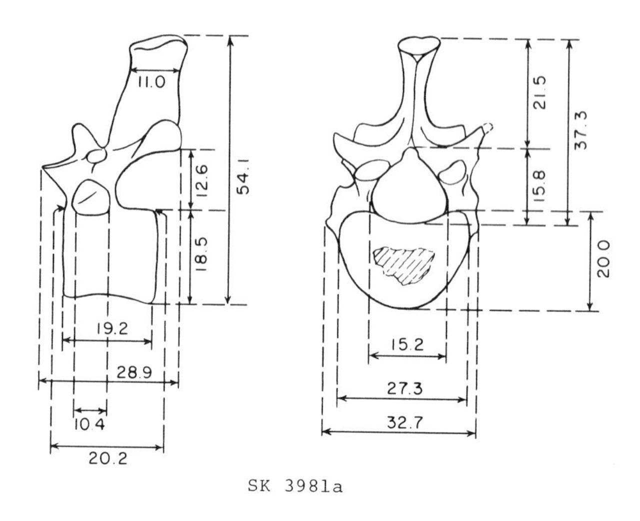

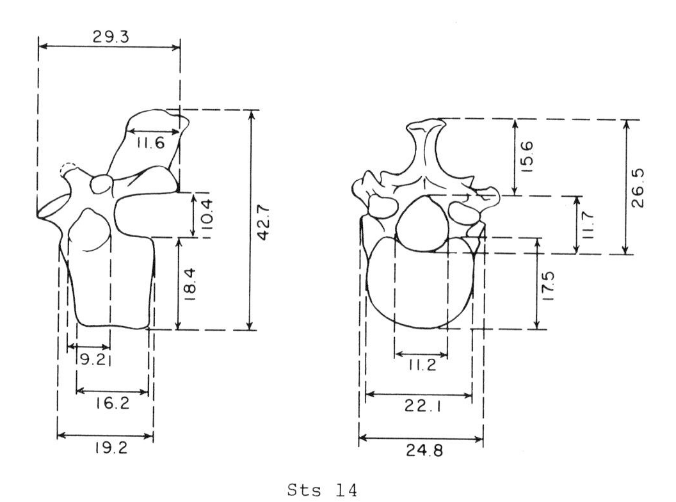

Fig. 66. Last thoracic vertebrae of *Paranthropus* (SK 3981a) and of *Homo africanus* (Sts 14). Dimensions in millimeters.

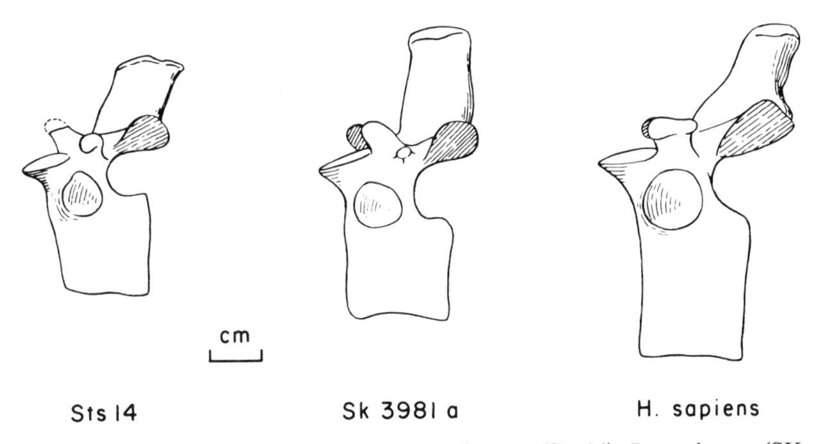

Fig. 67. Last thoracic vertebrae of *Homo africanus* (Sts 14), *Paranthropus* (SK 3981a), and *H. sapiens* (Bush) in lateral view.

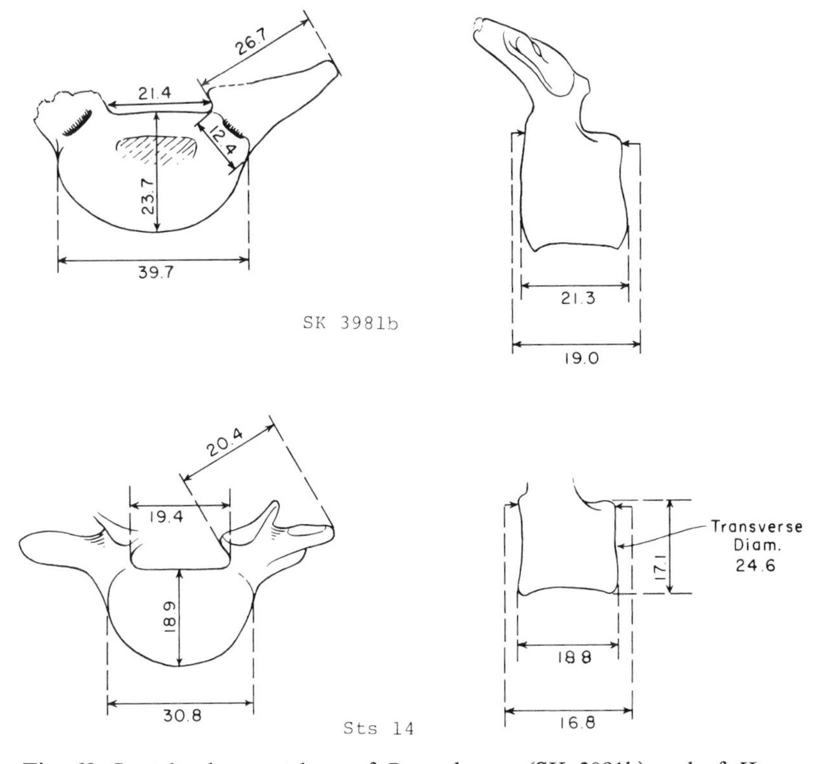

Fig. 68. Last lumbar vertebrae of *Paranthropus* (SK 3981b) and of *Homo africanus* (Sts 14). Dimensions in millimeters.

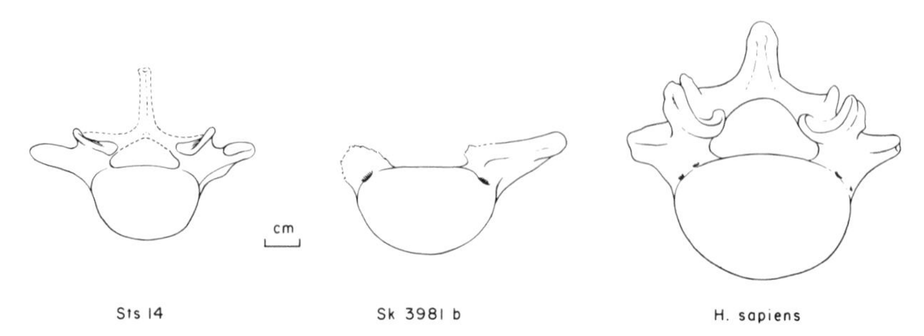

Fig. 69. Last lumbar vertebrae of *Homo africanus* (Sts 14), *Paranthropus* (SK 3981b), and *H. sapiens*.

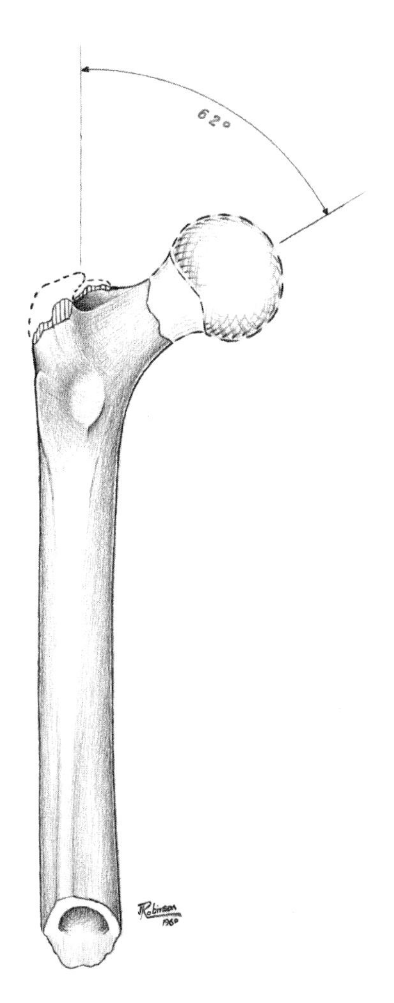

Fig. 70. Restoration of posterior view of the proximal portion of the left femur of *Homo africanus* (Sts 14). Neck length and head size are known within close limits. The greater trochanter has been conservatively reconstructed. The gluteal tuberosity, spiral line, and linea aspera are observable on the somewhat damaged original (see also fig. 71). Owing to the strong mottling of the specimen by manganese dioxide staining, detail does not show well in photographs. Approximately half natural size.

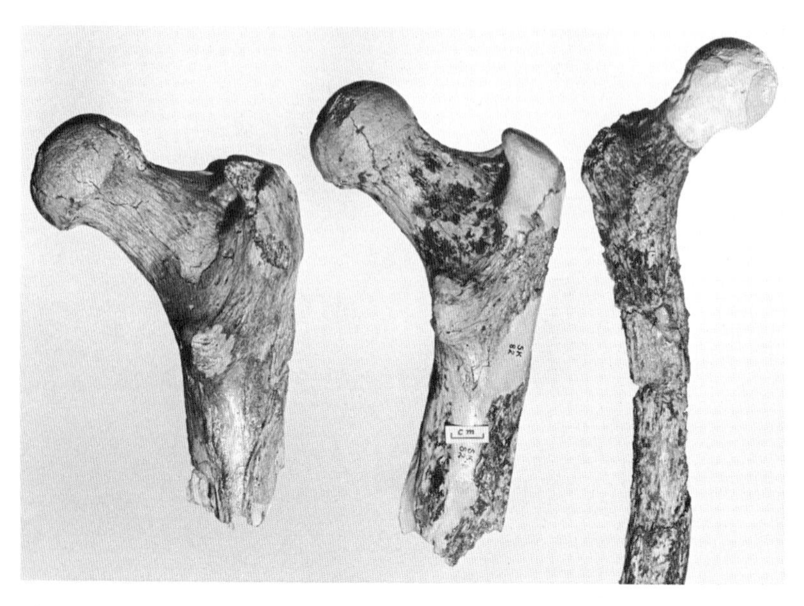

Fig. 71. Posterior views of the proximal ends of femora of *Paranthropus* (left, SK 97; right, SK 82) and of *Homo africanus* (Sts 14). About two-thirds natural size.

Fig. 72. Posterior views of the proximal ends of the femora of, from left to right, *Homo africanus* (Sts 14); chimpanzee; *Paranthropus* SK 82 and SK 97; *H. sapiens* (Bush female).

Fig. 73. Anterior views of the proximal ends of femora of, from left to right, *Homo africanus* (Sts 14); *Paranthropus* SK 82 and SK 97; *H. sapiens* (Bush female).

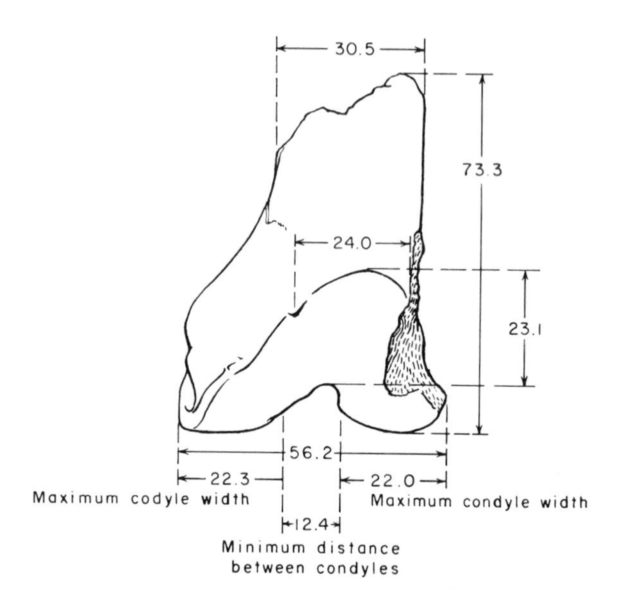

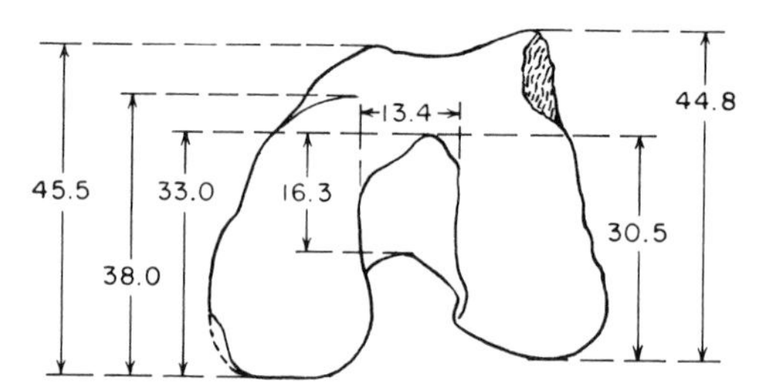

Fig. 74. Distal end of femur of *Homo africanus* (TM 1513). Dimensions in millimeters.

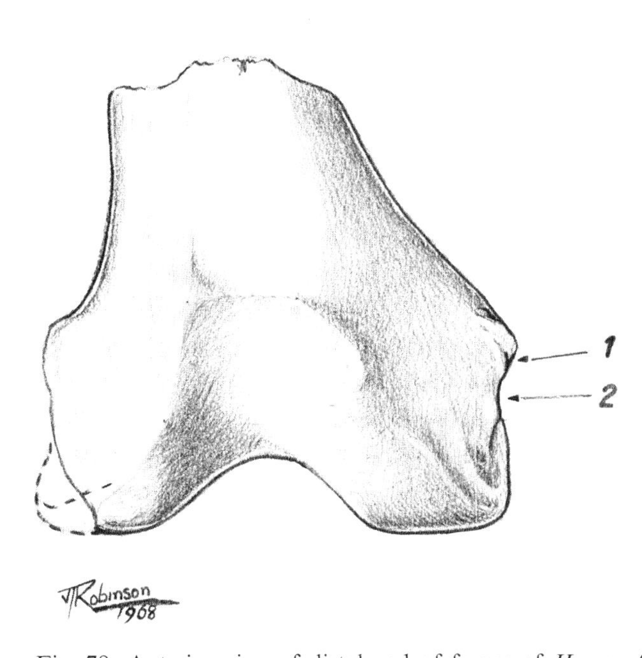

Fig. 78. Anterior view of distal end of femur of *Homo africanus* (Sts 34): 1, adductor tubercle; 2, medial epicondyle.

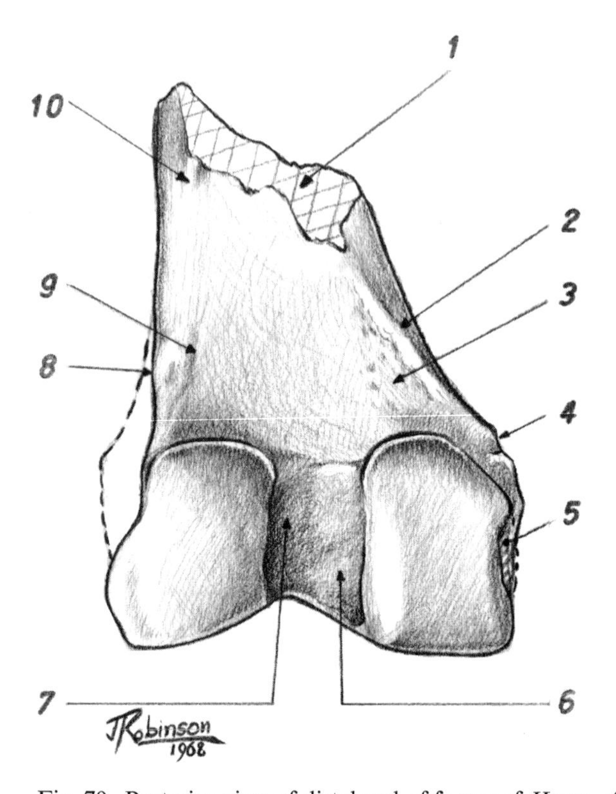

Fig. 79. Posterior view of distal end of femur of *Homo africanus* (TM 1513): 1, plaster reconstruction; 2, medial supracondylar ridge; 3, attachment site for medial head of gastrocnemius; 4, adductor tubercle; 5, small area of damage; 6, attachment for posterior cruciate ligament; 8, tubercle at end of intermuscular septum; 9, (?) attachment area for plantaris; 10, lateral supracondylar ridge.

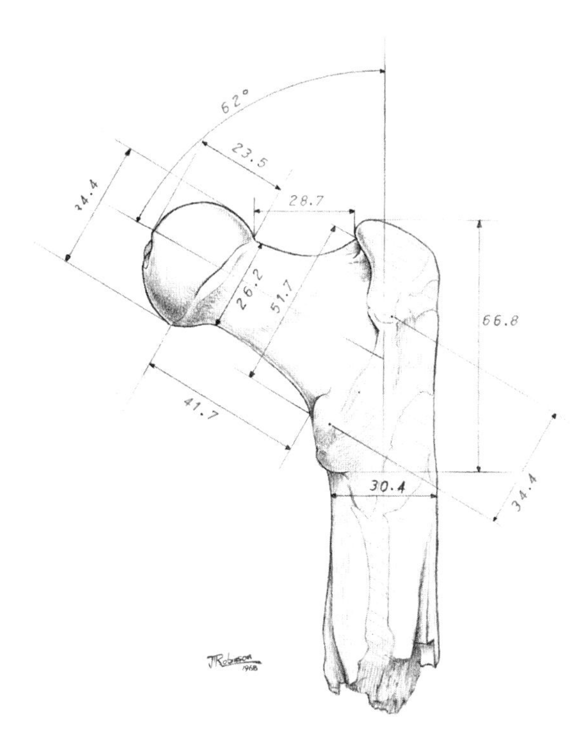

Fig. 80. Posterior aspect of proximal end of *Paranthropus* femur SK 82 indicating some dimensions. Note obturator externus groove. Lined areas have surface bone missing. Approximately half natural size.

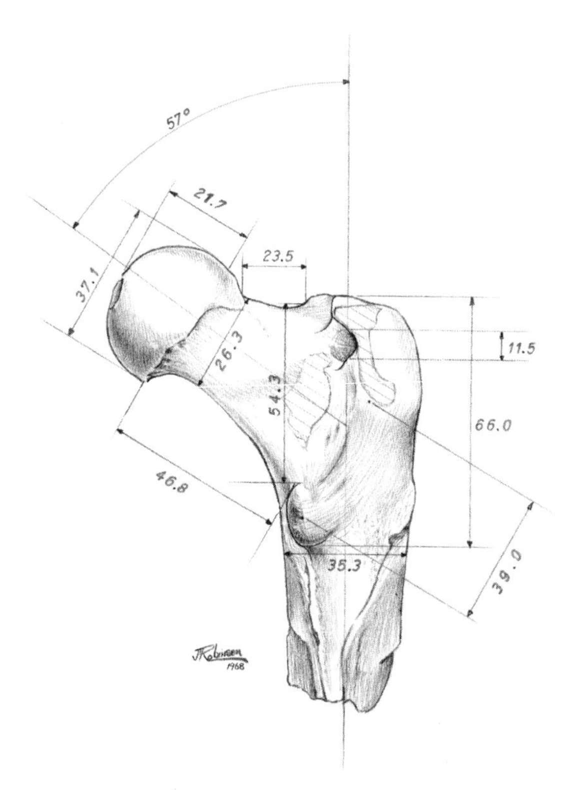

Fig. 81. Posterior aspect of proximal end of *Paranthropus* femur SK 97 indicating some dimensions. Note obturator externus groove. Lined areas have surface bone missing. Approximately half natural size.

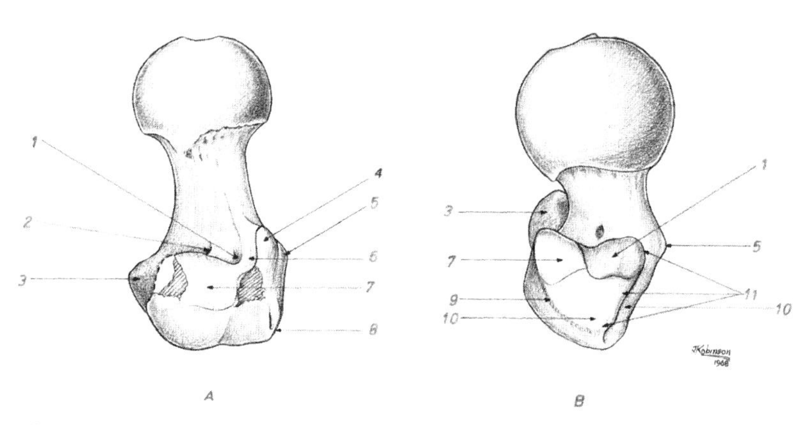

Fig. 82. Proximal ends of femora of, A, *Paranthropus* SK 97, and B, *Homo sapiens* (Bush), seen in superior view: 1, area of attachment for obturator externus and gemelli; 2, attachment area for obturator externus; 3, lesser trochanter; 4, anteriormost attachment for gluteus minimus; 5, femoral tubercle; 6, attachment for pyriformis and anterior retinaculum; 7, attachment area for pyriformis; 8, most posterior attachment for gluteus minimus; 9, attachment for gluteus medius; 10, bursal area; 11, attachment for gluteus minimus.

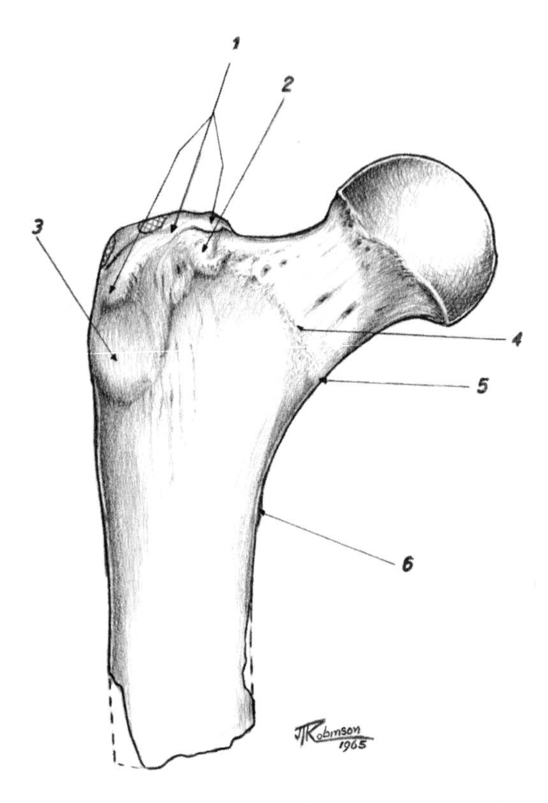

Fig. 83. Anterior view of proximal end of *Paranthropus* femur SK 97: 1, attachment for gluteus minimus; 2, femoral tubercle; 3, bursal area; 4, trochanteric line; 5, small tubercle; 6, lesser trochanter.

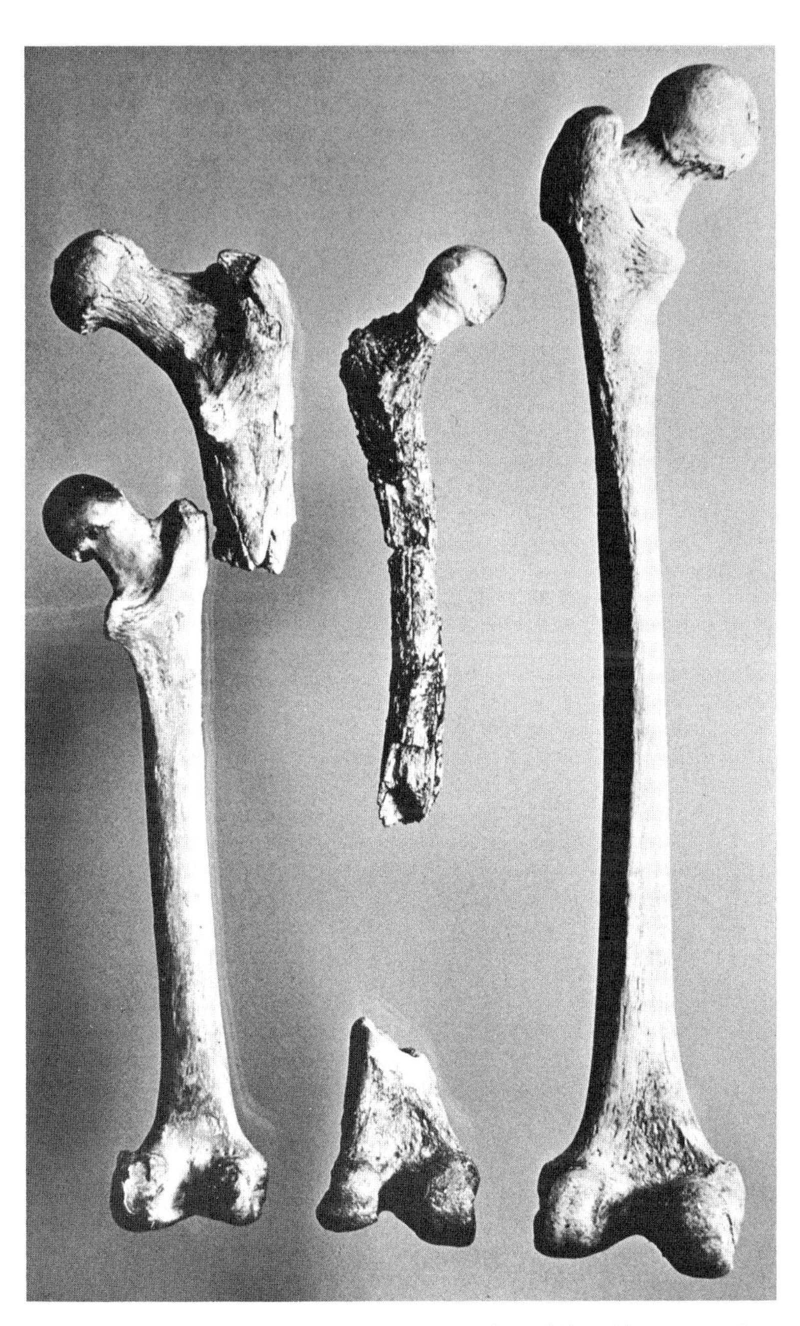

Fig. 84. Femora of some hominoids. From left to right: chimpanzee, *Paranthropus*, *Homo africanus* and *H. sapiens* (Bush).

Fig. 85. Femora of *Homo sapiens* (left) and *H. africanus*.

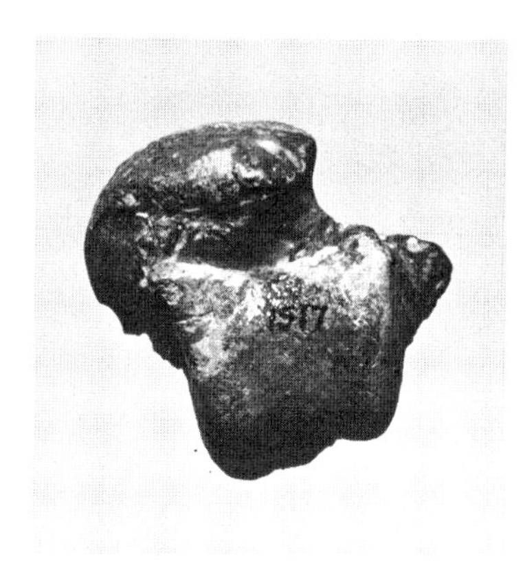

Fig. 86. Right talus of *Paranthropus* from Kromdraai. Approximately natural size.

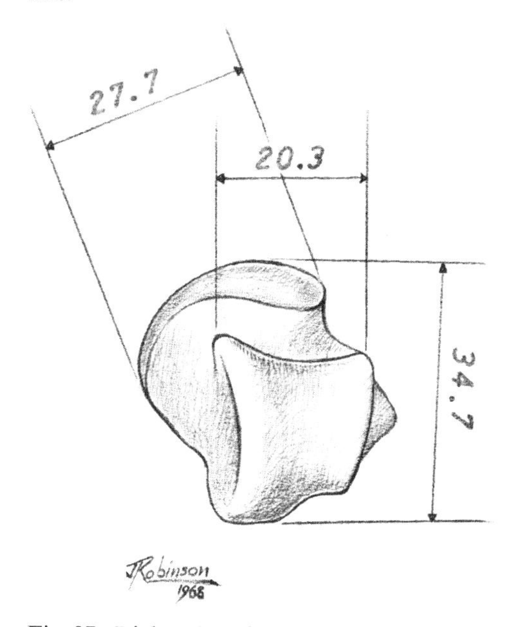

Fig. 87. Right talus of *Paranthropus* from Kromdraai. Approximately natural size.

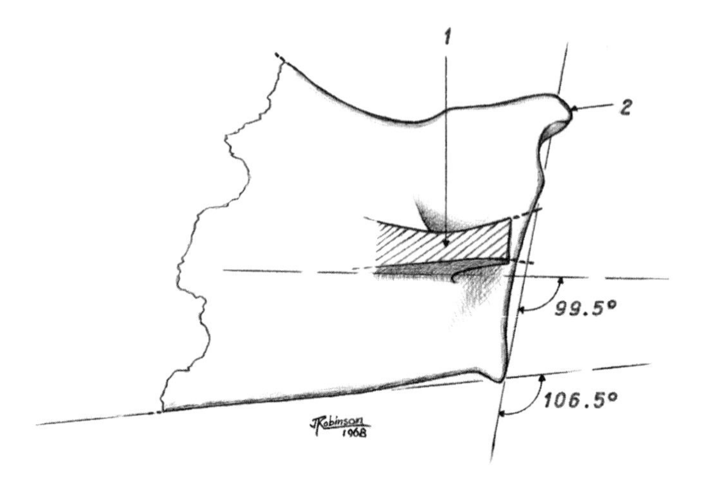

Fig. 88. Proximal end of scapula of *Homo africanus* (Sts 7).

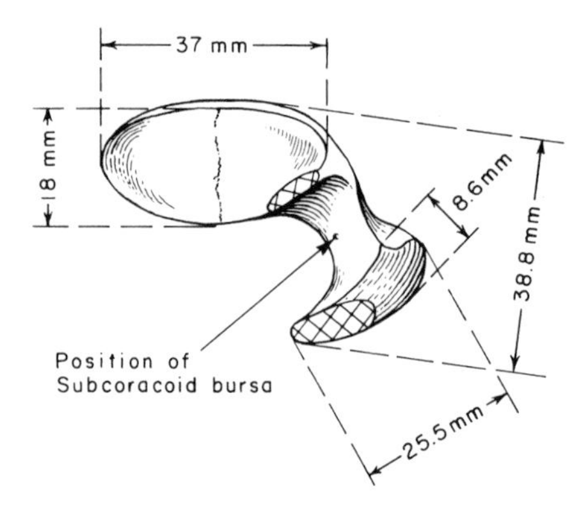

Fig. 89. Glenoid and coracoid of Sts 7 (*Homo africanus*).

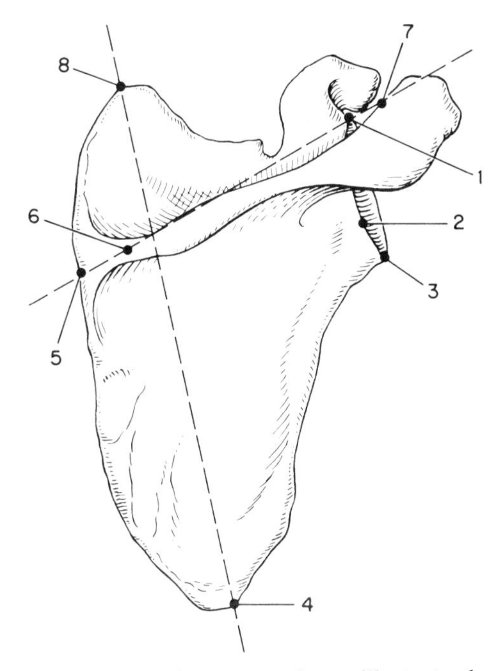

Fig. 90. A primate scapula to illustrate the angle *e* used by Ashton and Oxnard. Redrawn from Ashton and Oxnard.

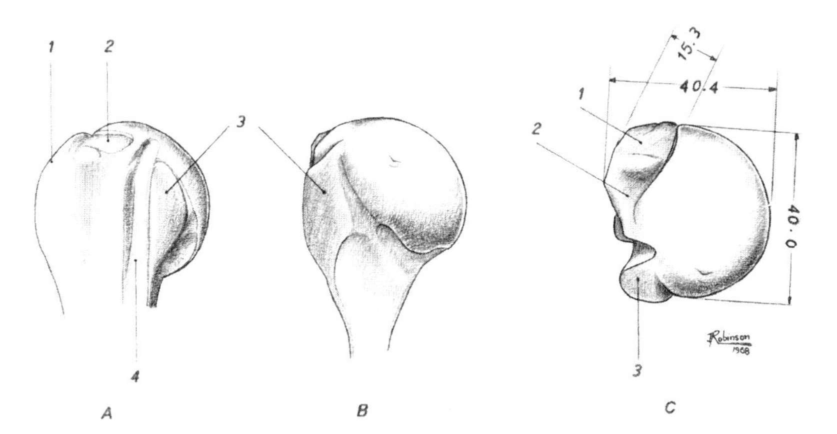

Fig. 91. Proximal end of humerus of *Homo africanus* Sts 7: 1, facet for infraspinatus; 2, facet for supraspinatus; 3, lesser tuberosity.

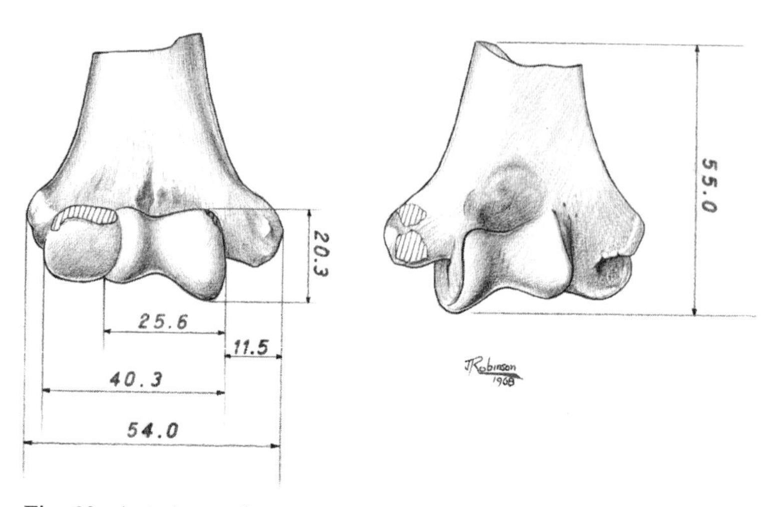

Fig. 92. Anterior and posterior views of the distal end of a *Paranthropus* humerus (TM 1517) from Kromdraai.

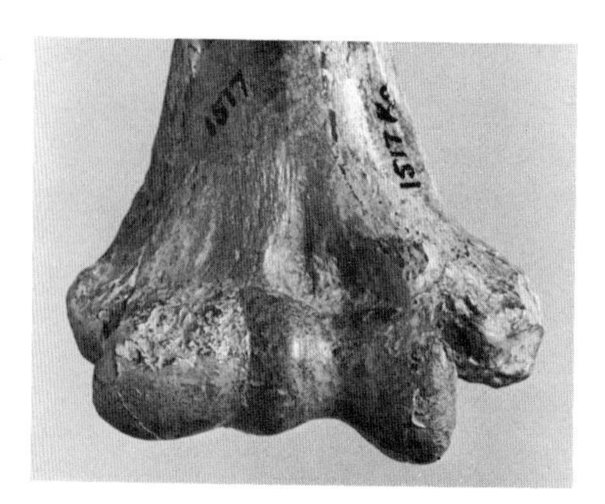

Fig. 93. Anterior view of *Paranthropus* humerus fragment TM 1517. Approximately natural size.

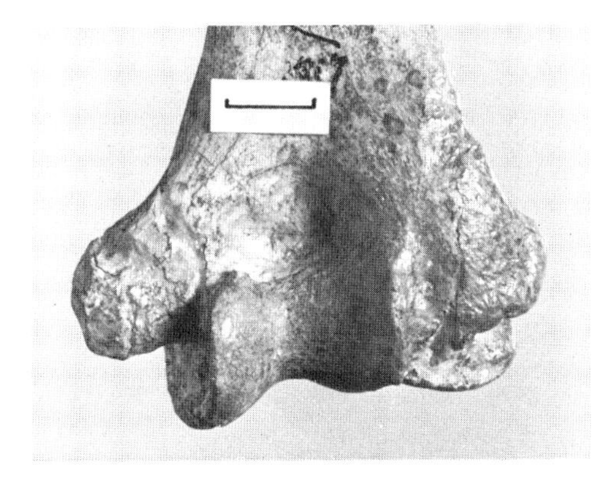

Fig. 94. Posterior view of *Paranthropus* humerus fragment TM 1517. Approximately natural size.

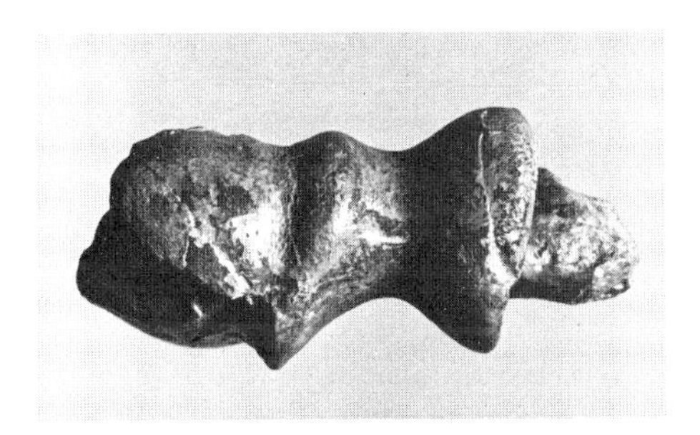

Fig. 95. End view of the articular surfaces of *Paranthropus* humerus TM 1517. Approximately natural size.

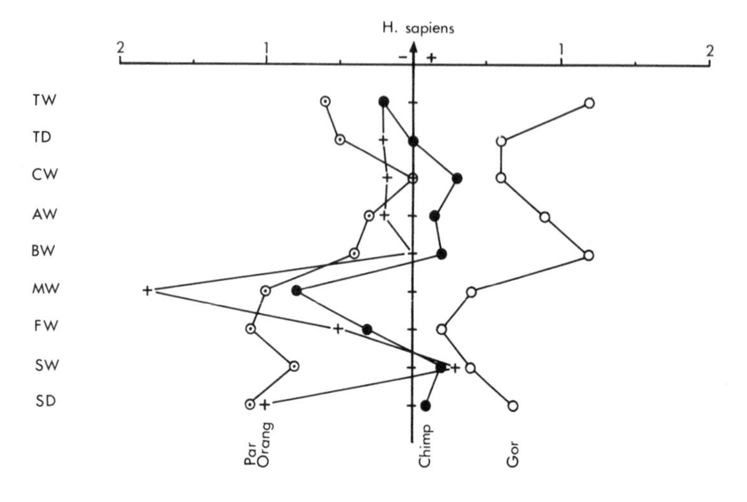

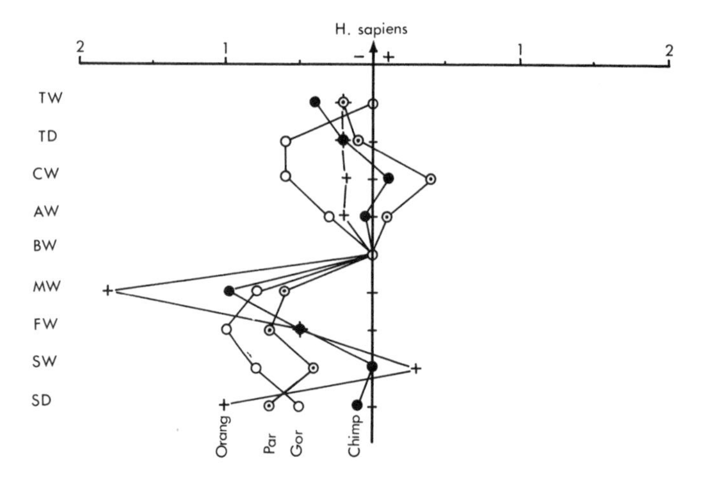

Fig. 96. Ratio diagram comparing various dimensions of the distal end of the humerus of some hominoids. *Homo sapiens* is used as the standard for comparison. TW, trochlear width; TD, trochlear depth; CW, capitulum width; AW, articular surface width; BW, bi-epicondylar width; MW, width of medial epicondyle; FW, width of fossa olecrani; SW, greatest width of lower humerus shaft; SD, greatest depth (thickness) of lower humerus shaft. Original data from Straus, 1948.

Fig. 97. Modified ratio diagram comparing the same dimensions used in fig. 96 but using bi-epicondylar width for coincidence.

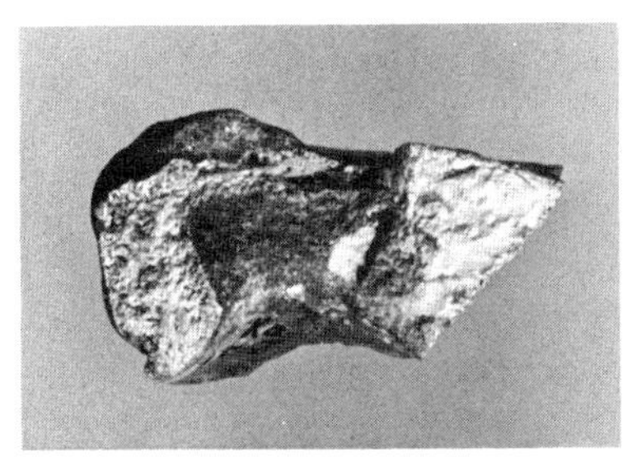
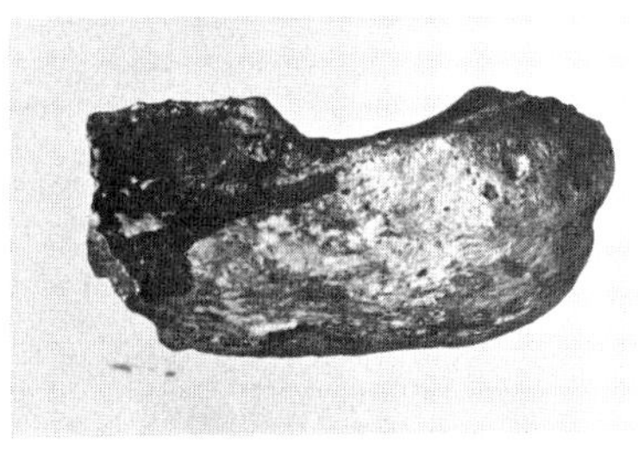

Fig. 98. Fragment of ulna of *Paranthropus* (TM 1517) from Kromdraai. Approximately natural size.

Fig. 99. Fragment of ulna of *Paranthropus* (TM 1517) from Kromdraai. Approximately natural size.

Fig. 100. Fragment of ulna of *Paranthropus* (TM 1517) from Kromdraai. Approximately natural size.

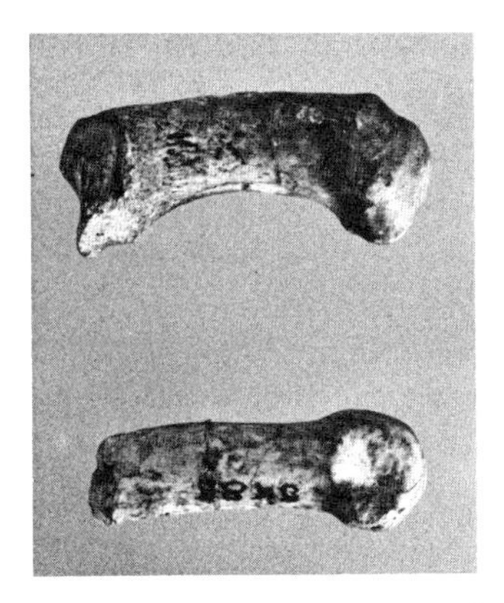

Fig. 101. Thumb metacarpal of *Paranthropus* (SK 84) from Swartkrans, above, and incomplete fourth metacarpal of *Homo erectus* (SK 85) from the same site. Approximately natural size.

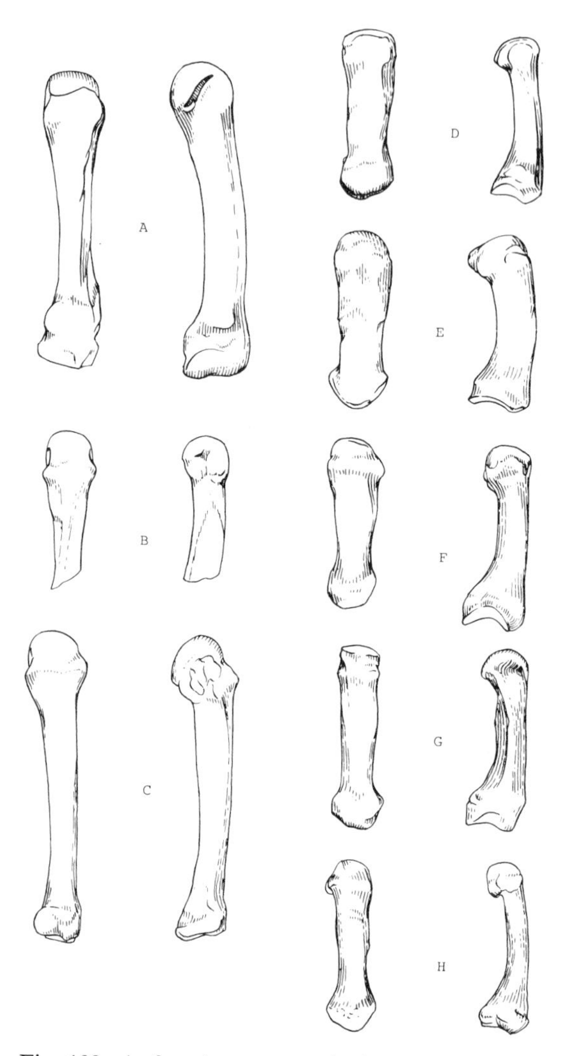

Fig. 102. A, fourth metacarpal of *Homo sapiens* $\times \frac{3}{4}$; B, fourth metacarpal of *H. erectus* (= "Telanthropus") SK 85 $\times \frac{5}{8}$; C, fourth metcarpal of *Gorilla* $\times \frac{1}{2}$; D, first metacarpal of *H. sapiens* $\times \frac{1}{2}$; E, first metacarpal of *Paranthropus* SK 84 $\times \frac{3}{4}$; F, first metacarpal of *Gorilla* $\times \frac{1}{2}$; G, first metacarpal of *Pongo* $\times \frac{5}{8}$; H, first metacarpal of *Pan* $\times \frac{1}{2}$. Redrawn from Napier (1959).

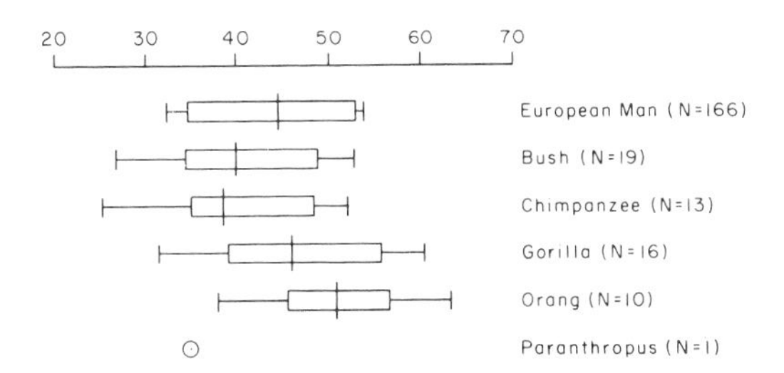

Fig. 103. Comparison of maximum absolute length of the thumb metacarpal in a number of hominoids. Scale is in millimeters. Based on Napier 1959.

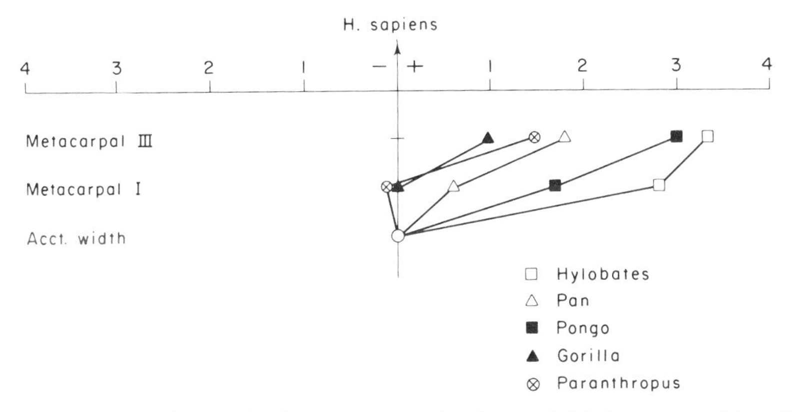

Fig. 104. Modified ratio diagram comparing first and third metacarpal length and using acetabular width for comparison.

Fig. 105. Second metacarpal of *Paranthropus* (TM 1517) from Kromdraai.

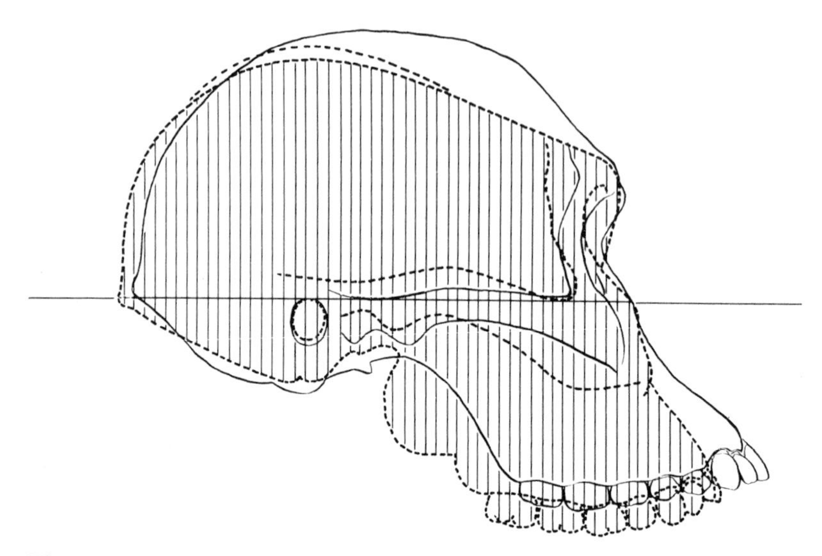

Fig. 106. Superimposed drawings of *Paranthropus* (lined) and *Homo africanus*. Original drawing was done natural size.

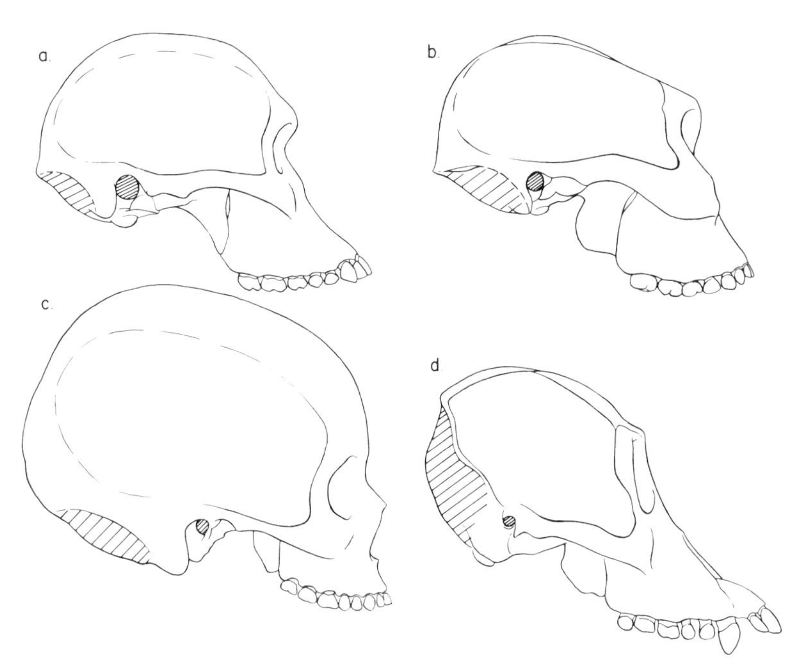

Fig. 107. Lateral view of skulls of: a, *Homo africanus* from Sterkfontein; b, *Paranthropus* from Swartkrans; c, *Homo sapiens*; d, *Pongo*. The nuchal portion of the occiput is shaded in each case. All specimens to same scale.

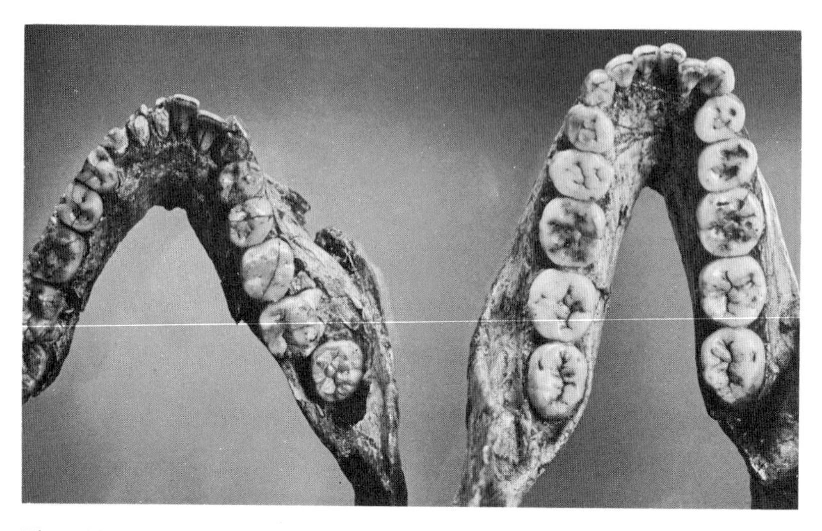

Fig. 108. Mandibular dentition of *Homo africanus* (left) and *Paranthropus*. The misplaced right lateral incisor of the latter occupied the present position during life. Compression during fossilization has narrowed the distance between the two halves of the *Paranthropus* mandible.

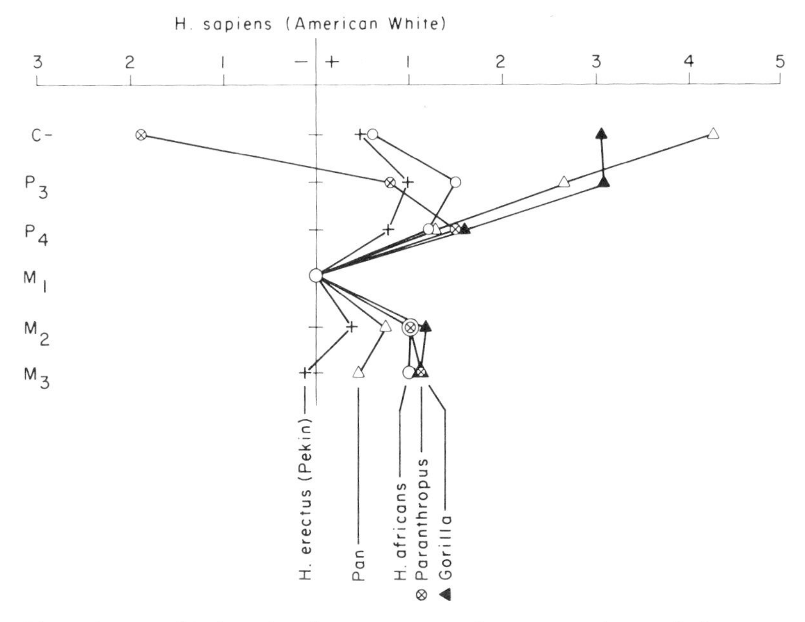

Fig. 109. Modified ratio diagram comparing proportions of the mean robustness (length × breadth of crown) of the mandibular teeth (excluding incisors) of some hominoids. All "curves" are related to a standard size of first molar.

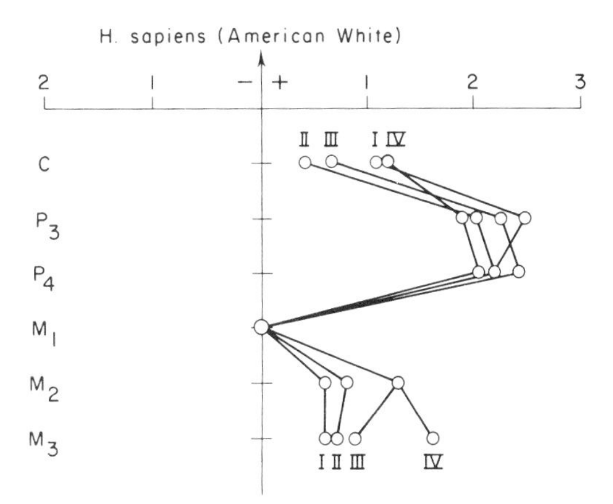

Fig. 110. Modified ratio diagram comparing proportionate robustness of the crowns of some mandibular teeth in four specimens of *Gigantopithecus*. Specimens I–III are from China (Woo's data) and IV is from India (data from Simons and Chopra). Compare with fig. 109.

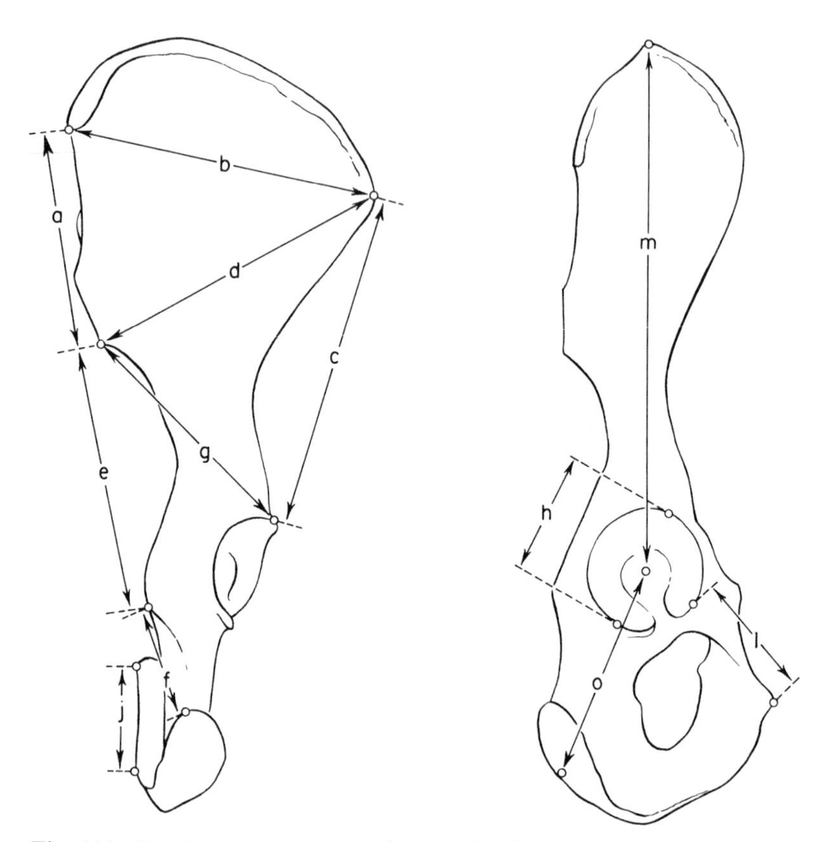

Fig. 111. Key to measurements taken on the innominate bone. For some reason not now obvious to me, pubis length was measured in all cases from the acetabular margin rather than from the center of the acetabulum. The point is not of great moment since pubic length is not especially important in the discussions and how it is measured is not nearly so critical as in the case of ischial length. The conversion to true pubic length, however, easily can be made by those who wish to do so by adding $\frac{1}{2}h$ to l. It may not be immediately clear from the key that j is pubic symphysis length.

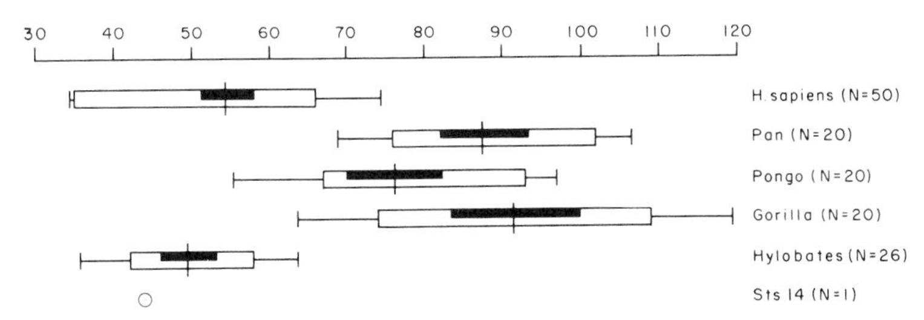

Fig. 112. Comparison of absolute dimension *e* of the innominate bone of some hominoids. The central bar represents the mean, the box represents the observed range, the black bar represents the confidence interval for the mean where $p = .001$, and maximum length represents the standard population range (mean ± 3 s.d.).

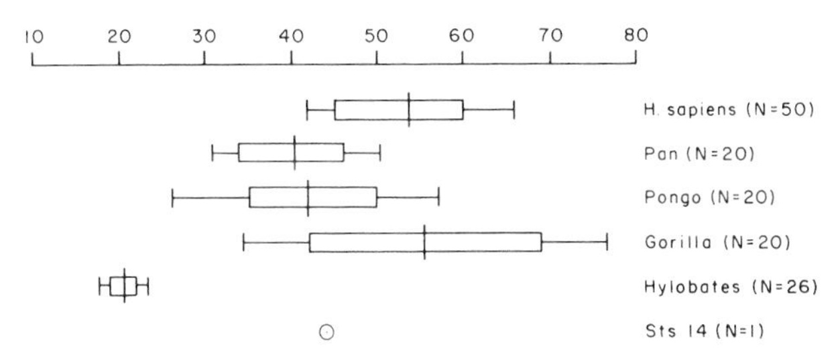

Fig. 113. Comparison of absolute dimension *h* (acetabular width) in some hominoids. Conventions as for fig. 112.

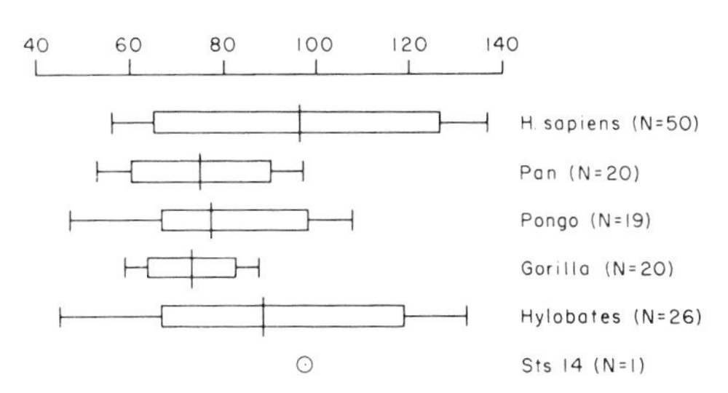

Fig. 114. Comparison of the ratio $100e/c$ for some hominoids. Conventions as for fig. 112.

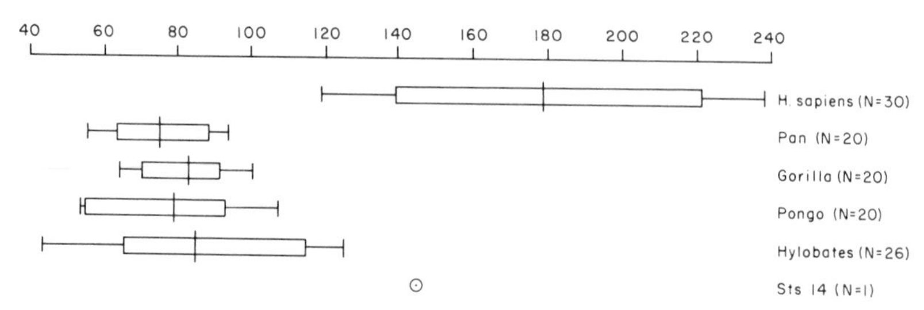

Fig. 115. Comparison of the ratio 100*g*/*c* for some hominoids. Conventions as for fig. 112.

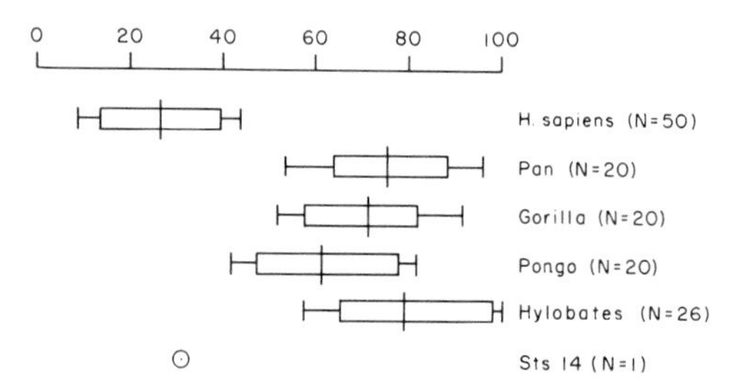

Fig. 116. Comparison of the ratio 100*a*/*d* for some hominoids. Conventions as in fig. 112.

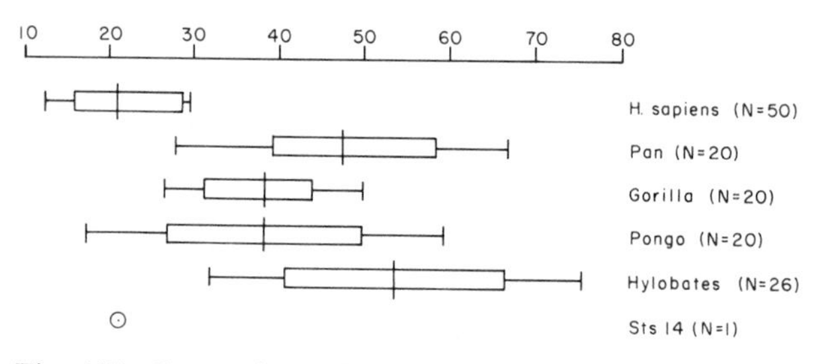

Fig. 117. Comparison of the ratio 100*f*/*b* for some hominoids. Conventions as in fig. 112.

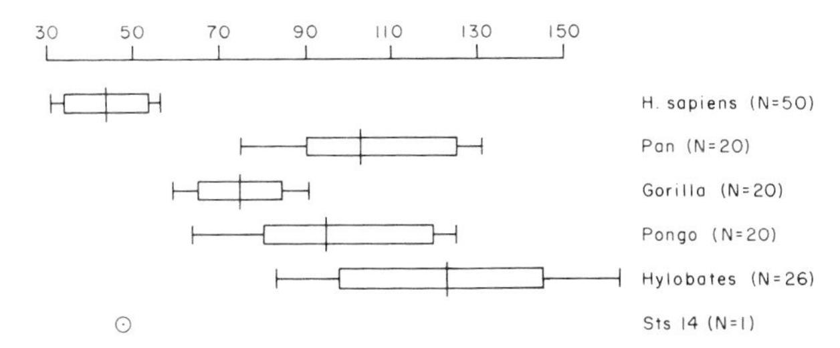

Fig. 118. Comparison of the ratio 100*c*/*d* for some hominoids. Conventions as in fig. 112.

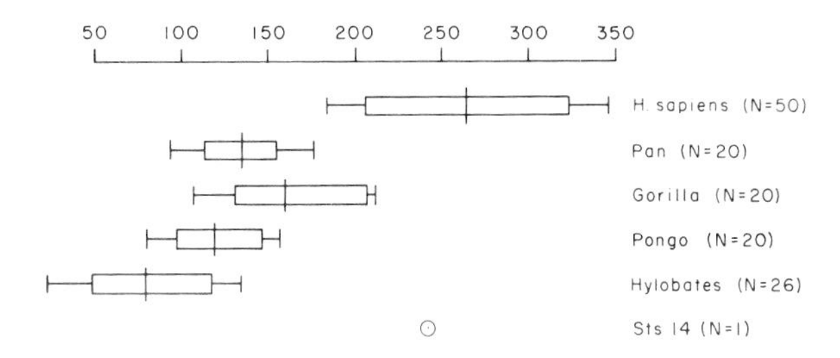

Fig. 119. Comparison of the ratio 100*b*/*c* for some hominoids. Conventions as in fig. 112.

Index